One Dimensional
Spline
Interpolation
Algorithms

One Dimensional Spline Interpolation Algorithms

Helmuth Späth
Universität Oldenburg
Oldenburg, Germany

CRC Press
Taylor & Francis Group
Boca Raton London New York

CRC Press is an imprint of the
Taylor & Francis Group, an **informa** business

AN A K PETERS BOOK

First published 1995 by A K Peters, Ltd.

Published 2018 by CRC Press
Taylor & Francis Group
6000 Broken Sound Parkway NW, Suite 300
Boca Raton, FL 33487-2742

© 1995 by Taylor & Francis Group, LLC
CRC Press is an imprint of Taylor & Francis Group, an Informa business

First issued in paperback 2019

No claim to original U.S. Government works

ISBN 13: 978-0-367-44907-0 (pbk)
ISBN 13: 978-1-56881-016-4 (hbk)

Visit the Taylor & Francis Web site at
http://www.taylorandfrancis.com

and the CRC Press Web site at
http://www.crcpress.com

Library of Congress Cataloging-in-Publication Data

Späth, Helmuth.
    [Eindimensionale Spline-Interpolations Algorithmen. English]
    One dimensional spline interpolation algorithms / H. Späth.
      p. cm,
    Includes bibliographical references and index.
    ISBN 1-56881-016-4
    1. Spline theory--Data processing. 2. Functions--Data processing.
I. Title.
QA224.S59513   1995   95-40858
    511'.42--dc20   CIP

Contents

Preface

This is the result, after almost two decades of research, of bringing up-to-date and recapitulating (first for one dimension) my little book *Spline Algorithms for Curves and Surfaces*, of which almost five thousand copies in four editions have been sold, and which has also been translated into English.

Our intention, as it was previously, is to provide an elementary and directly applicable introduction to the computation of those (as simple as possible) spline functions, which are determined by the requirement of smooth and shape-preserving interpolation and (in two cases) the smoothing of measured or collected data.

By elementary, we mean in particular that we have chosen to give explicit and easily evaluated forms of the spline interpolants (instead of in terms of recursive B-splines) and that in general existence and uniqueness can be decided, since we can demonstrate strict diagonal dominance of tridiagonal and (in two cases) five-diagonal coefficient matrices of linear systems of equations in appropriate unknowns.

This book should also be useful for applications, since not only do we derive the formulas and algorithms as such, but we also give efficient Fortran-77 subroutines. These are used to calculate numerous examples that in turn allow the reader to assess how the various spline interpolants perform depending on the configuration of the data.

Since the earlier book, much is new, especially reasearch that has appeared in the literature in the last two decades. In this regard, local

Hermite quadratic and cubic C^1-splines should especially be mentioned. For the required purposes, these seem to be superior to other polynomial spline interpolants. Also, the numerous variants of simplest possible rational spline interpolants are especially emphasized.

Just as for my last book, *Mathematical Software for Linear Regression* (1987), the implementation of most, and the testing of all, the subroutines was carried out by Mr. Jörg Meier (Dipl. Math.), scientific assistant at the Department of Mathematics; without him this book would not have been possible. Students J. Haschen, R. Obst, and A. Stark contributed to the literature searches as well as to various preliminary studies. The non-trivial task of text preparation was carried out with care and patience by Mrs. Büsselmann, also of the department.

Oldenburg, May 1989   H. Späth

Preface to the English Edition

A number of typographical errors and small discrepancies have been corrected from the German edition. Most of these were discovered by Prof. Len Bos during the translation, which could not have been carried out in a more congenial manner. Many thanks! The handling of publication matters, in this case by Alice Peters, was very supportive and extremely reliable.

H. Späth

1

Polynomial Interpolation

1.1. The Lagrange Form of the Interpolating Polynomial

Suppose that we are given n points (x_k, y_k), $k = 1, \cdots, n$ with pairwise distinct x_k. Equivalently, by renumbering if necessary, we may assume that $x_1 < \cdots < x_n$. Then there is a unique polynomial, p_{n-1}, of degree $n - 1$, which interpolates this data. Indeed, the *Lagrange form* of p_{n-1} may be explicitly given by means of the *fundamental polynomials* or *cardinal functions*, L_i, defined by

$$L_i(x) = \prod_{\substack{k=1 \\ k \neq i}}^{n} \frac{x - x_k}{x_i - x_k}. \tag{1.1}$$

Some plotted examples of such L_i are given, for example, in [121], p. 83. Then, since $L_i(x_k) = \delta_{ik}$, we have

$$p_{n-1}(x) = \sum_{i=1}^{n} y_i L_i(x). \tag{1.2}$$

This representation requires $O(n^2)$ arithmetic operations for each evaluation of p_{n-1}. A more economical and also numerically more stable form can be obtained as follows. Set

$$\lambda_i = \prod_{\substack{k=1 \\ k \neq i}}^{n} \frac{1}{x_i - x_k}. \tag{1.3}$$

Notice that these factors are independent of the point of evaluation, v, and thus need only be computed once. Also, the special case of (1.2) with $y_i = 1$, $i = 1, \cdots, n$, yields the relation,

$$\sum_{i=1}^{n} L_i(x) = 1.$$

Together, these may be used to rewrite (1.2) in the so-called *barycentric representation*,

$$p_{n-1}(x) = \frac{\sum_{i=1}^{n} y_i \frac{\lambda_i}{x - x_i}}{\sum_{i=1}^{n} \frac{\lambda_i}{x - x_i}}, \tag{1.4}$$

of the Lagrange interpolating polynomial. This formula is well-defined for $x \neq x_i$ and may be extended continuously by setting $p_{n-1}(x_i) := y_i$, $i = 1, \cdots, n$. Using (1.4) requires only a further $O(n)$ operations per evaluation. For numerical reasons ([171]) it is good policy to renumber the *interpolation nodes* x_i so that

$$|\bar{x} - x_1| \geq |\bar{x} - x_2| \geq \cdots \geq |\bar{x} - x_n| \tag{1.5}$$

holds, where $\bar{x} = \frac{1}{n} \sum_{i=1}^{n} x_i$. Since the values (1.3) are independent of the y_i, the barycentric representation is especially recommended when several polynomials with different y_i but the same nodes x_i are to be evaluated.

1.2. The Newton Form of the Interpolating Polynomial

If this is not the case, or if the intention is to add to the number of given points one by one, then the Newton form of the interpolating polynomial is preferred. We write

$$\begin{aligned} p_{n-1}(x) \quad = \quad & a_1 + a_2(x - x_1) + a_3(x - x_1)(x - x_2) + \\ & \cdots + a_n(x - x_1)(x - x_2) \cdots (x - x_{n-1}), \end{aligned} \tag{1.6}$$

where the coefficients a_k are denoted by

$$a_k = f[x_1, x_2, \cdots, x_k]. \tag{1.7}$$

```
      SUBROUTINE NEWDIA(N,X,Y,A,IFLAG)
      DIMENSION X(N),Y(N),A(N)
      IFLAG=0
      IF (N.LT.1) THEN
          IFLAG=1
          RETURN
      END IF
      DO 10 I=1,N
          A(I)=Y(I)
10    CONTINUE
      DO 30 K=N,2,-1
          DO 20 I=K,N
          A(I)=(A(I)-A(I-1))/(X(I)-X(K-1))
20        CONTINUE
30    CONTINUE
      RETURN
      END
```

Calling sequence:

CALL NEWDIA(N,X,Y,A,IFLAG)

Purpose:
The determination of the coefficients of the Newton interpolating polynomial of degree N−1.

Description of the parameters:

N   Number of given points. N must be at least 1.
X   ARRAY(N): Upon calling must contain the abscissas
                    $x_k, k = 1, \cdots, n$, with $x_i \neq x_j$ for $i \neq j$.
Y   ARRAY(N): Upon calling must contain the
                    ordinates $y_k, k = 1, \cdots, n$.
A   ARRAY(N): Upon successful execution (IFLAG=0)
                    contains the required polynomial coefficients.
IFLAG   =0:   Normal execution.
        =1:   N< 1 not permitted.

Remark: The difference scheme is worked out in a diagonal fashion.

Figure 1.1. Subroutine NEWDIA and its description.

```
SUBROUTINE NEWSOL(N,X,A,T,F,IFLAG)
DIMENSION X(N),A(N)
IFLAG=0
IF (N.LT.1) THEN
      IFLAG=1
      RETURN
END IF
F=A(N)
DO 10 K=N-1,1,-1
      F=F*(T-X(K))+A(K)
10    CONTINUE
RETURN
END
```

Calling sequence:

CALL NEWSOL(N,X,A,T,F,IFLAG)

Purpose:
The calculation of the function value of the Newton interpolating polynomial at the point T.

Description of the parameters:

N	Number of given points. $N \geq 1$ is required.	
X	ARRAY(N):	Upon calling must contain the abscissas x_k, $k = 1, \cdots, n$, with $x_i \neq x_j$ for $i \neq j$.
A	ARRAY(N):	Upon calling must contain the polynomial coefficients $a_1, a_2, \cdots, a_n$.
T	Point at which the polynomial is to be evaluated.	
F	Value of the polynomial at the point T.	
IFLAG	=0:	Normal execution.
	=1:	N< 1 not permitted.

Figure 1.2. Subroutine NEWSOL and its description.

The *divided differences* $f[x_1, x_2, \cdots, x_n]$ may be calculated recursively
from

$$f[x_1, x_2, \cdots, x_n] = \frac{f[x_2, \cdots, x_k] - f[x_1, \cdots, x_{k-1}]}{x_k - x_1}, \qquad (1.8)$$

where $f[x_i] := y_i$, $i = 1, \cdots, n$. (A nice, expository derivation may be found,
for example, in [8].) According to [164], it is recommended for numerical
reasons that the x_i be renumbered so that

$$|v - x_1| \leq |v - x_2| \leq \cdots \leq |v - x_n|, \qquad (1.9)$$

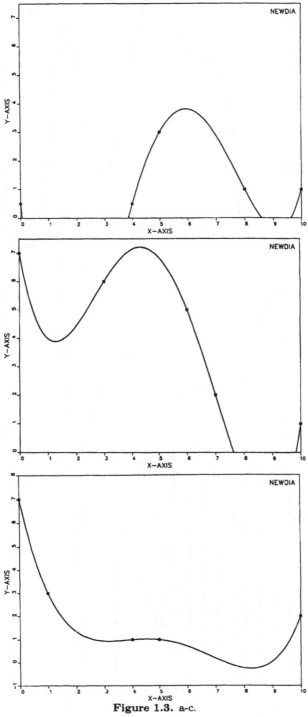

Figure 1.3. a-c.

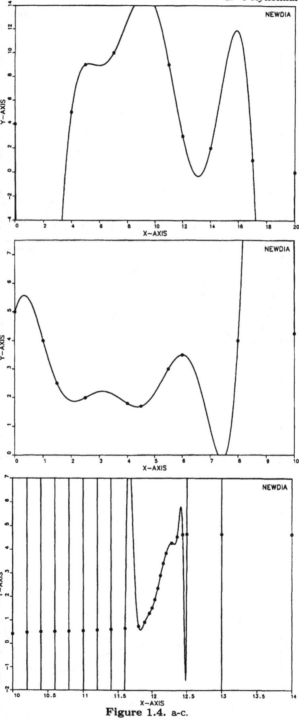

Figure 1.4. a-c.

where v is the point of evaluation. The Newton form of the interpolating polynomial may be efficiently evaluated by means of Horner's rule; i.e.,

$$
\begin{aligned}
p_{n-1}(x) &= a_1 + (x - x_1)(a_2 + (x - x_2)(a_3 + \cdots (x - x_{n-1})a_n)\cdots) \\
&= (\cdots((a_n(x - x_{n-1}) + a_{n-1})(x - x_{n-2}) + a_{n-2}\cdots) \\
&\quad \times (x - x_1) + a_1.
\end{aligned}
\tag{1.10}
$$

By this method, the Newton form without the renumbering is about half as expensive ([171]) as the Lagrange form without the renumbering (1.5).

Although the rearrangement of the given points for reasons of numerical stability, (1.5), is independent of the point of evaluation, while that for the Newton form, (1.9), is not, it is still in general preferable to use the representation (1.10) with the calculation of the coefficients by means of (1.8) (without renumbering). Therefore, we only give the subroutines NEWDIA (Fig. 1.1) for the calculation of divided differences and NEWSOL (Fig. 1.2) for the evaluation by a polynomial by Horner's rule, (1.10). NEWDIA uses (1.8) and the diagonal scheme of [121].

These routines do not involve the renumbering of (1.9). In general, this is not worthwhile, as interpolating polynomials of higher degree ($n \geq 4$), where it might be relevant, are not recommended for other reasons. This will be clear from the examples computed with NEWDIA and NEWSOL given in this section. In each of the first three, five points were given. These, together with the corresponding polynomial interpolants of fourth degree, are shown in Figs. 1.3a, b and c. This sequence of plots shows that polynomial interpolation preserves neither positivity nor monotonicity nor convexity of the data. In contrast, a simple *polygonal path* does possess these *shape-preserving* properties. The examples of Figs. 1.4a ($n = 9$) and b ($n = 10$) are taken from [154, pp. 31 and 105]. We will often encounter them later. Finally, the example of Fig. 1.4c ([121, p. 109]), involving 24 points, shows an interpolating polynomial of degree 23, which is completely ill-behaved. Although in special cases higher-degree polynomial interpolants can be useful, in general the results are such that this type of interpolation is not, in practice, applicable.

2

Polygonal Paths as Linear Spline Interpolants

2.1. General Spline Interpolants

For the given abscissas, we will now almost always suppose that

$$x_1 < x_2 < \cdots < x_n. \tag{2.1}$$

In general, by a spline interpolant $s \in C^m[x_1, x_n]$ with knots x_k, we mean a set of $n - 1$ functions, s_k, defined on $[x_k, x_{k+1}]$, respectively, $k = 1, \cdots, n - 1$, that are stitched together so as to be m-times $(m \geq 0)$ continuously differentiable at the knots and that satisfy the interpolation conditions,

$$s_k(x_k) = y_k, \quad s_k(x_{k+1}) = y_{k+1}, \quad k = 1, \cdots, n - 1. \tag{2.2}$$

For a polygonal path through the points (x_k, y_k), $k = 1, \ldots, n$, we have $m = 0$ and the s_k are all line segments with endpoints (x_i, y_i), $i = k, k + 1$. For $m = 1$, we will be connecting parabolic segments, and for $m = 2$, cubic polynomials as well as other functions. As we shall see with polynomials of degree five and $m = 4$, $m > 2$ is in general unsuitable, since, as we saw in Chapter 1, the unacceptable properties of polynomials of higher degrees again take effect. One could, in principle, choose a different function type on each interval for s_k, but we avoid this for practical considerations.

2.2.   Various Representations of a Polygonal Path

The line segments s_k can be represented in a number of ways. For example,

$$s_k(x) \quad = \quad A_k + B_k x, \tag{2.3}$$

$$s_k(x) \quad = \quad A_k + B_k(x - x_k), \tag{2.4}$$

$$s_k(x) \quad = \quad A_k + B_k t, \tag{2.5}$$

$$s_k(x) \quad = \quad A_k u + B_k t \tag{2.6}$$

are all posssible. Here,

$$t \quad = \quad \frac{x - x_k}{h_k}, \quad h_k = \Delta x_k = x_{k+1} - x_k,$$

$$u \quad = \quad 1 - t = \frac{x_{k+1} - x}{h_k}. \tag{2.7}$$

For each form, we may obtain the values of the corresponding $2(n-1)$ parameters A_k and B_k, $k = 1, \cdots, n-1$ from the interpolation conditions (2.2). The computational expense differs in each case. For the forms (2.4), (2.5), and (2.6), it follows immediately from $s_k(x_k) = y_k$ that $A_k = y_k$. For (2.4), it follows from $s_k(x_{k+1}) = y_{k+1}$ that the slope, B_k, of s_k is given by $B_k = \Delta y_k / \Delta x_k$.

The form (2.6) appears to be the most elegant, as then $B_k = y_{k+1}$ and thus,

$$s_k = y_k u + y_{k+1} t. \tag{2.8}$$

Hence, other than the given data (x_k, y_k), $k = 1, \cdots, n$, no new parameters are introduced and consequently no additional storage locations are required. (Note, however, that for (2.4), the y_k could be also be overwritten by the B_k.) The form (2.3) is unsuitable, as x does not vary intrinsically with respect to the interval $[x_k, x_{k+1}]$. In (2.4), $x - x_k$ varies from 0 to Δx_k, and in (2.5), t varies from 0 to 1 (standardized interval length).

Up till now, we have discussed the piecewise representation of the polygonal path, s. It is reasonable to ask if there is also a closed-form representation that holds on all of $[x_1, x_n]$? We introduce the notation,

$$(x - x_i)_+ = \begin{cases} x - x_i & \text{for } x \geq x_i \\ 0 & \text{otherwise} \end{cases}. \tag{2.9}$$

If we set

$$s(x) = \alpha + \beta_1(x - x_1) + \sum_{i=2}^{n-1} \beta_i(x - x_i)_+, \tag{2.10}$$

then clearly, s restricts to a linear polynomial on each of the intervals $[x_k, x_{k+1}]$, $k = 1, \ldots, n-1$, namely,

$$s_k(x) = \alpha + \sum_{i=1}^{k} \beta_i (x - x_i).$$

The parameters $\alpha, \beta_1, \cdots, \beta_{n-1}$ can be successively calculated from the interpolation conditions (2.2). The representation (2.10) has the advantage over the previous forms that it is not necessary to always first determine in which interval the point of evaluation, v, lies. The disadvantage is that the computation of the $\beta_2, \cdots, \beta_{n-1}$ and the evaluation itself are both expensive.

A representation analogous to the Lagrange form of the interpolating polynomial is obtained from the introduction of the so-called *B-splines*. (B stands for basis). Those of first order are given by

$$N_i(x) = \begin{cases} 0 & \text{for } x \leq x_{i-1} \\ \frac{x - x_{i-1}}{\Delta x_{i-1}} & \text{for } x_{i-1} \leq x \leq x_i \\ \frac{x_{i+1} - x}{\Delta x_i} & \text{for } x_i \leq x \leq x_{i+1} \\ 0 & \text{for } x \geq x_{i+1} \end{cases} \qquad (2.11)$$

The points $x_0 < x_1$ and $x_{n+1} > x_n$ are otherwise arbitrary. Then clearly,

$$s(x) = \sum_{i=1}^{n} y_i N_i(x) \qquad (2.12)$$

is also a representation of the interpolating polygonal path, since s restricts to a linear on $[x_k, x_{k+1}]$ and it also satisfies the interpolation conditions. Further, from the definition (2.11), only N_k and N_{k+1} differ from zero on $[x_k, x_{k+1}]$. Hence,

$$\begin{aligned} s_k(x) &= y_k N_k(x) + y_{k+1} N_{k+1}(x) \\ &= y_k u + y_{k+1} t, \end{aligned}$$

and we recover the form (2.6). B-splines ([19,20]), for $m > 2$, are an important and indispensable tool for data smoothing ([67]) in one and two variables as well as for free-form curves and surfaces ([37]). For spline interpolation with $m \leq 2$, they are, however, in the case of polynomial segments, too complicated, and for non-polynomial segments, only explicitly available in exceptional cases. Hence, we will not pursue them further.

```
      SUBROUTINE INTONE(X,N,V,I,IFLAG)
      DIMENSION X(N)
      IFLAG=0
      IF (I.GE.N) I=1
      IF (V.LT.X(1).OR.V.GT.X(N)) THEN
          IFLAG=3
          RETURN
      END IF
      IF (V.LT.X(I)) GOTO 10
      IF (V.LE.X(I+1)) RETURN
      L=N
      GOTO 30
10    L=I
      I=1
20    K=(I+L)/2
      IF (V.LT.X(K)) THEN
          L=K
      ELSE
          I=K
      END IF
30    IF (L.GT.I+1) GOTO 20
      RETURN
      END
```

Calling sequence:

CALL INTONE(X,N,V,I,IFLAG)

Purpose:
Determination of an index I with $X(I) \leq V \leq X(I+1)$. $X(1) < X(2) < \cdots < X(N)$ is required.

Description of the parameters:

X	ARRAY(N): Abscissas of the given points.	
N	Number of given X-values.	
V	Abscissa of the point at which the spline function is to be evaluated.	
I	Input:	Upon calling I must contain a value between 1 and $n-1$.
	Output:	I with $X(I) \leq V \leq X(I+1)$.
IFLAG	=0:	Normal execution.
	=3:	$V < X(1)$ and $V > X(N)$ not allowed.

Figure 2.1. Subroutine INTONE and its description.

```
FUNCTION POLVAL(N,X,Y,V,IFLAG)
DIMENSION X(N),Y(N)
DATA I/1/
IFLAG=0
IF (N.LT.2) THEN
   IFLAG=1
   RETURN
END IF
CALL INTONE(X,N,V,I,IFLAG)
IF (IFLAG.NE.0) RETURN
XI=X(I)
T=(V-XI)/(X(I+1)-XI)
POLVAL=Y(I+1)*T+Y(I)*(1.-T)
RETURN
END
```

FUNCTION POLVAL(N,X,Y,V,IFLAG)

Purpose:
POLVAL is a FUNCTION subprogram for the calculation of a function
value of a polygonal path at a point $V \in [X(1),X(N)]$.

Description of the parameters:

N	Number of given points.
X	ARRAY(N): Abscissas.
Y	ARRAY(N): Ordinates.
V	Point at which the function is to be evaluated.
IFLAG	=0: Normal execution.
	=1: $N \geq 2$ is required.
	=3: Error in the interval determination (INTONE).

Required subroutines: INTONE.

Remark: The statement 'DATA I/1/' has the effect that I
is set to 1 at the first call to POLVAL.

Figure 2.2. Function POLVAL and its description.

2.3. Evaluation by Searching an Ordered List

As for all spline interpolants, before (2.4) or (2.6) can be used to evaluate a polygonal path $s(v)$ at an abscissa $v \in [x_1, x_n]$, there first arises the problem of finding that index i for which $x_i \leq v \leq x_{i+1}$. For this there are several solutions of varying efficiency ([63,109]). Here, we proceed from the reasonable assumption that function values at a monotonically increasing sequence of absicissas $v_1 < v_2 < \cdots < v_{\tilde{n}}$ are to be calculated. If $\tilde{n}$ is substantially larger than n, then it will frequently be the case that in passing from v_j to v_{j+1}, the new abscissa will lie in the same interval, with index i, as did v_j. Thus, we initialize $i = 1$ and store the (possibly changed) index i of the interval in which the last v_j lay. If v_{j+1} is not in $[x_i, x_{i+1}]$, then the new i is found by a binary search on $[x_1, x_i]$ if $v_{j+1} < x_i$ and on $[x_i, x_n]$ if $v_{j+1} \geq x_i$ (the regular case). This procedure is implemented by the subroutine INTONE of Fig. 2.1; the program description is also found in Fig. 2.1. The function POLVAL (Fig. 2.2) evaluates a polygonal path by making use of INTONE. An example showing the polygonal path interpolant to the data of Fig. 1.4c is given in Fig. 2.3. Although the "curve" appears to be very smooth, this is deceiving as there are jumps in the first derivative at the nodes. If the abscissas can be chosen to be sufficiently close together, however, then these jumps become arbitrarily small. This fact is at the heart of any plotter software.

2.4. Properties of Polygonal Paths

As opposed to spline interpolants of higher degrees, the polygonal path s has notable shape-preserving properties. Positivity in the data is preserved: if $y_k > 0$, $k = 1, \cdots, n$, then $s > 0$ on $[x_1, x_n]$. Monotonicity is also preserved: if for instance $y_1 < y_2 < \cdots < y_n$, then since $s'_k = \frac{\Delta y_k}{\Delta x_k} > 0$, $s' > 0$ on the whole interval $[x_1, x_n]$. Denote by d_k the slopes

$$d_k = \frac{\Delta y_k}{\Delta x_k}, \quad k = 1, \cdots, n - 1. \tag{2.13}$$

If the given data is convex in the sense that $d_1 < d_2 < \cdots < d_{n-1}$, then since $s'' = 0$, the polygonal path is also convex. Examples of these three properties are obtained from Figs. 1.3a-c by connecting the points by line segments. Now if $y_k = f(x_k)$, $k = 1, \cdots, n$, for $f \in C^1[x_1, x_n]$, then the

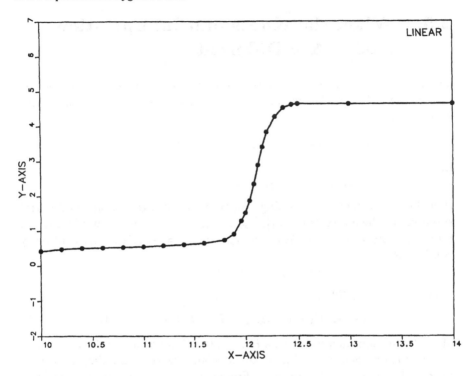

Figure 2.3.

interpolating polygonal path s has the property that

$$\int_{x_1}^{x_n} [s'(x)]^2 dx \le \int_{x_1}^{x_n} [f'(x)]^2 dx, \qquad (2.14)$$

i.e., among all interpolating functions, the polygonal path minimizes the aggregate slope squared. For the proof, we show that in

$$0 \le \int_{x_1}^{x_n} (f' - s')^2 dx = \int_{x_1}^{x_n} (f')^2 dx - 2\int_{x_1}^{x_n} (f' - s')s'dx - \int_{x_1}^{x_n} (s')^2 dx,$$

the middle term on the right disappears. In fact,

$$\int_{x_1}^{x_n} (f' - s')s'dx = \sum_{k=1}^{n-1} \int_{x_k}^{x_{k+1}} (f' - s')s'dx$$

$$= \sum_{k=1}^{n-1} \left[(f - s)s' \Big|_{x_k}^{x_{k+1}} - \int_{x_k}^{x_{k+1}} (f - s)s''dx \right]$$

by integration by parts. The first term is zero, since $f(x_k) = y_k = s(x_k)$, and the second, since $s'' = 0$.

2.5.   When the Knots and Interpolation Nodes Are Different

The interpolating line segments need not necessarily be joined continuously just at the nodes x_k. We could also use parameters $\alpha_k \in (0,1)$ and define knots z_k by

$$z_k = \alpha_k x_k + (1 - \alpha_k) x_{k+1}, \quad k = 1, \cdots, n - 1.$$

Let

$$s_k(x) = y_k + B_k(x - x_k), \quad k = 1, \cdots, n,$$

then be n line segments passing respectively through (x_k, y_k) and corresponding to the intervals $[x_1, z_1], [z_k, z_{k+1}], k = 1, \cdots, n-2$, and $[z_{n-1}, x_n]$. Since $z_k - x_k = (1 - \alpha_k) \Delta x_k$ and $z_k - x_{k+1} = -\alpha_k \Delta x_k$, the continuity conditions

$$s_k(z_k) = s_{k+1}(z_k), \quad k = 1, \cdots, n - 1,$$

yield the linear system,

$$(1 - \alpha_k) B_k + \alpha_k B_{k+1} = d_k, \quad k = 1, \cdots, n - 1, \qquad (2.15)$$

of $n-1$ equations in n unknowns $B_1, \cdots, B_n$. If, for example, B_1 is specified ahead of time, then it is uniquely solvable and the corresponding polygonal path is thus uniquely determined. From experience, the appearance of this interpolant depends strongly on the choice of B_1.

This dependency can be eliminated in the case of an odd number of points $n = 2m + 1$ and symmetric data, i.e.,

$$y_k = y_{n+1-k}, \Delta x_k = \Delta x_{n-k}, \alpha_k = 1 - \alpha_{n-k}, \quad k = 1, \cdots, m,$$

by the reasonable requirement that $B_1 = -B_n$. The system (2.15) becomes

$$\begin{aligned}
\alpha_1 B_2 \qquad\qquad\qquad -(1 - \alpha_1) B_n \quad &= d_1 \\
& \quad \vdots \\
(1 - \alpha_m) B_m \quad +\alpha_m B_{m+1} \qquad\qquad\qquad &= d_m \quad . \\
& \quad \vdots \\
\alpha_1 B_{n-1} \quad +(1 - \alpha_1) B_n \quad &= -d_1
\end{aligned}$$

By adding the first and last equations, we see that $B_2 = -B_{n-1}$. Then, using this in the addition of the second and second from last equations, we obtain $B_3 = -B_{n-2}$. Continuing in this way, we finally obtain that $B_m = -B_{m+2}$ and from the addition of the mth and $(m + 1)$st equations, that $B_{m+1} = 0$. Thus, starting with the mth equation, we may successively compute

$$B_k = -B_{n+1-k} = \frac{d_k - \alpha_k B_{k+1}}{1 - \alpha_k}, \quad k = m, m - 1, \cdots, 2.$$

```
      SUBROUTINE POLSYM(N,X,Y,W,B,IFLAG)
      DIMENSION X(N),Y(N),W(N),B(N)
      IFLAG=1
      IF (MOD(N,2).EQ.0.OR.N.LT.3) RETURN
      IFLAG=0
      M=(N-1)/2
      B(M+1)=0.
      DO 10 K=M,1,-1
         K1=K+1
         B(K)=((Y(K1)-Y(K))/(X(K1)-X(K))-W(K)*B(K1))/(1.-W(K))
         B(N-K+1)=-B(K)
10    CONTINUE
      RETURN
      END
```

Calling sequence:

CALL POLSYM(N,X,Y,W,B,IFLAG)

Purpose:

Calculation of a polygonal path with knots differing from interpolation nodes for data symmetric with respect to the y-axis.

Description of the parameters:

N	Number of given points n. N must be odd.
X	ARRAY(N): Vector of abscissas.
Y	ARRAY(N): Vector of ordinates.
W	ARRAY(N): Upon calling must contain the parameters α_k, $k = 1, \cdots, n-1$, with $0 < \alpha_k < 1$.
B	ARRAY(N): Upon completion with IFLAG=0 contains the slopes of the required polygonal path.
IFLAG =0:	Normal execution.
=1:	N odd and N$\geq$3 are required.

Figure 2.4. Subroutine POLSYM and its description.

This procedure is implemented in the subroutine POLSYM (Fig. 2.4). An example with $\alpha_k = 1/2$, $k = 1, \cdots, 5$, is given in Fig. 2.5.

2.6. Parametric Polygonal Paths

Suppose for the moment that the general assumption (2.1) does not hold and that the numbering of the points is to correspond to the order in which

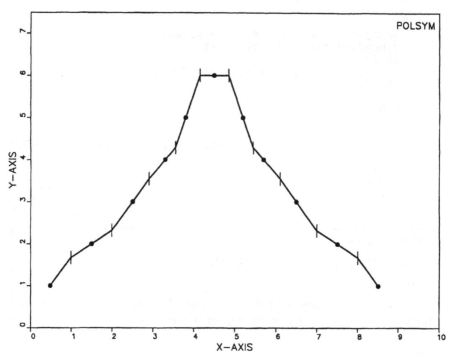

Figure 2.5.

the interpolant is to pass through them. Then, in general, the corresponding polygonal path in the plane can no longer be described by a function, and so, instead, we will make use of the parametric representation of a curve. Choose $v_1 < v_2 < \cdots < v_n$ arbitrarily and set

$$s_k(v) = \begin{cases} \xi(v) = x_k + \frac{\Delta x_k}{\Delta v_k}(v - v_k) \\ \eta(v) = y_k + \frac{\Delta y_k}{\Delta v_k}(v - v_k) \end{cases} \qquad (2.16)$$

in the interval $[v_k, v_{k+1}]$, $k = 1, \cdots, n - 1$. Then, as desired,

$$s_k(v_k) = \begin{pmatrix} x_k \\ y_k \end{pmatrix}$$

and

$$s_k(v_{k+1}) = \begin{pmatrix} x_{k+1} \\ y_{k+1} \end{pmatrix}.$$

Moreover, s_k does represent the straight line between these two points, since we may eliminate $(v - v_k)/\Delta v_k$ from the equations,

$$\xi - x_k = \frac{\Delta x_k}{\Delta v_k}(v - v_k),$$

$$\eta - y_k = \frac{\Delta y_k}{\Delta v_k}(v - v_k),$$

to obtain the usual equation of a line,

$$\frac{\eta - y_k}{\xi - x_k} = \frac{\Delta y_k}{\Delta x_k},$$

in case $\Delta x_k \neq 0$, or its reciprocal if $\Delta y_k \neq 0$. If $(x_1, y_1) = (x_n, y_n)$, then we obtain in this manner a closed polygonal path.

The magnitude of Δv_k has, in the case of the polgonal path, no effect on the appearance of the curve. It is suggested that one choose the v_k as the cumulative arclength along the curve, i.e,

$$v_1 = 0, \quad v_{k+1} = v_k + \sqrt{\Delta x_k^2 + \Delta y_k^2}, \quad k = 1, \cdots, n - 1. \qquad (2.17)$$

Here, as well as in the general case, this is called the canonical parameterization of the curve. If the curve segments are not straight lines, as we will be using later, then in general the arclength cannot be computed explicitly but (2.17) is the basis of a first approximation.

2.7. Smoothing with Polygonal Paths I

We now again assume that (2.1) holds. Further, we suppose that there are measurement errors in the y_k so that the desired

$$s_k(x) = A_k + B_k(x - x_k), \quad k = 1, \cdots, n - 1, \qquad (2.18)$$

is not to pass through the given points (x_k, y_k) themselves but through points (x_k, A_k) with yet to be determined "exact" ordinates A_k, $k = 1, \cdots, n$. In order to be as flexible as possible in the choice of these A_k, we introduce variable control parameters p_k and ask that the differences in the ordinates be proportional to the jumps in the first derivative of the polygonal path, i.e.,

$$p_k(A_k - y_k) = B_k - B_{k-1}, \quad k = 1, \cdots, n. \qquad (2.19)$$

Here we have set $B_0 = B_n = 0$. (A similar requirement is made for cubic splines in [153]; we will return to this later. More precise reasons for this type of model can be motivated by the theory of nonlinear optimization ([163]).)

The interpolation conditions (2.2) yield

$$A_k + B_k \Delta x_k = A_{k+1}, \quad k = 1, \cdots, n - 1.$$

Solving these for B_k and substituting in (2.19) with $B_0 = B_n = 0$ gives the linear system of equations,

$$M \cdot A = PY, \tag{2.20}$$

where

$$M = \begin{bmatrix} p_1 + \frac{1}{h_1} & -\frac{1}{h_1} & & & \\ -\frac{1}{h_1} & p_2 + \frac{1}{h_1} + \frac{1}{h_2} & -\frac{1}{h_2} & & \\ & -\frac{1}{h_2} & p_3 + \frac{1}{h_2} + \frac{1}{h_3} & -\frac{1}{h_3} & \\ & & & \ddots & \\ & & & -\frac{1}{h_{n-1}} & p_n + \frac{1}{h_{n-1}} \end{bmatrix},$$

$$A = \begin{bmatrix} A_1 \\ A_2 \\ A_3 \\ \cdot \\ A_n \end{bmatrix}, \quad PY = \begin{bmatrix} p_1 y_1 \\ p_2 y_2 \\ p_3 y_3 \\ \cdot \\ p_n y_n \end{bmatrix}.$$

The $n \times n$ coefficient matrix M is symmetric and, for $p_k > 0$, strictly diagonally dominant. Hence, the system of equations is always uniquely solvable for arbitrary control parameters $p_k > 0$. In this case then, the corresponding smoothing polygonal path also exists and is unique. (We will often show the existence of spline interpolants by an argument of strict diagonal dominance. The reader not confident with this material should consult either the appendix or an appropriate textbook.)

The subroutine POLSM1 (Figs. 2.6 and 2.7) sets up the linear system (2.20) for given $p_k > 0$ and obtains the solution by calling the symmetric tridiagonal matrix solver TRIDIS (see the appendix). (TRIDIS does not use pivoting, as this is not necessary for strictly diagonally dominant matrices ([8]).) Examples computed with POLSM1 are illustrated in Figs. 2.8a and b. In Fig. 2.8a the control parameters p_k were chosen to be $p_k = 1$, $k = 1, \cdots, 7$, and in 2.8b, $p_1 = p_3 = 10$, $p_2 = p_4 = 1$, $p_5 = p_6 = p_7 = 5$. In the limit as $p_k \to \infty$, $A_k = y_k$, and thus we recover the interpolating polygonal path. This can be seen by dividing the kth row of (2.20) by p_k and then passing to the limit.

2.8.   Smoothing with Polygonal Paths II

In statistics, there arises the problem ([39, 69]) of fitting a polygonal path with prescribed knots x_k, $k = 1, \cdots, n \geq 2$, to a set of points (u_i, v_i), $i = 1, \cdots, m \geq 3$, in the sense of least squares. Typically, m is substantially larger than n. In this, the assumptions that (2.1) holds and that $x_1 \leq u_i \leq x_n$, $i = 1, \cdots, m$, are also made. Abscissas u_i and x_k could be the same.

Using the B-spline representation (2.12), we wish to determine y_k corresponding to x_k, $k = 1, \cdots, n$, which minimize

$$S(y_1, \cdots, y_n) = \sum_{i=1}^{m} \left[\sum_{k=1}^{n} y_k N_k(u_i) - v_i \right]^2 \qquad (2.21)$$

(see also [67], p. 71). The conditions necessary for a minimum of (2.21)

```
      SUBROUTINE POLSM1(N,X,Y,P,EPS,A,B,IFLAG,F,G)
      DIMENSION X(N),Y(N),P(N),A(N),B(N),F(N),G(N)
      IFLAG=0
      IF (N.LT.2) THEN
          IFLAG=1
          RETURN
      END IF
      H1=0.
      DO 10 K=1,N-1
          PK=P(K)
          IF (PK.LE.0.) THEN
              IFLAG=4
              RETURN
          END IF
          H2=1./(X(K+1)-X(K))
          B(K)=H2
          F(K)=PK+H1+H2
          G(K)=-H2
          A(K)=PK*Y(K)
          H1=H2
 10   CONTINUE
      F(N)=P(N)+H1
      A(N)=P(N)*Y(N)
      CALL TRIDIS(N,F,G,A,EPS,IFLAG)
      IF (IFLAG.NE.0) RETURN
      DO 20 K=1,N-1
          B(K)=(A(K+1)-A(K))*B(K)
 20   CONTINUE
      RETURN
      END
```

Figure 2.6. Program listing of POLSM1.

Calling sequence:

CALL POLSM1(N,X,Y,P,EPS,A,B,IFLAG,F,G)

Purpose:
Determination of the coefficients A_k and B_k of a smoothing linear spline function (knots same as nodes).

Description of the parameters:

N	Number of given points. $N \geq 2$ is required.
X	ARRAY(N): Upon calling must contain the abscissa values x_k, $k = 1, \cdots, n$, with $x_1 < x_2 < \cdots < x_n$.
Y	ARRAY(N): Upon calling must contain the ordinate values y_k, $k = 1, \cdots, n$.
P	ARRAY(N): Upon calling must contain the values of the weights p_k, $k = 1, \cdots, n$.
EPS	see TRIDIS.
A,B	ARRAY(N): Upon completion with IFLAG=0 contain the desired spline coefficients, $k = 1, \cdots, n$.
IFLAG	=0: Normal execution.
	=1: $N \geq 2$ is required.
	=2: Error in solving the system (TRIDIS).
	=3: $p_k > 0$ is required.
F,G	ARRAY(N): Work space.

Required subroutine: TRIDIS.

Figure 2.7. Description of Subroutine POLSM1.

are

$$\frac{\partial S}{\partial y_j} = 2 \sum_{i=1}^{m} N_j(u_i) \left[\sum_{k=1}^{n} N_k(u_i) y_k - v_i \right] = 0, \qquad (2.22)$$

which yield the linear system of equations,

$$\sum_{i=1}^{m} \sum_{k=1}^{n} N_j(u_i) N_k(u_i) y_k = \sum_{i=1}^{m} N_j(u_i) v_i, \quad j = 1, \cdots, n. \qquad (2.23)$$

Its coefficient matrix C can be written as

$$C = A^T A,$$

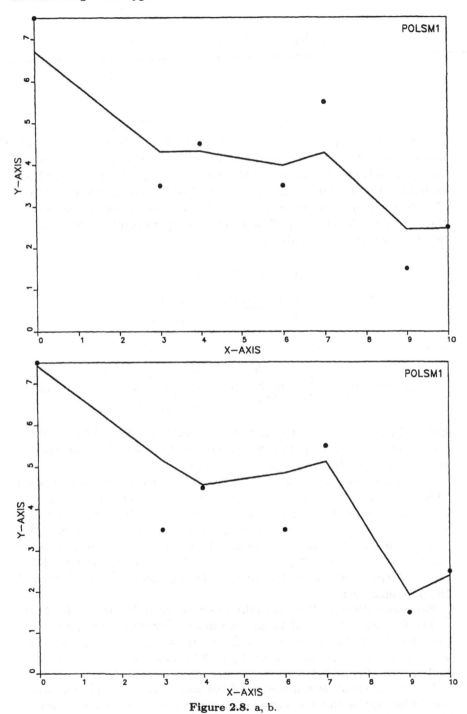

Figure 2.8. a, b.

with

$$A = \begin{bmatrix} N_1(u_1) & N_2(u_1) & \cdots & N_n(u_1) \\ N_1(u_2) & N_2(u_2) & \cdots & N_n(u_2) \\ \cdot & \cdot & \cdot & \cdot \\ \cdot & \cdot & \cdot & \cdot \\ \cdot & \cdot & \cdot & \cdot \\ N_1(u_m) & N_2(u_m) & \cdots & N_n(u_m) \end{bmatrix},$$

and so, in particular, it is symmetric. In order for C to be nonsingular, it is necessary that the rank of A be n, which necessarily presupposes $m \geq n$ and an appropriate distribution of the abscissas u_i and x_k. (Sufficient conditions do not seem to be known.) Further, since $N_k(u_i)N_{k+2}(u_i) = 0$ (see (2.11)), C is tridiagonal. The summands in the sums over i are determined by those elements of the diagonal and sub- and super-diagonals, which are in general nonzero. Specifically, these are

$$(N_k(u_i))^2 = \begin{cases} \left(\frac{u_i - x_{k-1}}{\Delta x_{k-1}}\right)^2 & \text{for } x_{k-1} \leq u_i \leq x_k \\ \left(\frac{x_{k+1} - u_i}{\Delta x_k}\right)^2 & \text{for } x_k \leq u_i \leq x_{k+1} \\ 0 & \text{otherwise} \end{cases} \qquad (2.24)$$

and

$$N_k(u_i)N_{k+1}(u_i) = \begin{cases} 0 & \text{for } x_{k-1} \leq u_i \leq x_k \\ \frac{x_{k+1} - u_i}{\Delta x_k}\frac{u_i - x_k}{\Delta x_k} & \text{for } x_k \leq u_i \leq x_{k+1} \\ 0 & \text{for } x_{k+1} \leq u_i \leq x_{k+2} \end{cases} \qquad . \qquad (2.25)$$

Evidently, the elements of C are also nonnegative. For arbitrary distributions of the u_i and x_k, this system of equations cannot, unfortunately, easily seen to be strictly diagonally dominant.

The subroutine POLSM2 (Figs. 2.9 and 2.10) forms the linear system (2.23) by means of (2.24) and (2.25) and attempts to solve it with TRIDIS (thereby assuming that no pivoting is necessary). If TRIDIS is not able to run till completion (see its description), execution is terminated. This is never the case in examples of practical importance. The resulting $A_k = y_k$, $k = 1, \cdots, n$, and B_k, $k = 1, \cdots, n - 1$, are the coefficients of the representation (2.4), and so B-splines need not be involved in evaluation of the polygonal path.

Four examples with the same initial data are given in Figs. 2.11a–2.12b.

The following choices of knots were made. For all of them, $x_1 = u_1$, and then for 2.11a, $x_2 = u_9$, for 2.11b, $x_2 = u_4$, $x_3 = u_9$, for 2.12a, $x_2 = (u_6 + u_7)/2$, $x_3 = u_9$, and finally for 2.12b, $x_2 = u_4$, $x_3 = (u_6 + u_7)/2$, and $x_4 = u_9$. Figure 2.11a shows the smoothing straight line.

For the practical determination of the knots x_k, the number of which should be kept as small as possible for practical reasons, one can proceed

```
         SUBROUTINE POLSM2(M,N,U,V,X,EPS,A,B,IFLAG,F,G)
         DIMENSION U(M),V(M),X(N),A(N),B(N),F(N),G(N)
         IFLAG=0
         IF (M.LT.3.OR.N.LT.2.OR.M.LT.N) THEN
             IFLAG=1
             RETURN
         END IF
         DO 10 K=1,N-1
             B(K)=X(K+1)-X(K)
10       CONTINUE
         DO 30 K=1,N
             K1=K+1
             K2=K-1
             T=0.
             B2=0.
             RS=0.
             DO 20 I=1,M
                 R1=0.
                 R2=0.
                 UH=U(I)
                 IF (K.GT.1) THEN
                     IF (UH.GE.X(K2).AND.UH.LE.X(K)) THEN
                         R1=(UH-X(K2))/B(K2)
                     END IF
                 END IF
                 IF (K.LT.N) THEN
                     IF (UH.GE.X(K).AND.UH.LE.X(K1)) THEN
                         R1=(X(K1)-UH)/B(K)
                         R2=(UH-X(K))/B(K)
                     END IF
                 END IF
                 T=T+R1*R2
                 B2=B2+R1*R2
                 RS=RS+R1*V(I)
20           CONTINUE
             F(K)=T
             A(K)=RS
             IF (K.LT.N) G(K)=B2
30       CONTINUE
         CALL TRIDIS(N,F,G,A,EPS,IFLAG)
         IF (IFLAG.NE.0) RETURN
         DO 40 K=1,N-1
             B(K)=(A(K+1)-A(K))/B(K)
40       CONTINUE
         RETURN
         END
```

Figure 2.9. Program listing of POLSM2.

Calling sequence:

CALL POLSM2(M,N,U,V,X,EPS,A,B,IFLAG,F,G)

Purpose: Determination of a smoothing polygonal path with fewer knots than interpolation points.

Description of the parameters:

M	Number of given points.
N	Number of knots.
U	ARRAY(M): Upon calling must contain the abscissa values u_k, $k = 1, \cdots, m$, with $u_1 \leq u_2 \leq \cdots \leq u_m$.
V	ARRAY(M): Upon calling must contain the ordinate values v_k, $k = 1, \cdots, m$.
X	ARRAY(N): Upon calling must contain the values x_k, $k = 1, \cdots, n$.
EPS	see TRIDIS.
A,B	ARRAY(N): Upon execution with IFLAG=0 contain the desired spline coefficients, $k = 1, \cdots, n - 1$.
IFLAG	=0: Normal execution.
	=1: M$\geq$3 and N$\geq$2 and M$\geq$N are required.
	=2: Error in solving the linear system (TRIDIS).
F,G	ARRAY(N): Work space.

Required subroutines: TRIDIS.

Figure 2.10. Description of Subroutine POLSM2.

in a manner analogous to that for cubic splines ([58]). Initially, choose $n = 2$, and $x_1 = u_1$ and $x_2 = u_m$, then $n = 3$, with $x_1 = u_1$, $x_3 = u_m$, and x_2 chosen so that about the same number of abscissas u_i lie on either side of it. The interval $[x_1, x_2]$ or $[x_2, x_3]$ for which the sum of the squares of the errors is largest is then again so subdivided and so on until a prescribed maximum value of $n \leq m$ is attained.

It would be very difficult to fix n and determine the x_k so as to also minimize (2.20) in these variables. We will not consider such so-called *free knot* problems; such do not arise in the interpolation problems constituting our main object of study.

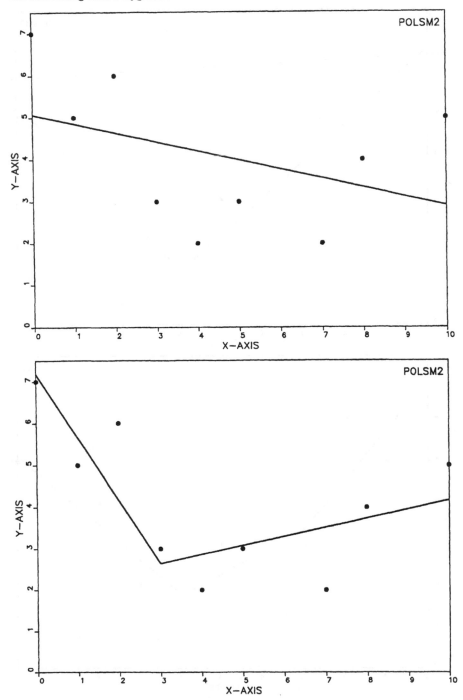

Figure 2.11. a, b.

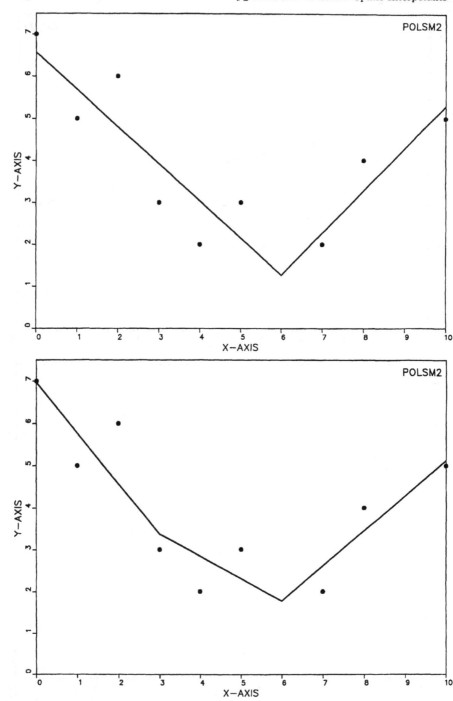

Figure 2.12. a, b.

3

Quadratic Spline Interpolants

3.1. Knots the Same as Nodes

Suppose once more that (2.1) holds. We wish now to join together parabolic segments,

$$s_k(x) = A_k + B_k(x - x_k) + C_k(x - x_k)^2, \quad k = 1, \cdots, n-1, \qquad (3.1)$$

at the nodes x_k so as to form a *once* continuously differentiable quadratic spline interpolant s. The interpolation conditions (2.2) become

$$\begin{aligned} A_k &= y_k, \\ A_k + h_k B_k + h_k^2 C_k &= y_{k+1}, \end{aligned} \qquad (3.2)$$

from which we obtain the relation,

$$B_k + h_k C_k = d_k, \quad k = 1, \cdots, n-1. \qquad (3.3)$$

Since

$$s_k'(x) = B_k + 2\, C_k(x - x_k), \qquad (3.4)$$

the C^1 conditions yield the equations,

$$B_{k-1} + 2\, h_{k-1} C_{k-1} = B_k, \quad k = 2, \cdots, n-1. \qquad (3.5)$$

Now solve (3.3) for C_k to obtain

$$C_k = \frac{1}{h_k}(d_k - B_k), \quad k = 1, \cdots, n-1. \tag{3.6}$$

Moreover, if we introduce the additional unknown B_n, then (3.5) holds also for $k = n$, and so substituting (3.6) into (3.5), we obtain ([87]) the linear system of equations,

$$B_{k-1} + B_k = 2\,d_{k-1}, \quad k = 2, \cdots, n, \tag{3.7}$$

for the determination of $B_1, \cdots, B_n$. Unfortunately, this system consists of n unknowns but only $n-1$ equations, and thus we require one extra condition. If, for example, we fix a value for B_1, then (3.7) can easily be solved recursively. It can be shown by complete induction that

$$B_k = (-1)^{k+1}B_1 + 2\sum_{j=1}^{k-1}(-1)^{k+j+1}d_j, \quad k = 2, \cdots, n. \tag{3.8}$$

The coefficients of (3.1) are then uniquely determined by (3.8), (3.2), and (3.6).

From (3.8), one readily sees a *shape-preserving* property of s ([87]). Suppose that $y_k \geq y_{k-1}$, $k = 2, \cdots, n$, and $d_k \geq d_{k-1}$, $k = 2, \cdots, n-1$, as well as that $0 \leq B_1 \leq 2\,d_1$. Then it follows that $B_k \geq 0$ for $k = 1, \cdots, n$. But then $s'(x)$ is continuous piecewise linear and, by (3.4), nonnegative at the x_k. Hence, $s'(x) \geq 0$ and we see that a certain kind of *monotonicity* is preserved. By substituting (3.8) in (3.6) and using the fact that $s_k''(x) = 2\,C_k$, we can obtain similar conditions for *convexity* preservation ([87]).

3.2.   Optimal Initial Slope

For the choice of the value B_1 of the slope at x_1, $B_1 = d_1$, for example, suggests itself. But we could also ask that B_1 be, in a certain sense, optimal. For example, the minimality property (2.14) of polygonal paths suggests that we distinguish an s among all quadratic spline interplolants by choosing B_1 to minimize

$$\tilde{F}_1 = \int_{x_1}^{x_n} [s'(x)]^2 dx.$$

We may calculate

$$\tilde{F}_1 = \sum_{k=1}^{n-1} \int_{x_k}^{x_{k+1}} [s_k'(x)]^2 dx$$

$$= \sum_{k=1}^{n-1} \int_{x_k}^{x_{k+1}} [B_k + 2C_k(x - x_k)]^2 dx$$

$$= \sum_{k=1}^{n-1} (B_k^2 h_k + 2B_k C_k h_k^2 + \frac{4}{3} C_k^2 h_k^3)$$

$$= \sum_{k=1}^{n-1} h_k (B_k^2 + 2B_k(d_k - B_k) + \frac{4}{3}(d_k - B_k)^2)$$

$$= \sum_{k=1}^{n-1} h_k (\frac{1}{3} B_k^2 - \frac{2}{3} B_k d_k + \frac{4}{3} d_k^2)$$

$$= \frac{1}{3} \sum_{k=1}^{n-1} h_k (B_k - d_k)^2 + \sum_{k=1}^{n-1} h_k d_k^2.$$

As the second term in the last line is independent of the B_k, it suffices to minimize

$$F_1(B_1, \cdots, B_{n-1}) = \frac{1}{3} \sum_{k=1}^{n-1} h_k (B_k - d_k)^2. \qquad (3.9)$$

A second possibility that offers itself is ([75])

$$F_2 = \int_{x_1}^{x_n} [s''(x)]^2 dx,$$

as the integral on the right side, as we shall see later, is minimized by certain cubic spline interpolants. It is easily seen that

$$F_2 = \sum_{k=1}^{n-1} \int_{x_k}^{x_{k+1}} [s_k''(x)]^2 dx$$

$$= 4 \sum_{k=1}^{n-1} h_k C_k^2,$$

and thus by (3.6),

$$F_2(B_1, \cdots, B_{n-1}) = 4 \sum_{k=1}^{n-1} \frac{1}{h_k} (B_k - d_k)^2. \qquad (3.10)$$

A third choice is to take a *convex combination*,

$$F_3(B_1, \cdots, B_{n-1}) = \lambda F_1(B_1, \cdots, B_{n-1}) + (1 - \lambda) F_2(B_1, \ldots, B_{n-1})$$

$$= \sum_{k=1}^{n-1} \left(\frac{\lambda}{3} h_k + \frac{4(1 - \lambda)}{h_k} \right) (B_k - d_k)^2, \qquad (3.11)$$

where $\lambda \in [0, 1]$ is a freely chosen parameter.

As a fourth possibility ([75]), we may consider the integral of the *curvature* squared,

$$
\begin{aligned}
\tilde{F}_4 &= \int_{x_1}^{x_n} \frac{[s''(x)]^2}{(1 + [s'(x)]^2)^3} \, dx \\
&= 4 \sum_{k=1}^{n-1} \int_{x_k}^{x_{k+1}} \frac{C_k^2}{(1 + [B_k + 2C_k(x - x_k)]^2)^3} \, dx.
\end{aligned}
$$

Although these integrals may be computed explicitly, they are to a high degree nonlinear in B_k and C_k. Thus, $\tilde{F}_4$ may only with difficulty be minimized. If, as suggested in [75], we approximate s'_k by d_k, then we obtain the approximation,

$$
F_4 = 4 \sum_{k=1}^{n-1} \frac{h_k}{[1 + d_k^2]^3} C_k^2,
$$

to $\tilde{F}_4$. By (3.6), this reduces to

$$
F_4(B_1, \cdots, B_{n-1}) = 4 \sum_{k=1}^{n-1} \frac{1}{h_k[1 + d_k^2]^3} (B_k - d_k)^2. \tag{3.12}
$$

(Only a slight difference results when the exponent 3 is replaced by 5 in the denominator of the expression for $\tilde{F}_4$.) Now since the functions F_1, F_2, F_3, and F_4 are all of the form,

$$
F_5(B_1, \cdots, B_{n-1}) = \sum_{k=1}^{n-1} w_k(B_k - d_k)^2, \tag{3.13}
$$

with $w_k > 0$, a fifth and final possibility is for arbitrary $w_k > 0$ in (3.13).

Now, by the chain rule,

$$
\frac{dF_5}{dB_1} = \frac{\partial F_5}{\partial B_1} + \frac{\partial F_5}{\partial B_2} \frac{dB_2}{dB_1} + \cdots + \frac{\partial F_5}{\partial B_{n-1}} \frac{dB_{n-1}}{dB_1}. \tag{3.14}
$$

Thus, since by (3.8), $dB_k/dB_1 = (-1)^{k-1}$, $k = 2, \cdots, n-1$, we see that

$$
\sum_{k=1}^{n-1} (-1)^{k-1} w_k(B_k - d_k) = 0 \tag{3.15}
$$

is a condition necessary for an extremum. Now substitute (3.8) to obtain

$$
B_1 = \frac{\sum_{j=1}^{n-1}(-1)^{j-1} \left[w_j + 2\sum_{k=j+1}^{n-1} w_k \right] d_j}{\sum_{k=1}^{n-1} w_k}. \tag{3.16}
$$

```
      SUBROUTINE QUAOPT(N,X,Y,ICASE,LAMBDA,W,A,B,C,IFLAG)
      DIMENSION X(N),Y(N),W(N),A(N),B(N),C(N)
      REAL LAMBDA
      IFLAG=0
      IF (N.LT.3) THEN
          IFLAG=3
          RETURN
      END IF
      IF (ICASE.LT.1.OR.ICASE.GT.5) THEN
          IFLAG=5
          RETURN
      END IF
      IF (ICASE.EQ.1) LAMBDA=1.
      IF (ICASE.EQ.2) LAMBDA=0.
      N1=N-1
      DO 10 K=1,N1
          K1=K+1
          A(K)=X(K1)-X(K)
          C(K)=(Y(K1)-Y(K))/A(K)
10    CONTINUE
      IF (ICASE.EQ.1) LAMBDA=1.
      IF (ICASE.EQ.2) LAMBDA=0.
      IF (ICASE.LE.3) THEN
          P1=LAMBDA/3.
          P2=4.*(1.-LAMBDA)
          DO 20 K=1,N1
              AK=A(K)
              W(K)=P1*AK+P2/AK
20        CONTINUE
      END IF
      IF (ICASE.EQ.4) THEN
          DO 30 K=1,N1
              CK=C(K)
              H=1.+CK*CK
              W(K)=4./(A(K)*H*H*H)
30        CONTINUE
      END IF
      WSUM=0.
      DO 40 K=1,N1
          WSUM=WSUM+W(K)
40    CONTINUE
      HSUM=WSUM
      ZSUM=0.
      VZ=1.
      DO 50 J=1,N1
          WJ=W(J)
          HSUM=HSUM-WJ
          ZSUM=ZSUM+VZ*(WJ+2.*HSUM)*C(J)
          VZ=-VZ
50    CONTINUE
      B(1)=ZSUM/WSUM
      DO 60 K=1,N1
          K1=K+1
          BK=B(K)
          CK=C(K)
          B(K1)=2.*CK-BK
          C(K)=(CK-BK)/A(K)
          A(K)=Y(K)
60    CONTINUE
      RETURN
      END
```

Figure 3.1. Program listing of QUAOPT.

Calling sequence:

CALL QUAOPT(N,X,Y,ICASE,LAMBDA,W,A,B,C,IFLAG)

Purpose:
Determination of the coefficients A_k, B_k, and C_k of a quadratic spline interpolant with knots the same as nodes. The value for $B_1 = y_1'$ is determined optimally by one of five optional criteria.

Description of the parameters:

N		Number of given points. N$\geq$3 is required.
X	ARRAY(N):	Upon calling must contain the abscissa values x_k, $k = 1, \cdots, n$, with $x_1 < x_2 < \cdots < x_n$.
Y	ARRAY(N):	Upon calling must contain the ordinate values y_k, $k = 1, \cdots, n$.
A,B,C	ARRAY(N):	Upon completion with IFLAG=0 will contain the required coefficients in the first N$-$1 locations.
ICASE		1$\leq$ICASE$\leq$5. Number of chosen criteria $F_1, \cdots, F_5$.
LAMBDA	REAL:	The given parameter for ICASE=3: 0<LAMBDA<1.
W	ARRAY(N):	Output: corresponds to the w_k, $k = 1, \cdots, n - 1$. The weights $w_k > 0$ must be given for ICASE=5.
IFLAG	=0:	Normal execution.
	=1:	N$\geq$3 is required.
	=5:	ICASE<1 and ICASE>5 not permitted.

Figure 3.2. Description of QUAOPT.

If we differentiate the left side of (3.15) by similar means, we see that

$$\frac{d^2 F_5}{dB_1^2} = 2 \sum_{k=1}^{n-1} w_k > 0,$$

and hence this value of B_1 does indeed produce a minimum. The remaining values $B_2, \cdots, B_{n-1}$ and $C_1, \cdots, C_{n-1}$ can be calculated from (3.16) together with (3.6) and (3.7).

The five variations (ICASE=1,$\cdots$,5) are implemented in the subroutine QUAOPT (Figs. 3.1 and 3.2). Evaluation is carried out by QUAVAL (Fig. 3.3). The results of QUAOPT for ICASE=1,$\cdots$,4 and four given points with strictly decreasing ordinates are shown in Figs. 3.4a–3.5b.

```
FUNCTION QUAVAL(N,X,A,B,C,V,IFLAG)
DIMENSION X(N),A(N),B(N),C(N)
DATA I/1/
IFLAG=0
IF (N.LT.2) THEN
    IFLAG=1
    RETURN
END IF
CALL INTONE(X,N,V,I,IFLAG)
IF (IFLAG.NE.0) RETURN
DX=V-X(I)
QUAVAL=A(I)+DX*(B(I)+C(I)*DX)
RETURN
END
```

FUNCTION QUAVAL(N,X,A,B,C,V,IFLAG)

Purpose:
QUAVAL is a FUNCTION subprogram for the calculation of a function value of a quadratic spline interpolant (knots same as nodes) at a point $V \in [X(1),X(N)]$.

Description of the parameters:

N	Number of given points.
X	ARRAY(N): X-values.
A,B,C	ARRAY(N): Vectors of the spline coefficients.
V	Point at which the spline function is to be evaluated.
IFLAG	=0: Normal execution.
	=1: $N \geq 2$ is required.
	=3: Error in the interval determination (INTONE).

Required subroutines: INTONE.

Remark: The statement 'DATA I/1/' has the effect that I is set to 1 at the first call to QUAVAL.

Figure 3.3. Function QUAVAL and its description.

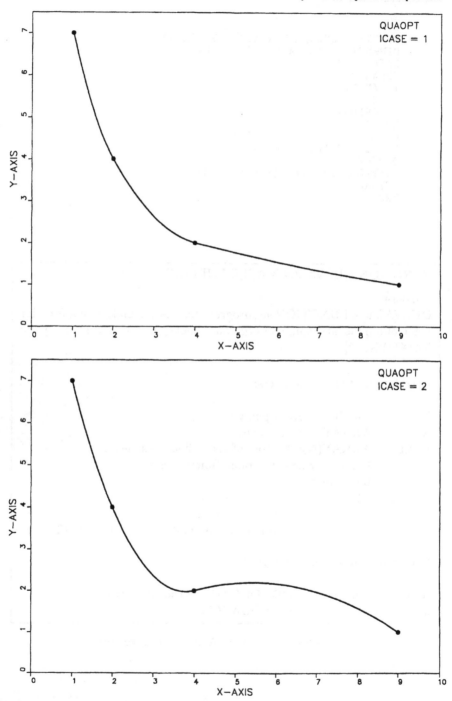

Figure 3.4. a, b.

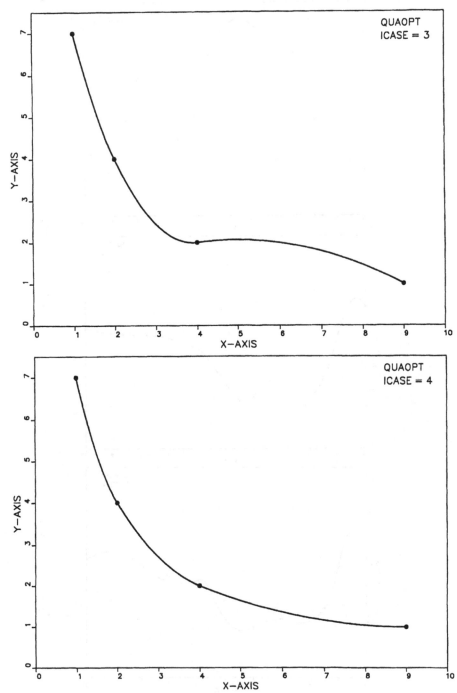

Figure 3.5. a, b.

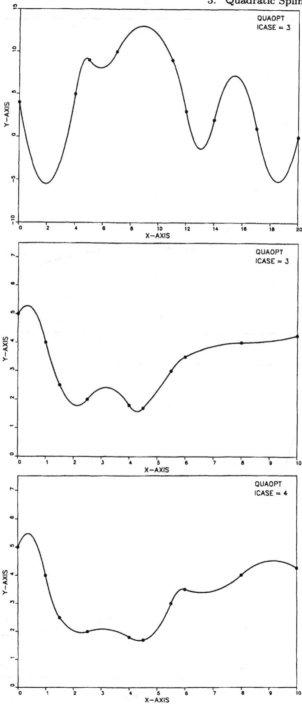

Figure 3.6. a-c.

ICASE=1 gives better results than does ICASE=2; ICASE=3 ($\lambda = 1/2$) gives the expected compromise. ICASE=4 gives the most usable solution. The weights in this case turned out to be: $w_1 = .0004$, $w_2 = .25$, and $w_3 = .7112$. Varying the weights to $w_1 = 1$, $w_2 = 5$, and $w_3 = 10$, or to $w_1 = 1$, $w_2 = 20$, and $w_3 = 100$, makes no significant difference to 3.5b. The results for the example of Fig. 1.4a are disappointing. Figure 3.6a shows the case ICASE=3 ($\lambda = 1/2$); ICASE=1,2 give practically the same graph. The oscillations at the edges could not be removed, even by varying the w_k with ICASE=5. Figures 3.6b and 3.6c show the results for the data of Fig. 1.4b for ICASE=3 ($\lambda = 1/2$) and ICASE=4.

Using QUAOPT is only worthwhile if the data is monotone or convex. In other cases, joining parabolic segments to be only once continuously differentiable rarely gives acceptable results.

3.3. Periodic Quadratic Spline Interpolants with Knots the Same as Nodes

The condition we now give for B_1 is

$$B_1 = \alpha B_n, \quad \alpha = \pm 1. \tag{3.17}$$

When $y_1 = y_n$, the choice of $\alpha = 1$ corresponds to periodic extension of the data. The case $\alpha = -1$ and symmetric data,

$$y_k = y_{n+1-k}, \quad \Delta x_k = \Delta x_{n-k},$$

and odd $n = 2m + 1$ has a meaning similar to that for a polygonal path. In both these cases, the coefficient matrix H of the linear system (3.7) has the appearance,

$$H = \begin{bmatrix} 1 & 1 & & & & & \\ & 1 & 1 & & & & \\ & & 1 & 1 & & & \\ & & & \cdot & \cdot & & \\ & & & & \cdot & \cdot & \\ & & & & & \cdot & \\ & & & & & 1 & 1 \\ -\alpha & & & & & & 1 \end{bmatrix}. \tag{3.18}$$

By expanding along the bottom row, we see that

$$det(H) = 1 + (-1)^n \alpha. \tag{3.19}$$

This means that for $\alpha = +1$, periodic spline interpolants exist only for even n, and only for odd n when $\alpha = -1$. In both cases, an explicit solution ([32]) can be given for symmetric data. We have refrained from implementing this, as periodic cubic and rational spline interpolants will produce more useful curves.

3.4. Knots at the Midpoints of the Nodes

Suppose again that we are given points (x_k, y_k), $k = 1, \cdots, n$, with $x_1 < \cdots < x_n$. In order to remove the preceding asymmetry in the choice of the slope B_1 or B_n at the left or right boundary point, we intend now to join ([66,97]) the parabolic segments (3.1) in a C^1 fashion at the midpoints,

$$z_k = \frac{1}{2}(x_k + x_{k+1}), \quad k = 1, \cdots, n-1. \tag{3.20}$$

(Later, we will allow the z_k to be variable between the x_k.) At the boundary, we require in addition the initial points,

$$z_0 = x_1 - \frac{h_1}{2}, \quad z_n = x_n + \frac{h_{n-1}}{2}. \tag{3.21}$$

Then we have

$$z_{k-1} - x_k = -\frac{h_{k-1}}{2}, \quad z_k - x_k = \frac{h_k}{2},$$

where $h_0 := h_1$ and $h_n := h_{n-1}$. We will use these relations shortly. Also, the interpolation condition for (3.1) immediately yields $A_k = y_k$, $k = 1, \cdots, n$. (Note that we are now considering n parabolic segments s_k. Previously there were $n - 1$.) Now introduce as unknowns the function values f_k at the points z_k, $k = 0, \cdots, n$. Then

$$s_k(z_{k-1}) = f_{k-1} = y_k - \frac{h_{k-1}}{2}B_k + \frac{h_{k-1}^2}{4}C_k,$$

$$s_k(z_k) = f_k = y_k + \frac{h_k}{2}B_k + \frac{h_k^2}{4}C_k. \tag{3.22}$$

These equations can be solved by Cramer's rule for B_k and C_k as functions of f_{k-1} and f_k :

$$B_k = \frac{2}{h_{k-1} + h_k}\left[\frac{h_{k-1}}{h_k}(f_k - y_k) - \frac{h_k}{h_{k-1}}(f_{k-1} - y_k)\right],$$

$$C_k = \frac{4}{h_{k-1} + h_k}\left[\frac{1}{h_k}(f_k - y_k) + \frac{1}{h_{k-1}}(f_{k-1} - y_k)\right]. \tag{3.23}$$

The f_k are then characterized by the C^1 conditions,

$$s'_k(z_k) = s'_{k+1}(z_k), \quad k = 1, \cdots, n-1,$$

that is,

$$B_k + h_k C_k = B_{k+1} - h_k C_{k+1}, \quad k = 1, \cdots, n-1.$$

By substituting (3.23) and dividing through by h_k^2, we obtain

$$\frac{1}{h_{k-1}h_k(h_{k-1}+h_k)}f_{k-1} + \frac{1}{h_k^2}\left(\frac{1}{h_{k-1}+h_k} + \frac{2}{h_k} + \frac{1}{h_k+h_{k+1}}\right)f_k$$

$$+\frac{1}{h_k h_{k+1}(h_k+h_{k+1})}f_{k+1}$$

$$= \frac{1}{h_k^2}\left(\frac{h_{k-1}+h_k}{h_{k-1}h_k}y_k + \frac{h_k+h_{k+1}}{h_k h_{k+1}}y_{k+1}\right), \quad k = 1, \cdots, n-1. \quad (3.24)$$

If we assign values to f_0 at z_0 and to f_n at z_n, then we have a linear system of $n-1$ equations in $n-1$ unknowns $f_1, \cdots, f_{n-1}$. The coefficient matrix is symmetric and the existence of the desired spline is then most easily shown by showing that it is also strictly diagonally dominant. This property would follow if

$$\frac{1}{h^2}\left(\frac{1}{x+h} + \frac{1}{h}\right) > \frac{1}{xh(x+h)}$$

held for $h = h_k$ and $x = h_{k-1}$ as well as $x = h_{k+1}$. But if we multiply this inequality by h and cancel the positive common denominator $xh^2(x+h)$, we obtain the condition, sufficient for strict diagonal dominance,

$$x^h + 2hx - h^2 > 0.$$

This inequality is satisfied for $x > (\sqrt{2}-1)h$ and $x < -(\sqrt{2}+1)h$. The second case is of no relevance, as x cannot be negative. Using this result in the equivalent form $h < (\sqrt{2}+1)x$ as well as the fact that the inequality must hold for all k, we obtain

$$(\sqrt{2}-1) < \frac{h_k}{h_{k-1}} < (\sqrt{2}+1). \quad (3.25)$$

This means that the ratio of two consequtive abscissa intervals can be neither too large nor too small in order to be able to show existence by means of strict diagonal dominance.

In such situations, it is most helpful to introduce other unknowns. In our case, there remains the possibility of introducing values f'_k, $k = 0, \cdots, n$, for the derivatives at z_k. Analogously to (3.22), we then have

$$\begin{aligned} s'_k(z_{k-1}) &= f'_{k-1} = B_k - h_{k-1}C_k, \\ s'_k(z_k) &= f'_k = B_k + h_k C_k, \end{aligned} \quad (3.26)$$

```
          SUBROUTINE QUASKF(N,X,Y,Z,EPS,A,B,C,IFLAG)
          DIMENSION X(N),Y(N),Z(0:N),A(N+1),B(N+1),C(N+1)
          IFLAG=0
          IF (N.LT.2) THEN
              IFLAG=1
              RETURN
          END IF
          NP1=N+1
          NM1=N-1
          Y0=A(1)
          YN1=A(2)
          H1=X(2)-X(1)
          H2=H1
          Z(0)=X(1)-H1/2.
          Z(N)=X(N)+(X(N)-X(NM1))/2.
          R1=Y(1)
          H4=1./(H1+H2)
          A(1)=H4
          B(1)=3.*H4
          C(1)=4.(R1-Y0)/(H1*H1)
          DO 10 K=1,NM1
              K1=K+1
              IF (K.LT.NM1) H3=X(K+2)-X(K1)
              Z(K)=X(K)+H2/2.
              H4=1./(H2+H3)
              R2=Y(K1)
              A(K1)=H4
              B(K1)=(2.+H1/(H1+H2)+H3*H4)/H2
              C(K1)=4.*(R2-R1)/(H2*H2)
              H1=H2
              H2=H3
              R1=R2
10        CONTINUE
          B(NP1)=3.*H4
          C(NP1)=4.*(YN1-R1)/(H3*H3)
          CALL TRIDIS(NP1,B,A,C,EPS,IFLAG)
          IF (IFLAG.NE.0) RETURN
          H1=X(2)-X(1)
          DO 20 K=1,N
              K1=K+1
              IF (K.LT.N) H2=X(K1)-X(K)
              H=1./(H1+H2)
              A(K)=Y(K)
              B(K)=H*(H1*C(K1)+H2*C(K))
              C(K)=H*(C(K1)-C(K))
              H1=H2
20        CONTINUE
          RETURN
          END
```

Figure 3.7. Program listing of QUASKF.

Calling sequence:

CALL QUASKF(N,X,Y,Z,EPS,A,B,C,IFLAG)

Purpose:
Determination of the coefficients A_k, B_k, and C_k of a quadratic spline interpolant $s(x)$ for given points with abscissa values x_k and ordinate values y_k, $k = 1, \cdots, n \geq 2$, and knots at the points $z_k = (x_k + x_{k+1})/2$, $k = 1, \cdots, n - 1$, $z_0 = x_1 - h_1/2$, and $z_n = x_n + h_{n-1}/2$. In addition, values for $s(z_0)$ and $s(z_n)$ must be given in A(1) and A(2), respectively.

Description of the parameters:

N,X,Y   The same as for QUAOPT.

Z   ARRAY(0:N): Upon completion contains the abscissa values of the knots z_k, $k = 0, \cdots, n$.

EPS   see TRIDIS.

A,B,C   ARRAY(N):   Upon completion with IFLAG=0 contain the required spline coefficients, K=1,...,N.

IFLAG =0:   Normal execution.

       =1: $N \geq 2$ required.

       =2:   Error in solving the system (TRIDIS).

Required subroutines: TRIDIS.

Figure 3.8. Description of QUASKF.

from which we obtain

$$B_k = \frac{1}{h_{k-1} + h_k}(h_{k-1}f'_k + h_k f'_{k-1}),$$

$$C_k = \frac{1}{h_{k-1} + h_k}(f'_k - f'_{k-1}). \tag{3.27}$$

The f'_k are then determined from the C^0 conditions,

$$s_k(z_k) = s_{k+1}(z_k), \quad k = 1, \cdots, n - 1.$$

By substituting (3.27) and dividing through by $h_k^2/4$, we obtain

$$\frac{1}{h_{k-1} + h_k}f'_{k-1}$$
$$+ \left[\frac{1}{h_{k-1} + h_k}\left(1 + 2\frac{h_{k-1}}{h_k}\right) + \frac{1}{h_k + h_{k+1}}\left(1 + 2\frac{h_{k+1}}{h_k}\right)\right]f'_k$$
$$+ \frac{1}{h_k + h_{k+1}}f'_{k+1} = \frac{4}{h_k}d_k, \quad k = 1, \cdots, n - 1. \tag{3.28}$$

```
FUNCTION QUAVAM(N,X,Z,A,B,C,V,IFLAG)
DIMENSION X(N),Z(N+1),A(N),B(N),C(N)
DATA I/1/
IFLAG=0
IF (N.LT.1) THEN
    IFLAG=1
    RETURN
END IF
CALL INTONE(Z,N+1,V,I,IFLAG)
IF (IFLAG.NE.0) RETURN
DX=V-X(I)
QUAVAM=A(I)+DX*(B(I)+C(I)*DX)
RETURN
END
```

Figure 3.9. Program listing of QUAVAM.

FUNCTION QUAVAM(N,X,Z,A,B,C,V,IFLAG)

Purpose:
QUAVAM is a FUNCTION subprogram for the calculation of a function value of a quadratic spline interpolant (knots either at the midpoints or variable between the nodes) at $V \in [Z(1),Z(N+1)]$.

Remarks: The statement 'DATA I/1/' has the effect that I is set to 1 at the first call to QUAVAM. The vector Z must be dimensioned from 0 to N in the main program when calling the subroutines QUASKF, QUASKV, QUAPKV, and QUASKG. If now QUAVAM (here Z is dimensioned from 1 to N+1) is called with this convention to evaluate a spline function, then the elements of the vector, as intended, are shifted $Z(0)$ to $Z(1)$, $Z(1)$ to $Z(2)$,...,$Z(N)$ to $Z(N+1)$.

Description of the parameters:
N,X,A,B,C,V,IFLAG as in QUAVAL.
Z   ARRAY(N+1): Vector with the abscissa values of the knots
                $z_k, \ k = 1, \cdots, n - 1$.

Required subroutines: INTONE.

Figure 3.10. Description of QUAVAM.

For programming purposes, we will make use of the algebraically equivalent but simpler expression,

$$\frac{1}{h_k}\left[2 + \frac{h_{k-1}}{h_{k-1} + h_k} + \frac{h_{k+1}}{h_k + h_{k+1}}\right],$$

for the factor of f'_k. The form (3.28) has, however, the advantage that from it the *strict diagonal dominance* of the system of equations is more easily recognized. This is the case when f'_0 and f'_n are assigned fixed values.

If, as previously, it is still desired to prescribe function values f_0 and f_n, then some calculations are needed to determine f'_0 and f'_n. First of all, the additional condition $s_1(z_0) = f_0$ gives

$$A_1 - \frac{h_1}{2}B_1 + \frac{h_1^2}{4}C_1 = f_0,$$

and then the substitution of $A_1 = y_1$ and the formulas for B_1 and C_1 from (3.27) yields

$$\frac{3}{2h_1}f'_0 + \frac{1}{2h_1}f'_1 = \frac{4}{h_1^2}(y_1 - f_0). \tag{3.29}$$

Similarly, from the condition $s_n(z_n) = f_n$, we obtain

$$\frac{1}{2h_{n-1}}f'_{n-1} + \frac{3}{2h_{n-1}}f'_n = \frac{4}{h_{n-1}^2}(f_n - y_n). \tag{3.30}$$

The addition of these two equations to (3.28) produces a system of $n + 1$ equations in $n + 1$ unknowns $f'_0, \cdots, f'_n$. Clearly, (3.29) and (3.30) do not disturb the strict diagonal dominance of the system. Hence, we may easily conclude the existence of quadratic spline interpolants for any distribution of abscissas with $x_1 < \cdots < x_n$. That there are two more equations than in (3.24) is of little import.

The example in Fig. 3.11 was calculated with the subroutine QUASKF (Knots Fixed) (Figs. 3.7 and 3.8) and the evaluation carried out by QUA-VAM (Figs. 3.9 and 3.10).

3.5.   Knots Variable between the Nodes

One thing we might try, in order to better control the shape of the resulting quadratic spline interpolant, is to not to place the knots z_k at the midpoints of the nodes but allow them to vary between them ([175]). In other words, we replace (3.20) by

$$\begin{aligned}
z_k &= \alpha_k x_k + (1 - \alpha_k)x_{k+1} \\
&= x_k + (1 - \alpha_k)h_k \\
&= x_{k+1} - \alpha_k h_k, \quad k = 1, \cdots, n - 1.
\end{aligned} \tag{3.31}$$

We again add two additional knots

$$\begin{aligned}
z_0 &= x_1 - \alpha_0 h_0 \quad (h_0 := h_1), \\
z_n &= x_n + (1 - \alpha_n)h_n, \quad (h_n := h_{n-1}).
\end{aligned} \tag{3.32}$$

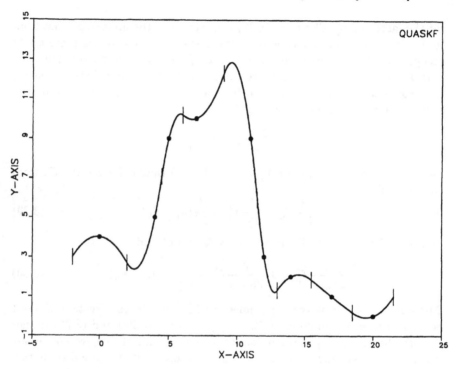

Figure 3.11.

Of course, $0 < \alpha_k < 1$, $k = 0, \cdots, n$. Having learned from the experience of the last section, we introduce immediately the values f'_k of the derivative at z_k, $k = 0, \cdots, n$, of the desired function. Just as for (3.26), we obtain the conditions,

$$s'_k(z_{k-1}) = f'_{k-1} = B_k - 2\alpha_{k-1}h_{k-1}C_k,$$
$$s'_k(z_k) = f'_k = B_k + 2(1 - \alpha_k)h_kC_k,$$

from which result

$$B_k = \frac{1}{\alpha_{k-1}h_{k-1} + (1 - \alpha_k)h_k}(\alpha_{k-1}h_{k-1}f'_k + (1 - \alpha_k)h_kf'_{k-1}),$$
$$C_k = \frac{1}{2(\alpha_{k-1}h_{k-1} + (1 - \alpha_k)h_k)}(f'_k - f'_{k-1}). \qquad (3.33)$$

We have used the facts that $z_{k-1} - x_k = -\alpha_{k-1}h_{k-1}$ and that $z_k - x_k = (1 - \alpha_k)h_k$. The C^0 conditions then yield

$$\frac{(1 - \alpha_k)^2 h_k^2}{\alpha_{k-1}h_{k-1} + (1 - \alpha_k)h_k} f'_{k-1} + \left[\frac{(1 - \alpha_k)h_k(2\alpha_{k-1}h_{k-1} + (1 - \alpha_k)h_k)}{\alpha_{k-1}h_{k-1} + (1 - \alpha_k)h_k} \right.$$

$$\left. + \frac{\alpha_k h_k(2(1 - \alpha_{k+1})h_{k+1} + \alpha_k h_k)}{\alpha_k h_k + (1 - \alpha_{k+1})h_{k+1}} \right] f'_k$$

$$+ \frac{\alpha_k^2 h_k^2}{\alpha_k h_k + (1 - \alpha_{k+1})h_{k+1}} f'_{k+1} = 2(y_{k+1} - y_k),$$

from which the strict diagonal dominance of the coefficient matrix is easily seen. For $\alpha_k = 1/2$, $k = 0, \cdots, n$, we recover (3.28). Dividing through by $(1 - \alpha_k)\alpha_k h_k^2$ produces a form convenient for programming:

$$\frac{1 - \alpha_k}{\alpha_k[\alpha_{k-1}h_{k-1} + (1 - \alpha_k)h_k]} f'_{k-1} + \frac{1}{(1 - \alpha_k)\alpha_k h_k}$$

$$\times \left[1 + \frac{(1 - \alpha_k)\alpha_{k-1}h_{k-1}}{\alpha_{k-1}h_{k-1} + (1 - \alpha_k)h_k} + \frac{\alpha_k(1 - \alpha_{k+1})h_{k+1}}{\alpha_k h_k + (1 - \alpha_{k+1}h_{k+1})} \right] f'_k$$

$$+ \frac{\alpha_k}{(1 - \alpha_k)[\alpha_k h_k + (1 - \alpha_{k+1})h_{k+1}]} f'_{k+1} \qquad (3.34)$$

$$= \frac{2}{(1 - \alpha_k)\alpha_k h_k^2}(y_{k+1} - y_k), \quad k = 1, \cdots, n - 1.$$

Just as before, we obtain the boundary conditions,

$$\frac{\alpha_0}{h_1} \left[1 + \frac{(1 - \alpha_1)h_1}{\alpha_0 h_1 + (1 - \alpha_1)h_1} \right] f'_0 + \frac{\alpha_0^2}{\alpha_0 h_1 + (1 - \alpha_1)h_1} f'_1$$

$$= \frac{2}{h_1^2}(y_1 - f_0) \qquad (3.35)$$

and

$$\frac{(1 - \alpha_n)^2}{\alpha_{n-1}h_{n-1} + (1 - \alpha_n)h_{n-1}} f'_{n-1} \qquad (3.36)$$

$$+ \frac{1 - \alpha_n}{h_{n-1}} \left[1 + \frac{\alpha_{n-1}h_{n-1}}{\alpha_{n-1}h_{n-1} + (1 - \alpha_n)h_{n-1}} \right] f'_n = \frac{2}{h_{n-1}^2}(f_n - y_n).$$

With $\alpha_0 = \alpha_1 = \alpha_{n-1} = \alpha_n = 1/2$, we recover (3.29) and (3.30). The equations (3.35) and (3.36) are consistent with strict diagonal dominance. To be sure, the coefficient matrix of the system (3.35), (3.34), and (3.36) for the determination of $f'_0, \cdots, f'_n$ is no longer symmetric. Hence, the appropriate subroutine TRIDIU (see the appendix) must be used by QUASKV (Knots Variable) (Fig. 3.12).

```
      SUBROUTINE QUASKV(N,X,Y,Z,ALPHA,EPS,A,B,C,D,IFLAG)
      DIMENSION X(N),Y(N),Z(0:N),ALPHA(0:N),
     &          A(N+1),B(N+1),C(N+1),D(N+1)
      IFLAG=0
      IF (N.LT.2) THEN
         IFLAG=1
         RETURN
      END IF
      NP1=N+1
      NM1=N-1
      DO 10 K=0,N
         H=ALPHA(K)
         IF (H.LE.0.OR.H.GE.1.) THEN
         IFLAG=10
         RETURN
      END IF
10    CONTINUE
      YO=A(1)
      YN1=A(2)
      AK=ALPHA(0)
      AN=ALPHA(N)
      Z(0)=X(1)-AK*(X(2)-X(1))
      Z(N)=X(N)+(1.-AN)*(X(N)-X(NM1))
      H1=X(2)-X(1)
      H2=H1
      AK1=1.-ALPHA(1)
      H=AK1*H1
      H4=1./(AK*H1+H)
      R1=Y(1)
      C(1)=AK*AK*H4
      B(1)=AK*(1.+H*H4)/H1
      D(1)=2.*(R1-YO)/(H1*H1)
      DO 20 K=1,NM1
         KP1=K+1
         KM1=K-1
         IF (K.LT.NM1) H3=X(K+2)-X(KP1)
         AK=ALPHA(K)
         AK2=1.-ALPHA(KP1)
         Z(K)=X(KP1)-AK*H2
         R2=Y(KP1)
         H5=1./(AK*H2+AK2*H3)
         H6=1./(AK1*AK*H2)
         A(K)=AK1/AK*H4
         C(KP1)=AK/AK1*H5
         B(KP1)=H6*(1.+AK1*ALPHA(KM1)*H1*H4
     &          +AK2*AK*H3*H5)
         D(KP1)=2.*H6/H2*(R2-R1)
         H1=H2
         H2=H3
         H4=H5
         AK1=AK2
         R1=R2
20    CONTINUE
      A(N)=AK2*AK2*H4
      B(NP1)=AK2*(1.+AK*H1*H4)/H1
      D(NP1)=2.*(YN1-R2)/(H1*H1)
      CALL TRIDIU(NP1,A,B,C,D,EPS,IFLAG)
```

(*cont.*)
```
        IF (IFLAG.NE.0) RETURN
        H1=X(2)-X(1)
        DO 30 K=1,N
            KP1=K+1
            KM1=K-1
            IF (K.LT.N) H2=X(KP1)-X(K)
            AK=ALPHA(KM1)
            AK1=1.-ALPHA(K)
            H=1./(AK1*H2+AK*H1)
            A(K)=Y(K)
            B(K)=H*(AK*H1*D(KP1)+AK1*H2*D(K))
            C(K)=H/2.*(D(KP1)-D(K))
            H1=H2
30      CONTINUE
        RETURN
        END
```

Calling sequence:

CALL QUASKV(N,X,Y,Z,ALPHA,EPS,A,B,C,D,IFLAG)

Purpose:
Determination of the the coefficients A_k, B_k, and C_k of a quadratic spline interpolant s at given points (x_k, y_k), $k = 1, \cdots, n \geq 2$, for given factors $\alpha_k \in (0, 1)$, $k = 0, \cdots, n$, which specify the knots $z_k = x_k + \alpha_k h_k$, $k = 1, \cdots, n - 1$, $z_0 = x_1 - \alpha_0 h_1$, and $z_n = x_n + (1 - \alpha_n)h_{n-1}$. Here, $h_k := x_{k+1} - x_k$. In addition, values for $s(z_0)$ and $s(z_n)$ must be given in $A(1)$ and $A(2)$, respectively.

Description of the parameters:
N,X,Y,EPS,A,B,C as in QUASKF.

Z   ARRAY(0:N): Upon execution contains the abscissa values
        of the knots z_k, $k = 0, \cdots, n$.
ALPHA   ARRAY(N): Upon calling must contain the factors
        α_k, $k = 0, \cdots, n$.
D   ARRAY(N+1): Work space.
IFLAG   =0:   Normal execution.
        =1:   N$\geq$2 is required.
        =2:   Error in solving the system (TRIDIU).
        =10: $\alpha_k \leq 0$ and $\alpha_k \geq 1$ not permitted.

Required subroutines: TRIDIU.

Figure 3.12. QUASKV and its description.

The special case of $\alpha_0 = 0$ and $\alpha_n = 1$, i.e., $z_0 = x_1$ and $z_n = x_n$ with the natural choice additionally of $f_0 = y_1$ and $f_n = y_n$, is indeed, for given values f_0' and f_n' at x_1 and x_n, covered by the system (3.34). However, its solution cannot be calculated by QUASKV, as the equations (3.35) and (3.36) are then of the form $0 = 0$.

If we compare Figs. 3.13a and 3.13b with 3.11 ($\alpha_k = 1/2$), we see that the shape of the curves really can be favorably influenced by the choice of the knots z_k (indicated by vertical bars). In Fig. 3.13a, α=(.5, .25, .5, .75, .25, .5, .75, .5,. 5, .5); in 3.13b, α=(.5, .1, .5, .9, .1, .5, .9, .5, .5, .5); and in 3.13c, α=(.5, .2, .5, .8, .5, .5, .5, .5, .5, .5, .5, .5). However, just as before, only once continuous differentiability sometimes causes an erratic-looking transition at the knots. The user should be warned that using values of α_k with (approximately) $\alpha_k < .05$ or $\alpha_k > .95$ can result in wild oscillations in the resulting quadratic spline interpolant. As $\alpha_k \to 1$ or $\alpha_k \to 0$, the problem is ill-conditioned in the sense that small changes in the α_k can produce large changes in the result.

3.6.   Periodic Quadratic Spline Interpolants with Variable Knots between the Nodes

The quadratic spline interpolant of the previous section is said to be *periodic* when

$$s_1^{(j)}(x_1) = s_n^{(j)}(x_n), \quad j = 0, 1, 2,$$

or, in other words, when the graphs of the segments s_1 and s_n agree. This is attained formally by setting $\alpha_0 = 0$ and $\alpha_n = 1$, i.e., $z_0 = x_1$ and $z_n = x_n$. The derivatives f_0' and f_n' would then refer to x_1 and x_n. This agreement is also guaranteed when the coefficients agree, i.e., when $A_1 = A_n$, $B_1 = B_n$, and $C_1 = C_n$. It is easy to see that then

$$y_1 = y_n,$$
$$f_0' = f_n',$$
$$\frac{1}{2(1 - \alpha_1)h_1}(f_1' - f_0') = \frac{1}{2\alpha_{n-1}h_{n-1}}(f_n' - f_{n-1}').$$

It follows that the $n - 1$ equations (3.34) then hold, with y_n replaced by y_1 in the last equation, and f_0' by f_n' in the first. Moreover, we also have

$$\frac{1}{2(1 - \alpha_1)h_1}f_1' + \frac{1}{2\alpha_{n-1}h_{n-1}}f_{n-1}' - \left(\frac{1}{2\alpha_{n-1}h_{n-1}} + \frac{1}{2(1 - \alpha_1)h_1}\right)f_n' = 0.$$
$$(3.37)$$

The coefficient matrix of the equations (3.34) with f_n' instead of f_0' and

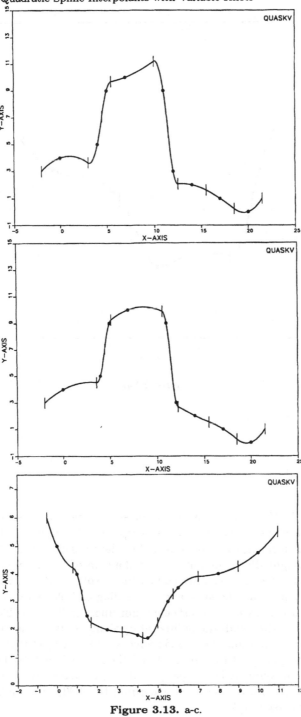

Figure 3.13. a-c.

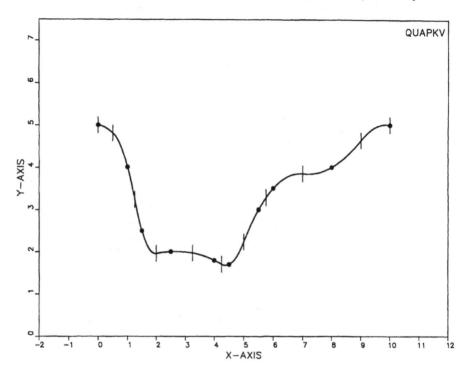

Figure 3.14.

the additional nth equation (3.37) is not symmetric. It is, however, *cyclically tridiagonal*, i.e., it is tridiagonal except that the uppermost right and lowermost left elements are nonzero. All of the rows except for the last one are strictly diagonally dominant. There the sum of the absolute values of the off-diagonal elements exactly equals the absolute value of the diagonal element. But since all the sub- and super-diagonal elements are nonzero, the matrix, in this case, is nevertheless nonsingular ([166, p. 23]). Hence, the desired periodic quadratic spline interpolant exists and is unique.

The system of equations described above is set up by QUAPKV (P̲eriodic, K̲nots V̲ariable) (Figs. 3.15 and 3.16) and then solved by TRIPEU (see the appendix for nonsymmetric cyclically tridiagonal coefficient matrices). The coefficients of (3.1) are calculated according to (3.33). Figure 3.14 shows an example for $\alpha_k = 1/2$.

```
      SUBROUTINE QUAPKV(N,X,Y,X,ALPHA,EPS,A,B,C,D,IFLAG)
      DIMENSION X(N),Y(N),Z(0:N),ALPHA(0:N),A(N),B(N),
     &          C(N),D(N)
      IFLAG=0
      IF (N.LT.3) THEN
          IFLAG=1
          RETURN
      END IF
      NP1=N+1
      NM1=N-1
      ALPHA(0)=0.
      ALPHA(N)=0.
      DO 10 K=1,NM1
          H=ALPHA(K)
          IF (H.LE.0.OR.H.GE.1.) THEN
              IFLAG=10
              RETURN
          END IF
10    CONTINUE
      Y(N)=Y(1)
      Z(0)=X(1)
      Z(N)=X(N)
      H1=0.
      H2=X(2)-X(1)
      AK1=1.-ALPHA(1)
      H4=1./(AK1*H2)
      R1=Y(1)
      DO 20 K=1,NM1
          KP1=K+1
          KM1=K-1
          IF (K.LT.NM1) THEN
              H3=X(K+2)-X(KP1)
          ELSE
              H3=0.
          END IF
          AK=ALPHA(K)
          AK2=1.-ALPHA(KP1)
          Z(K)=X(KP1)-AK*H2
          R2=Y(KP1)
          H5=1./(AK*H2+AK2*H3)
          H6=1./(AK1*AK*H2)
          A(K)=AK1/AK*H4
          C(K)=AK/AK1*H5
          B(K)=H6*(1.+AK1*ALPHA(KM1)*H1*H4+AK2*AK*H3*H5)
          D(K)=2.*H6/H2*(R2-R1)
          H1=H2
          H2=H3
          H4=H5
          AK1=AK2
          R1=R2
20    CONTINUE
      A(N)=1./(2.*AK*H1)
      C(N)=1./(2.*(1.-ALPHA(1))*(X(2)-X(1)))
      B(N)=-C(N)-A(N)
      D(N)=0.
      CALL TRIPEU(N,A,B,C,D,EPS,IFLAG)
      IF (IFLAG.NE.0) RETURN
      H1=0.
      DK=D(N)
      DO 30 K=1,N
          KP1=K+1
```

(*cont.*)

```
            KM1=K-1
            IF (K.LT.N) THEN
                H2=X(KP1)-X(K)
            ELSE
                H2=0.
            END IF
            AK=ALPHA(KM1)
            AK1=1.-ALPHA(K)
            H=1./(AK1*H2+AK*H1)
            A(K)=Y(K)
            B(K)=H*(AK*H1*D(K)+AK1*H2*DK)
            C(K)=H/2.*(D(K)-DK)
            DK=D(K)
            H1=H2
30          CONTINUE
            RETURN
            END
```

Figure 3.15. Program listing of QUAPKV.

Calling sequence:

CALL QUAPKV(N,X,Y,Z,ALPHA,EPS,A,B,C,D,IFLAG)

Purpose:

Determination of the coefficients A_k, B_k, and C_k of a periodic quadratic spline interpolant for given points (x_k, y_k), $k = 1, \cdots, n \geq 3$, given factors $\alpha_k \in (0,1)$, $k = 1, \cdots, n-1$, and knots $z_k = x_k + \alpha_k h_k$, $k = 1, \cdots, n-1$. Here, $h_k := x_{k+1} - x_k$. QUAPKV assigns the following values: ALPHA(0)=ALPHA(N)=0, Y(N)=Y(1), Z(0)=X(1), and Z(N)=X(N).

Description of the parameters:

N,X,Y,EPS as in QUASKF.

Z	ARRAY(0:N):	Upon completion contains the abscissa values of the knots z_k, $k = 0, \cdots, n$.
ALPHA	ARRAY(0:N)	Upon calling must contain the factors α_k, $k = 1, \cdots, n-1$.
A,B,C	ARRAY(N):	Upon completion with IFLAG=0 will contain the desired spline coefficients, K=1,...,N.
D	ARRAY(N):	Work space.
IFLAG	=0:	Normal execution.
	=1:	N<3 not permitted.
	=2:	Error in solving the linear system (TRIPEU).
	=10:	$\alpha_k \leq 0$ and $\alpha_k \geq 1$ not permitted.

Required subroutines: TRIPEU.

Figure 3.16. Description of QUAPKV.

3.7.   Nodes Variable between Knots

In [142,175], a problem is discussed that at first glance appears to be different from that of the previous section. Knots $z_0 < z_1 < \cdots < z_n$ are given together with factors $\beta_k \in (0,1)$, $1 \leq k \leq n$, and from these, interpolation points are calculated according to the formula,

$$x_k = \beta_k z_{k+1} + (1 - \beta_k) z_k, \quad k = 1, \cdots, n.$$

(Given values of x_k can be prescribed by setting $\beta_k = (z_k - x_k)/(z_k - z_{k-1})$.) But from the z_k, $k = 1, \cdots, n$, and the x_k, $k = 1, \cdots, n$, we can easily calculate $\alpha_0, \cdots, \alpha_n$ of (3.31) and (3.32). In fact,

$$
\begin{aligned}
\alpha_0 &= \frac{x_1 - z_0}{h_1}, \\
\alpha_k &= \frac{x_{k+1} - z_k}{h_k} \quad k = 1, \cdots, n-1, \\
\alpha_n &= 1 - \frac{z_n - x_n}{h_{n-1}}.
\end{aligned}
\tag{3.38}
$$

It is true that $\alpha_k \in (0,1)$, $k = 1, \cdots, n-1$, but it may not be the case that $0 < \alpha_0 < 1$ or that $0 < \alpha_n < 1$. In this case then, a slight modification of QUASKV would be required before it could be called with the x_k and α_k to solve the problem posed above.

3.8.   Quadratic Histosplines

In the sciences, it is sometimes required to approximate a histogram by a smooth curve with areas above each interval matching those of the histogram ([144]). For example, in statistics there is the problem of the reconstruction of density functions ([6]), and in physics the problem of representing measured multi-group spectra by smooth curves. A *histogram* is defined by $x_1 < x_2 < \cdots < x_n$ and given heights y_k above the base lines running from x_k to x_{k+1}, $k = 1, \cdots, n-1$. We are looking for a quadratic spline s with segments s_k, $k = 1, \cdots, n-1$, as in (3.1), that satisfy both the *area conditions*,

$$h_k y_k = \int_{x_k}^{x_{k+1}} s_k(x)\,dx = A_k h_k + \frac{1}{2} B_k h_k^2 + \frac{1}{3} C_k h_k^3, \tag{3.39}$$

$k = 1, \cdots, n-1$, and the C^1 conditions,

$$
\begin{aligned}
s_k(x_{k+1}) &= s_{k+1}(x_{k+1}), \quad k = 1, \cdots, n-2, \\
s'_{k-1}(x_k) &= s'_k(x_k), \quad k = 1, \cdots, n-1.
\end{aligned}
$$

These may be expressed as

$$
\begin{aligned}
A_k + \frac{1}{2} h_k B_k + \frac{1}{3} h_k^2 C_k &= y_k, & k &= 1, \cdots, n-1, \\
A_k + h_k B_k + h_k^2 C_k &= A_{k+1}, & k &= 1, \cdots, n-2, \\
B_{k-1} + 2 h_{k-1} C_{k-1} &= B_k, & k &= 1, \cdots, n-1.
\end{aligned}
\tag{3.40}
$$

Further, following [98,155], we set *boundary conditions* in the form of given values y_0 and y_n at x_1 and x_n. We then have

$$
A_1 = y_0 \quad \text{and} \quad A_n = y_n.
\tag{3.41}
$$

```
      SUBROUTINE QUHIST(N,X,Y,EPS,A,B,C,IFLAG)
      DIMENSION X(N),Y(0:N),A(N),B(N),C(N)
      IFLAG=0
      IF (N.LT.2) THEN
         IFLAG=1
         RETURN
      END IF
      N1=N-1
      N2=N-2
      DO 20 K=1,N1
         KP1=K+1
         KM1=K-1
         H2=1./(X(KP1)-X(K))
         R2=H2*Y(K)
         IF(K.EQ.1) GOTO 10
         B(KM1)=2.*(H1+H2)
         C(KM1)=H2
         A(KM1)=3.*(R2+R1)
         IF(K.EQ.2) A(KM1)=A(KM1)-H1*Y(0)
         IF(K.EQ.N1) A(KM1)=A(KM1)-H2*Y(N)
10       H1=H2
         R1=R2
20    CONTINUE
      CALL TRIDIS(N2,B,C,A,EPS,IFLAG)
      IF(IFLAG.NE.0) RETURN
      DO 30 K=N2,1,-1
         A(K+1)=A(K)
30    CONTINUE
      A(1)=Y(0)
      A(N)=Y(N)
      DO 40 K=1,N1
         K1=K+1
         DX=X(K1)-X(K)
         B(K)=(6.*Y(K)-4.*A(K)-2.*A(K1))/DX
         C(K)=(3.*A(K)+3.*A(K1)-6.*Y(K))/(DX*DX)
40    CONTINUE
      RETURN
      END
```

Figure 3.17. Program listing of QUHIST.

Calling sequence:

CALL QUHIST(N,X,Y,EPS,A,B,C,IFLAG)

Purpose: Determination of a quadratic histospline with coefficients A_k, B_k, and C_k, $k = 1, \cdots, n - 1$ for given abscissas $x_1 < x_2 < \cdots < x_n$, rectangle heights y_k, $k = 1, \cdots, n-1$, a left-end value, y_0, and a right-end value y_n. $N \geq 2$ is necessary.

Description of the parameters:

N, X, A, B, C as in QUAOPT.

Y   ARRAY(0:N): Y(0) must contain the left-end value y_0. Y(N) must contain the right-end value y_n. Y(1),Y(2),...Y(N−1) must contain the rectangle heights y_k, $k = 1, \cdots, n - 1$.

EPS   see TRIDIS.

IFLAG =0:   Normal execution.

       =1:   N≥2 is required.

       =2:   Error in solving the linear system (TRIDIS).

Required subroutine: TRIDIS.

Figure 3.18. Description of QUHIST.

Consequently, the second equation of (3.40) holds also for $k = n - 1$. Now solve the first two equations of (3.40) for B_k and C_k to obtain

$$B_k = \frac{2}{h_k}(3y_k - 2A_k + A_{k+1}),$$

$$C_k = \frac{3}{h_k^2}(A_k + A_{k+1} - 2y_k), \tag{3.42}$$

$k = 1, \cdots, n - 1$. The substitution of these in the third equation of (3.40) then yields

$$\frac{1}{h_{k-1}}A_{k-1} + 2(\frac{1}{h_{k-1}} + \frac{1}{h_k})A_k + \frac{1}{h_k}A_{k+1} = 3(\frac{y_{k-1}}{h_{k-1}} + \frac{y_k}{h_k}),$$
$$k = 2, \cdots, n - 1. \tag{3.43}$$

Together with (3.41), these again give a linear system of equations for the determination of the A_k, $k = 1, \cdots, n - 1$. The coefficient matrix is symmetric, tridiagonal, and strictly diagonally dominant. The desired histospline is thus uniquely determined and given by the A_k, $k = 2, \cdots, n - 1$, and (3.42).

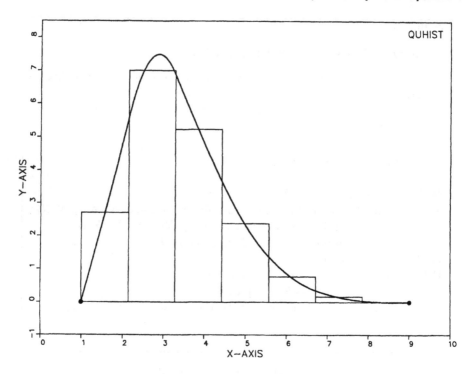

Figure 3.19.

   This procedure is implemented in the subroutine QUHIST (Figs. 3.17 and 3.18). As can be seen from the examples in Figs. 3.19–3.21b, it is, in practice, not always satisfactory: positivity, monotonicity, and unimodality of the histogram are not always preserved by the curve so obtained.

   The authors of [93,94] discuss how to improve quadratic splines in this regard. Later, we will study quartic and rational histosplines, which involve parameters for the flexible control of the resulting curve.

   If it is instead desired to prescribe boundary values y_1' and y_n' for the first derivative at x_1 and x_n ([98]), then we may proceed in one of two ways. One option is to maintain $A_1, \cdots, A_n$ as the unknowns. The two equations,

$$
\begin{aligned}
B_1 &= y_1', \\
B_{n-1} + 2h_{n-1}C_{n-1} &= y_n',
\end{aligned}
\tag{3.44}
$$

may be expressed, using (3.42), as functions of A_1 and A_2, and A_{n-1} and A_n, respectively. This gives two equations to be added to (3.43). A second option is to solve the first equation of (3.40) for A_k and the last for C_{k-1}

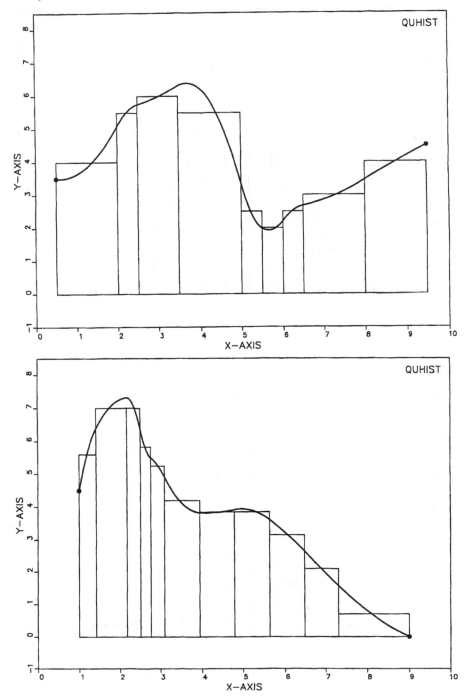

Figure 3.20. a, b.

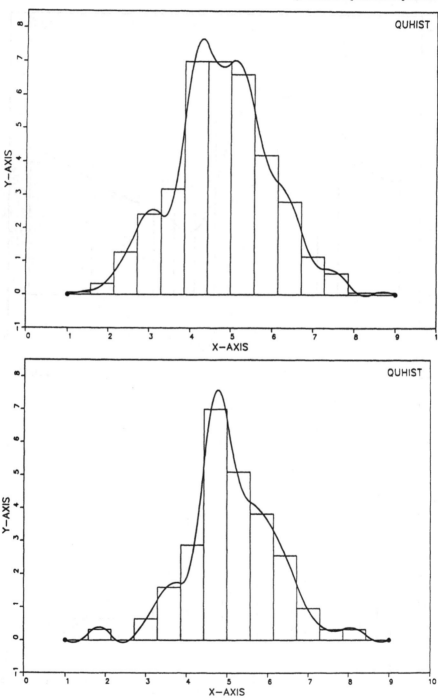

Figure 3.21. a, b.

and then substitute these expressions into the middle equation to obtain

$$h_{k-1}B_{k-1} + 2(h_{k-1} + h_k)B_k + h_kB_{k+1} = 6(y_{k+1} - y_k),$$
$$k = 2, \cdots, n-1, \tag{3.45}$$

with prescribed values $B_1 = y_1'$ and $B_n = y_n'$.

3.9. Quadratic Hermite Spline Interpolants

By Hermite interpolation, we mean that not only the function values y_k but also the first derivative values y_k' are to be interpolated at a given set of nodes $x_1 < x_2 < \cdots < x_n$. In contrast to usual quadratic and cubic spline interpolants, and even quadratic splines with knots intermediate to the nodes, quadratic and cubic Hermite spline interpolants do not require the solution of a system of linear equations. Hence, they have the advantage of being local in nature, i.e., they can be evaluated at any point without having to have processed all of the given data. The problem then lies in finding good estimates for the y_k' in terms of the x_k and the y_k, should these not already be given. For example, one might use divided differences for this purpose.

As in [145], we at first consider just a single interval $[x_k, x_{k+1}]$. Call this interval $[a, b]$ and the function values and first derivative values at a and b, y_a, y_b, y_a', and y_b', respectively. Does there exist a quadratic

$$p(x) = A + B(x - a) + C(x - a)^2$$

with

$$p(a) = y_a, p(b) = y_b, p'(a) = y_a', \text{and } p'(b) = y_b' \text{ ?}$$

These four conditions on the three coefficients A, B, and C yield (with $h = b - a$):

$$\begin{aligned}
A &= y_a, \\
A + hB + h^2C &= y_b, \\
B &= y_a', \\
B + 2hC &= y_b',
\end{aligned}$$

which on the one hand gives

$$C = \frac{1}{h^2}[y_b - y_a - hy_a'],$$

and on the other hand,

$$C = \frac{1}{2h}[y_b' - y_a'].$$

Hence a solution to this problem only exists when

$$\frac{y_b - y_a}{b - a} = \frac{y'_a + y'_b}{2}. \tag{3.46}$$

For example, if $y_a = y_b$, then we must have $y'_a = -y'_b$.

Now (3.46) almost never holds. In case it does not, we introduce an *additional knot* $z \in (a, b)$, fix two different quadratics,

$$\begin{array}{llll} p_1(x) & = & A_1 + B_1(x - a) + C_1(x - a)^2 & \text{in} \quad [a, z], \\ p_2(x) & = & A_2 + B_2(x - z) + C_2(x - z)^2 & \text{in} \quad [z, b], \end{array} \tag{3.47}$$

and determine the coefficients so that they join C^1 at the knot z and satisfy the interpolation conditions,

$$p_1(a) = y_a, \ \ p'_1(a) = y'_a, \ \ p_2(b) = y_b, \text{ and } p'_2(b) = y'_b.$$

From these, we immediately obtain

$$A_1 = y_a, \qquad B_1 = y'_a, \tag{3.48}$$

and

$$\begin{array}{rcl} A_2 + h_b B_2 + h_b^2 C_2 & = & y_b, \\ B_2 + 2h_b C_2 & = & y'_b, \end{array} \tag{3.49}$$

where $h_b = b - z$. The C^1 conditions at z, i.e.,

$$p_1(z) = p_2(z), \qquad p'_1(z) = p'_2(z),$$

yield

$$\begin{array}{rcl} A_1 + h_a B_1 + h_a^2 C_1 & = & A_2, \\ B_1 + 2h_a C_1 & = & B_2, \end{array} \tag{3.50}$$

where $h_a = z - a$. From these, we see that

$$\begin{array}{rcl} C_1 & = & \frac{1}{2h_a}(B_2 - y'_a), \\ A_2 & = & y_a + h_a y'_a + h_a^2 C_1. \end{array} \tag{3.51}$$

Now substitute (3.51) into the first equation of (3.49) to obtain

$$(\frac{h_a}{2} + h_b)B_2 + h_b^2 C_2 = y_b - y_a - \frac{h_a}{2}y'_a.$$

This and the second equation of (3.49) may be solved for

$$\begin{array}{rcl} B_2 & = & \frac{1}{h}[2(y_b - y_a) - h_a y'_a - h_b y'_b], \\ C_2 & = & \frac{1}{2h_b}(y'_b - B_2). \end{array} \tag{3.52}$$

The equations (3.48), (3.51), and (3.52) uniquely determine the desired coefficients.

Applying these results to each of $[x_k, y_k]$, $k = 1, \cdots, n - 1$, we see that for given y_k and y'_k, $k = 1, \cdots, n$, and knots,

$$z_k \in (x_k, x_{k+1}), \quad k = 1, \cdots, n - 1, \tag{3.53}$$

we may compute, using only local data, the two corresponding quadratics

```
      SUBROUTINE HEQUA(N,X,Y,GRAD,Z,EPS,T,A,B,C,IFLAG)
      DIMENSION X(N),Y(N),GRAD(N),Z(N-1),A(2*N-2),B(2*N-2),
     &          C(2*N-2),T(2*N-1)
      IFLAG=0
      IF (N.LT.2) THEN
          IFLAG=1
          RETURN
      END IF
      N2=2*N-3
      J=1
      GJ=GRAD(1)
      DO 10 K=1,N2,2
          J1=J+1
          IF ((X(J)+EPS).GT.Z(J).OR.Z(J).GT.(X(J1)-EPS)) THEN
              IFLAG=11
              RETURN
          END IF
          K1=K+1
          HA=Z(J)-X(J)
          HB=X(J1)-Z(J)
          GJ1=GRAD(J1)
          B(K)=GJ
          BK1=(2.*(Y(J1)-Y(J))-HA*GJ-HB*GJ1)/(X(J1)-X(J))
          B(K1)=BK1
          C(K)=(BK1-GJ)/(HA+HA)
          C(K1)=(GJ1-BK1)/(HB+HB)
          A(K)=Y(J)
          A(K1)=A(K)+HA*(GJ+HA*C(K))
          T(K)=X(J)
          T(K1)=Z(J)
          J=J1
          GJ=GJ1
 10   CONTINUE
      T(N2+2)=X(N)
      RETURN
      END
```

Figure 3.22. Program listing of HEQUA.

Calling Sequence:

CALL HEQUA(N,X,Y,GRAD,Z,EPS,T,A,B,C,IFLAG)

Purpose:
Given abscissas $x_1 < x_2 < \cdots < x_n$, corresponding ordinates $y_1, y_2, \cdots, y_n$, values $y_1', y_2', \cdots, y_n'$ for the derivatives, and intermediate points $z_1, z_2, \cdots, z_{n-1}$ with $x_k < z_k < x_{k+1}$ for $k = 1, \cdots, n-1$, HEQUA determines the coefficients A_k, B_k, and C_k, $k = 1, 2, \cdots, 2n-2$, of a local quadratic Hermite spline interpolant of the following form:

$$s(x) = \begin{cases} A_{2k-1} + B_{2k-1}(x - x_k) + C_{2k-1}(x - x_k)^2 & \text{if } x \in [x_k, z_k] \\ A_{2k} + B_{2k}(x - z_k) + C_{2k}(x - z_k)^2 & \text{if } x \in [z_k, x_{k+1}] \end{cases}$$

Description of the parameters:

N, X, Y as in QUAOPT.

GRAD   ARRAY(N): Upon calling must contain the values of the derivatives y_k', $k = 1, \cdots, n$.

Z   ARRAY(N−1): Upon calling must contain the values of the intermediate points z_k, $k = 1, \cdots, n-1$.

EPS   Accuracy test. If $(x_k + \text{EPS}) > z_k$ or $(x_{k+1} - \text{EPS}) < z_k$ for even one k, then execution is terminated with IFLAG=11. Recommendation: EPS=10^{-4}.

T   ARRAY(2*N−1): Upon completion contains the values $x_1 < z_1 < x_2 < z_2 < \cdots < x_{n-1} < z_{n-1} < x_n$.

A,B,C   ARRAY(2*N−2): Upon completion with IFLAG=0 contain the desired spline coefficients, K=1, 2, $\cdots$, 2*N−2.

IFLAG   =0:   Normal execution.
        =1: $N \geq 2$ is required.
        =11: $(x_k + \text{EPS}) > z_k$ or $(x_{k+1} - \text{EPS}) < z_k$ for some k.

Figure 3.23. Description of HEQUA.

to obtain, all together, $2n - 2$ such segments s_k, $k = 1, \cdots, 2n - 2$, with

$$\begin{aligned} s_{2k-1}(x) &= A_{2k-1} + B_{2k-1}(x - x_k) + C_{2k-1}(x - x_k)^2 & \text{in } [x_k, z_k], \\ s_{2k}(x) &= A_{2k} + B_{2k}(x - z_k) + C_{2k}(x - z_k)^2 & \text{in } [z_k, x_{k+1}], \end{aligned}$$

$$(3.54)$$

$k = 1, \cdots, n - 1$. Choosing z_k too near to x_k should be avoided for reasons of roundoff error. The subroutine HEQUA (Figs. 3.22 and 3.23), which implements the preceding procedure, takes this into consideration. The quadratic spline interpolant so obtained may again be evaluated by QUA-VAL.

In order to be able to use HEQUA, it is necessary to provide values for the y'_k. If these are not directly available, they will need to be approximated from the x_k and y_k. The next section is about several such procedures. We will also return to this kind of problem later.

3.10.   Approximation of First Derivative Values I

The simplest method is to use the slopes of the lines connecting the neighboring points as approximations for the y'_k, i. e.,

$$y'_k = \frac{y_{k+1} - y_{k-1}}{x_{k+1} - x_{k-1}}, \qquad k = 2, \cdots, n - 1, \tag{3.55}$$

and at the endpoints,

$$y'_1 = d_1, \qquad y'_n = d_{n-1}. \tag{3.56}$$

This is implemented in GRAD1 (Fig. 3.24).

In the case that the data come from an underlying function, a more accurate approximation can be made by taking y'_k to be the slope at x_k of the parabola through the $(k-1)$st, kth and $(k+1)$st points. For the endpoints, y'_1 and y'_n, we would use the slope at x_1 of the parabola passing through the first three points and at x_n of that through the last three points, respectively. From the Newton form of the interpolating polynomial, we obtain the explicit formula,

$$y'_k \;=\; \frac{\dfrac{y_k - y_{k-1}}{x_k - x_{k-1}}(x_{k+1} - x_k) + \dfrac{y_{k+1} - y_k}{x_{k+1} - x_k}(x_k - x_{k-1})}{x_{k+1} - x_{k-1}} \tag{3.57}$$

$$\;=\; \frac{h_k d_{k-1} + h_{k-1} d_k}{h_k + h_{k-1}}.$$

The formula for y'_1 can be obtained from (3.57) by switching indices 1 and 2 in the formula for y'_2. Similarly, that for y'_n is obtained by switching indices n and $n-1$ in the formula for y'_{n-1}. This is done in the subroutine GRAD2 (Fig. 3.25). Later, we will discuss a slightly modifed version, GRAD2B, and also replace the parabolas by simple rational functions with three parameters.

The subroutine GRAD3 (Figs. 3.26 and 3.27) uses NEWDIA to calculate the Newton interpolating polynomial of degree four through the points

```
      SUBROUTINE GRAD1(N,X,Y,B,IFLAG)
      DIMENSION X(N),Y(N),B(N)
      IFLAG=0
      IF (N.LT.2) THEN
          IFLAG=1
          RETURN
      END IF
      B(1)=(Y(2)-Y(1))/(X(2)-X(1))
      DO 10 K=2,N-1
          KP1=K+1
          KM1=K-1
          B(K)=(Y(KP1)-Y(KM1))/(X(KP1)-X(KM1))
10    CONTINUE
      B(N)=(Y(N)-Y(N-1))/(X(N)-X(N-1))
      RETURN
      END
```

Calling sequence:

CALL GRAD1(N,X,Y,B,IFLAG)

Purpose:
Calculation of the values for first derivatives by difference approximation.

Description of the parameters:

N	Number of given points.
X	ARRAY(N): Upon calling must contain the abscissas $x_k, k = 1, \cdots, n$.
Y	ARRAY(N): Upon calling must contain the ordinates $y_k, k = 1, \cdots, n$.
B	ARRAY(N): Upon completion with IFLAG=0 contains the values of the derivatives.
IFLAG	=0: Normal execution.
	=1: $N \geq 2$ is required.

Figure 3.24. Subroutine GRAD1 and its description.

```
      SUBROUTINE GRAD2(N,X,Y,B,IFLAG)
      DIMENSION X(N),Y(N),B(N)
      IFLAG=0
      IF (N.LT.3) THEN
         IFLAG=1
         RETURN
      END IF
      N1=N-1
      N2=N-2
      H1=X(2)-X(1)
      H2=X(3)-X(1)
      H3=(Y(2)-Y(1))/H1
      B(1)=(H3*H2-(Y(3)-Y(1))/H2*H1)/(X(3)-X(2))
      DO 10 K=2,N1
         KM1=K-1
         KP1=K+1
         H2=X(KP1)-X(K)
         H4=(Y(KP1)-Y(K))/H2
         B(K)=(H3*H2+H4*H1)/(X(KP1)-X(KM1))
         H1=H2
         H3=H4
10    CONTINUE
      H1=X(N)-X(N2)
      B(N)=((Y(N2)-Y(N))/H1*H2+H4*H1)/(X(N1)-X(N2))
      RETURN
      END
```

Calling sequence:

CALL GRAD2(N,X,Y,B,IFLAG)

Purpose:
Calculation of values for first derivatives (quadratic through three consecutive points).

Description of the parameters:

N,X,Y,B as in GRAD1.

IFLAG	=0:	Normal execution.
	=1:	$N \geq 3$ is required.

Figure 3.25. Subroutine GRAD2 and its description.

```
      SUBROUTINE GRAD3(N,X,Y,B,IFLAG)
      DIMENSION X(N),Y(N),B(N),A(5),TX(5),TY(5)
      INTEGER INDEX(5)
      IFLAG=0
      IF (N.LT.5) THEN
         IFLAG=1
         RETURN
      END IF
      CALL NEWDIA(4,X,Y,A,IFLAG)
      H=X(2)-X(1)
      H1=A(3)*H
      B(1)=A(2)-H1
      B(2)=A(2)+H1+A(4)*H*(X(2)-X(3))
      DO 30 K=3,N-2
         INDEX(1)=K
         INDEX(2)=K-1
         INDEX(3)=K+1
         INDEX(4)=K-2
         INDEX(5)=K+2
         IF (K.LE.2) INDEX(4)=K+3
         IF (K.EQ.1) INDEX(2)=K+4
         IF (K.GE.N-1) INDEX(5)=K-3
         IF (K.EQ.N) INDEX(3)=K-4
         DO 10 I=1,5
            TX(I)=X(INDEX(I))
            TY(I)=Y(INDEX(I))
10       CONTINUE
         CALL NEWDIA(5,TX,TY,A,IFLAG)
         XK=X(K)
         G=A(5)
         DO 20 I=4,2,-1
            G=G*(XK-TX(I))+A(I)
20       CONTINUE
         B(K)=G
30    CONTINUE
      DO 40 I=1,4
         IH=N-4+I
         TX(I)=X(IH)
         TY(I)=Y(IH)
40    CONTINUE
      CALL NEWDIA(4,TX,TY,A,IFLAG)
      H1=TX(3)-TX(1)
      H2=TX(3)-TX(2)
      B(N-1)=A(2)+A(3)*(H1+H2)+A(4)*H1*H2
      DO 50 I=1,3
         IH=N-3+I
         TX(I)=X(IH)
         TY(I)=Y(IH)
50    CONTINUE
      CALL NEWDIA(3,TX,TY,A,IFLAG)
      B(N)=A(2)+A(3)*(2.*TX(4)-TX(1)-TX(2))
      RETURN
      END
```

Figure 3.26. Program listing of GRAD3.

Calling sequence:

CALL GRAD3(N,X,Y,B,IFLAG)

Purpose:
Calculation of values for the first derivatives (quartic polynomial through five consecutive points (modifications at the boundary)).

Description of the parameters:

N,X,Y,B as in GRAD1.
IFLAG   =0:   Normal execution.
          =1:   N≥5 is required.

Required subroutine: NEWDIA.

Figure 3.27. Description of GRAD3.

indexed $k - 2, k - 1, k, k + 1$, and $k + 2$. This is then differentiated at x_k to obtain an estimate for y'_k. From experience, polynomials of degree four can behave badly near the endpoints. Hence, we use the following procedure: at x_1 and x_n the quartics are replaced by quadratics through points $1, 2, 3$, and $n - 2, n - 1, n$, respectively and at x_2 and x_{n-1} by cubics through points $1, 2, 3, 4$, and $n - 3, n - 2, n - 1, n$, respectively.

GRAD4 (Figs. 3.28 and 3.29) is based on [36]. The y'_k, $k = 3, \cdots, n - 2$, are approximated by the slope of the cubic, p_k, that passes through the three points indexed $k-1, k$, and $k+1$ and which, in addition, minimizes the weighted (by $1/h^2_{k-2}$ and $1/h^2_{k+1}$, respectively) sum of the square distances to y_{k-2} and y_{k+2} at x_{k-2} and x_{k+2}. Specifically, if we set

$$p_k(x) = A_k + B_k(x - x_k) + C_k(x - x_k)^2 + D_k(x - x_k)^3,$$

then

$$G(A_k, B_k, C_k, D_k) = \frac{1}{h^2_{k-2}}[p_k(x_{k-2}) - y_{k-2}]^2 + \frac{1}{h^2_{k+1}}[p_k(x_{k+2}) - y_{k+2}]^2$$

(3.58)

is to be minimized under the constraints that

$$p_k(x_j) = y_j, \qquad j = k - 1, k, k + 1.$$

(3.59)

From these constraints, it follows immediately that $A_k = y_k$, and moreover, B_k and C_k can be solved as functions of D_k. The substitution of

```
      SUBROUTINE GRAD4(N,X,Y,B,IFLAG)
      DIMENSION X(N),Y(N),B(N)
      IFLAG=0
      IF (N.LT.4) THEN
          IFLAG=1
          RETURN
      END IF
      DO 10 K=1,N
          KM1=K-1
          KP1=K+1
          IF (K.EQ.1) KM1=K+3
          IF (K.EQ.N) KP1=K-3
          X2=X(K)
          Y2=Y(K)
          XM1=X(KM1)
          XP1=X(KP1)
          X1=XM1-X2
          X3=XP1-X2
          PR1=(Y(KM1)-Y2)*X3
          PR2=(Y(KP1)-Y2)*X1
          DIV=(XM1-XP1)*X1*X3
          G=(X1*PR2-X3*PR1)/DIV
          CF=(PR1-PR2)/DIV
          IF (K.LT.3) THEN
              Z=0.0
              DIV=0.0
          ELSE
              KM2=K-2
              XM2=X(KM2)
              X0=XM2-X2
              XD=XM2-XM1
              XP=X0*XD*(XM2-XP1)
              W=XP/(XD*XD)
              Z=W*(Y(KM2)-Y2-X0*(G+X0*CF))
              DIV=W*XP
          END IF
          IF (K.LT.N-1) THEN
              KP2=K+2
              XP2=X(KP2)
              X4=XP2-X2
              XD=XP2-XP1
              XP=X4*XD*(XP2-XM1)
              W=XP/(XD*XD)
              Z=Z+W*(Y(KP2)-Y2-X4*(G+X4*CF))
              DIV=DIV+W*XP
          END IF
          B(K)=G+Z*X1*X3/DIV
10        CONTINUE
      RETURN
      END
```

Figure 3.28. Program listing of GRAD4.

Calling sequence:

CALL GRAD4(N,X,Y,B,IFLAG)

Purpose:
Calculation of values for first derivatives by the method of Ellis and McLain.

Description of the parameters:

N,X,Y,B as in GRAD1.
IFLAG   =0:   Normal execution.
         =1:   N$\geq$4 is required.

Figure 3.29. Description of GRAD4.

these expressions in the objective function, (3.58), yields a minimization problem of a single variable, which can be solved by the usual method of setting $dG/dD_k = 0$ and testing whether $d^2G/dD_k^2 > 0$ at the solution to verify that it is indeed a minimum. For the boundary points indexed 1 and 2, the y_k' are approximated by the slopes of the cubic through points 1, 2, 3, and 4 evaluated at x_1 and x_2, respectively. The right boundary points $n-1$ and n are treated analogously. In Figs. 3.30–3.31a, we see some examples illustrating the results of using HEQUA. To save space, we show only the outcomes of using GRAD1 and GRAD3 for the calculation of the approximate values of the y_k'.

In all of the plots, the z_k of (3.53) were chosen to be $z_k = (x_k + x_{k+1})/2$. For purposes of comparison, in Figs. 3.32a,b,c, we show what happens when the z_k are chosen, depending on the interval, so as to preserve local monotonicity and convexity/concavity.

3.11.   Shape Preservation through the Choice of Additional Knots

Now that we have become aquainted with several methods for obtaining values for the y_k', we can return to the problem of quadratic Hermite spline interpolation. We will show that, under certain conditions, the knots z_k of (3.54) can be chosen so as to preserve monotonicity and convexity/concavity. Since the method is local, we need only consider a general interval $[a, b]$. Now it can be shown that ([145]):

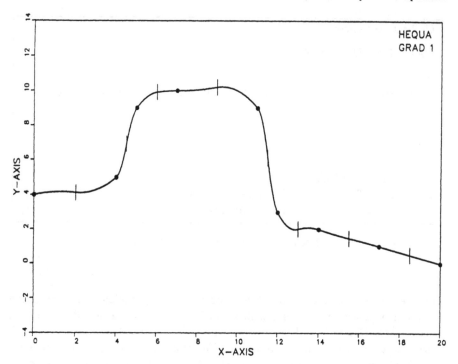

Figure 3.30.

1. If $y'_a y'_b \geq 0$, i.e., y'_a and y'_b have the same sign, then if $y'_z = B_2$
   also has this sign, then the C^1 quadratic spline formed by p_1 and p_2,
   (3.47), is monotone on $[a, b]$. This follows from the fact that then
   p'_1 and p'_2 are both line segments connecting points whose ordinates
   have the same sign.

2. If $y'_a \leq y'_z \leq y'_b$, then the Hermite spline interpolant is convex. This
   follows from the fact that then, by (3.51), $p''_1(x) = 2C_1 = (y'_z - y'_a)/h_a$
   and, by (3.52), $p''_2(x) = 2C_2 = (y'_b - y'_a)/h_b$ are both nonnegative.
   Similarly, if $y'_a \geq y'_z \geq y'_b$, then the interpolant is concave.

(These very simple conclusions are due to the fact that the polynomials
involved are only quadratics. For a *cubic* polynomial on an interval $[a, b]$,
it is still true that the Hermite interpolation problem can be solved locally,
but this cannot be shown in quite so elementary a fashion.)

   Conditions 1 and 2 would both be satisfied if we could choose $y'_z = B_2$
so that

$$y'_z = \gamma y'_a + (1 - \gamma) y'_b \qquad (3.60)$$

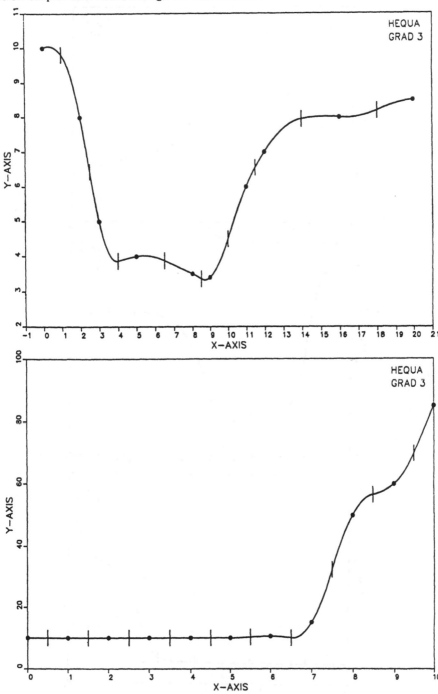

Figure 3.31. a, b.

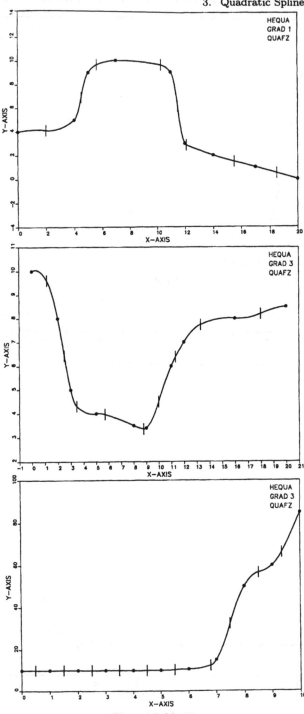

Figure 3.32. a-c.

for some $\gamma \in [0, 1]$. Then, fixing such a γ, the possibility of shape preservation by requiring conditions 1 and 2 reduces to being able to find a suitable knot z so that (3.60) holds.

From the conditions,

$$p_1'(z) = p_2'(z) = B_2 = y_z',$$

we see using (3.52) that for $y_a' \neq y_b'$,

$$z = \frac{1}{y_b' - y_a'}[-2(y_b - y_a) + b(y_b' + y_z') - a(y_a' + y_z')].$$

This, together with (3.60), yields

$$z = z(\gamma) = b - \gamma h + \frac{d}{y_b' - y_a'}, \tag{3.61}$$

where

$$d = -2(y_b - y_a) + h(y_a' + y_b'). \tag{3.62}$$

(d is actually the difference between the left and right sides of (3.46) times $2h$.) The condition,

$$a < z(\gamma) < b, \tag{3.63}$$

then reduces to

$$\alpha < \gamma < 1 + \alpha, \qquad \alpha = \frac{d}{h(y_b' - y_a')}. \tag{3.64}$$

Now the interval (3.64) has nonempty intersection with the interval $(0, 1)$ only if $-1 < \alpha < 1$. The possibilities for a knot z satisfying (3.60) are then $z \in (a + \alpha h, b)$ for $0 \leq \alpha < 1$ and $z \in (a, a + (1 + \alpha)h)$ for $-1 < \alpha < 0$. From experimental evidence, it seems that the best choice in either case is the midpoint,

$$z = a + \frac{1 + \alpha}{2}h. \tag{3.65}$$

In QUAFZ (Find Z, Fig. 3.33) α is computed by (3.64) and (3.62) if $| y_a' - y_b' | \geq \epsilon | y_a' |$. If $| \alpha | < 1 - \epsilon$, then z is computed by (3.65). If $| y_a' - y_b' | < \epsilon | y_a' |$ or $| \alpha | \geq 1 - \epsilon$, then z is set to $z = (a + b)/2$.

By applying QUAFZ, in sequence, to the intervals $[a, b] = [x_k, x_{k+1}]$ with $y_a = y_k$, $y_b = y_{k=1}$, $y_a' = y_k'$, and $y_b' = y_{k+1}'$, $k = 1, \cdots, n - 1$, we obtain knots $z_k = z$, which will either be the interval midpoint or a value chosen so that the resulting interpolant is locally shape-preserving in the sense of conditions 1 and 2 described previously.

Examples 3.32a,b,c were produced in this manner. In comparison with Figs. 3.30–3.31a, where the z_k were simply chosen to be the midpoints of

```
SUBROUTINE QUAFZ(A,B,YA,YB,Y1A,Y1B,EPS,Z)
H=B-A
Z=A+H/2.
IF (ABS(Y1A-Y1B).LE.EPS*ABS(Y1A)) RETURN
D=-2.*(YB-YA)+H*(Y1A+Y1B)
AL=D/(H*(Y1B-Y1A))
IF (ABS(AL).GE.(1.-EPS)) RETURN
Z=A+H*(1.+AL)/2.
RETURN
END
```

Calling sequence:

CALL QUAFZ(A,B,YA,YB,Y1A,Y1B,EPS,Z)

Purpose:
Determination when possible of shape-preserving knots for quadratic Hermite spline interpolation.

Description of the parameters:

A	Input:	Left endpoint of the interval.
B	Input:	Right endpoint of the interval.
YA	Input:	Ordinate value at A.
YB	Input:	Ordinate value at B.
Y1A	Input:	First derivative at A.
Y1B	Input:	First derivative at B.
EPS	Input:	Value for accuracy test.
		Recommendation: $EPS=10^{-4}$.
Z	Output:	Intermediate point with A<Z<B.

Figure 3.33. QUAFZ and its description.

the intervals $[x_k, x_{k+1}]$, there is evidently a distinct inprovement in the "critical" intervals. HEQUA should thus always be used with one of the derivative estimation routines (also those that we will give later) in conjunction with QUAFZ on each knot subinterval. As will become clear in Chapter 4, this method is a very serious competitor to cubic Hermite spline interpolation.

The method described in this section is similar to that of [145]. Another method is described in [78, 79], where up to two knots are inserted and slopes y'_k calculated appropriately. (See also Chapter 4.) A corresponding FORTRAN program may be found in [80]. One might also try to simultaneously calculate the y'_k and z_k in some suitable manner.

3.12. Quadratic Splines with Given Slopes at Intermediate Points

Suppose that we are given knots $x_1 < x_2 < \cdots < x_n$ and intermediate points,

$$z_k = \alpha_k x_k + (1 - \alpha_k)x_{k+1}, \qquad k = 1, \cdots, n-1, \qquad (3.66)$$

with

$$0 \le \alpha_k \le 1, \quad \alpha_k \ne \frac{1}{2}, \quad k = 1, \cdots, n-1, \qquad (3.67)$$

that are allowed to coincide with the knots but not with the midpoints. We look for quadratic segments s_k as in (3.1) forming a C^1 function with prescribed values, y'_k, of the first derivative at these intermediate points, i.e.,

$$s'_k(z_k) = y'_k, \qquad k = 1, \cdots, n-1. \qquad (3.68)$$

In addition, we prescribe boundary values y_0 and y_n with

$$s_1(x_1) = y_0, \qquad s_{n-1}(x_n) = y_n. \qquad (3.69)$$

This type of problem, but with $\alpha_k = \alpha, k = 1, \cdots, n-1, 0 \le \alpha \le 1, \alpha \ne \frac{1}{2}$, is treated in [127].

We will use the A_k of (3.1) as unknowns and set up a system of equations that these must satisfy. Now, from the continuity conditions,

$$A_k + h_k B_k + h_k^2 C_k = A_{k+1}, \qquad k = 1, \cdots, n-1, \qquad (3.70)$$

where, by (3.69),

$$A_1 = y_0, \qquad A_n = y_n. \qquad (3.71)$$

Furthermore, condition (3.68) yields

$$B_k + 2(z_k - x_k)C_k = y'_k, \qquad k = 1, \cdots, n-1. \qquad (3.72)$$

Then (3.70) and (3.72) may be solved for

$$B_k = \frac{2(z_k - x_k)(A_{k+1} - A_k)/h_k - h_k y'_k}{(z_k - x_k) - (x_{k+1} - z_k)} \qquad (3.73)$$

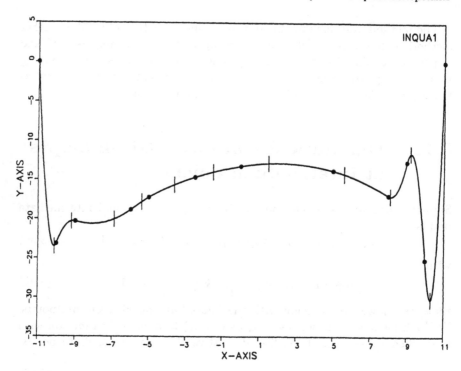

Figure 3.34.

and

$$C_k = \frac{y'_k - (A_{k+1} - A_k)/h_k}{(z_k - x_k) - (x_{k+1} - z_k)}, \tag{3.74}$$

$k = 1, \cdots, n-1$. This is always possible, since $\alpha_k \neq 1/2$. The C^1 conditions,

$$s'_{k-1}(x_k) = s'_k(x_k), \qquad k = 2, \cdots, n - 1,$$

then give the linear system of equations,

$$\frac{-(x_k - z_{k-1})}{h_{k-1}[(z_{k-1} - x_{k-1}) - (x_k - z_{k-1})]} A_{k-1} +$$

$$\left[\frac{x_k - z_{k-1}}{h_{k-1}[(z_{k-1} - x_{k-1}) - (x_k - z_{k-1})]} - \frac{z_k - x_k}{h_k[(z_k - x_k) - (x_{k+1} - z_k)]} \right] A_k$$

$$+ \frac{z_k - x_k}{h_k[(z_k - x_k) - (x_{k+1} - z_k)]} A_{k+1}$$

$$= \frac{1}{2} \left[\frac{h_{k-1}}{[(z_{k-1} - x_{k-1}) - (x_k - z_{k-1})]} y'_{k-1} + \frac{h_k}{[(z_k - x_k) - (x_{k+1} - z_k)]} y'_k \right],$$

$$k = 2, \cdots, n - 1. \tag{3.75}$$

```
      SUBROUTINE INQUA1(N,X,Y1,ALPHA,EPS1,EPS2,A,B,C,Z,
     &          IFLAG,D)
      DIMENSION X(N),Y1(N),ALPHA(N),A(N),B(N),C(N),Z(N),D(N)
      IFLAG=0
      IF (N.LT.3) THEN
         IFLAG=1
         RETURN
      END IF
      N1=N-1
      N2=N-2
      DO 10 K=1,N1
         H=ALPHA(K)
         IF (ABS(H-.5).LT.EPS1.OR.H.GT.1..OR.H.LT.0.) THEN
            IFLAG=22
            RETURN
         END IF
         Z(K)=H*X(K)+(1.-H)*X(K+1)
10    CONTINUE
      Y11=A(1)
      Y1N=A(N)
      DO 30 K=1,N1
         KP1=K+1
         KM1=K-1
         H2=X(KP1)-X(K)
         H3=Z(K)-X(K)
         H5=1./(H3-X(KP1)+Z(K))
         IF (K.EQ.1) GOTO 20
         H6=(X(K)-Z(KM1))*H4/H1
         H7=H3*H5/H2
         IF (K.GT.2) A(K-2)=-H6
         IF (K.LT.N1) C(KM1)=H7
         B(KM1)=H6-H7
         D(KM1)=(H1*H4*Y1(KM1)+H2*H5*Y1(K))/2.
         IF (K.EQ.2) D(KM1)=D(KM1)+H6*Y11
         IF (K.EQ.N1) D(KM1)=D(KM1)-H7*Y1N
20       H1=H2
         H4=H5
30    CONTINUE
      CALL TRIDU(N2,A,B,C,D,EPS2,IFLAG)
      IF(IFLAG.NE.0) RETURN
      DO 40 K=1,N2
         A(K+1)=D(K)
40    CONTINUE
      A(1)=Y11
      DO 50 K=1,N1
         K1=K+1
         H=X(K1)-X(K)
         G=(A(K1)-A(K))/H
         H1=Z(K)-X(K)
         H2=H1-X(K1)+Z(K)
         B(K)=(2.*H1*G-H*Y1(K))/H2
         C(K)=(Y1(K)-G)/H2
50    CONTINUE
      RETURN
      END
```

Figure 3.35. Program listing of INQUA1.

Calling sequence:

CALL INQUA1(N,X,Y1,ALPHA,EPS1,EPS2,A,B,C,Z,IFLAG,D)

Purpose:
Given abscissas x_k, $k = 1, \cdots, n$, ordinates y_1 and y_n corresponding to x_1
and x_n, and parameters α_k, $k = 1, \cdots, n-1$ with $0 \le \alpha_k \le 1$, $\alpha_k \ne 1/2$,
INQUA1 attempts to determine the coefficients A_k, B_k, and C_k of a
quadratic spline interpolant s with $s_k(x) = A_k + B_k(x-x_k) + C_k(x-x_k)^2$,
which has the given values y_k' for the first derivative at the points $z_k = \alpha_k x_k + (1 - \alpha_k)x_{k+1}$.

Description of the parameters:

N	Number of given points. N≥3 is required.
X	ARRAY(N): Upon calling must contain the abscissas x_k, $k = 1, \cdots, n$, with $x_1 < x_2 < \cdots < x_n$.
Y1	ARRAY(N): Input: Upon calling must contain the values of the first derivatives y_k', $k = 1, \cdots, n$.
EPS1	Value for accuracy test. Recommendation: EPS1=10^{-t} (t : number of available digits). A number is set to zero if it is less than EPS1.
EPS2	see TRIDIU.
ALPHA	ARRAY(N): Upon calling must contain the factors α_k, $k = 1, \cdots, n$.
A	ARRAY(N): Input: Upon calling A(1) must contain the value y_1 and A(N) the value y_n.
A,B,C	ARRAY(N): Upon completion with IFLAG=0 contain the desired spline coefficients.
Z	ARRAY(N): Upon completion contains the values z_k, $k = 1, \cdots, n - 1$.
IFLAG	=0: Normal execution.
	=1: N<3 not allowed.
	=2: Error in solving the linear system (TRIDIU).
	=22: $\alpha_k \ne 1/2$ and $0 \le \alpha_k \le 1$ required.
D	ARRAY(N): Work space.

Required subroutine: TRIDIU.

Figure 3.36. Description of INQUA1.

Because of (3.71), this is a system of $n - 2$ equations in the $n - 2$ unknowns $A_2, \cdots, A_{n-1}$. The coefficient matrix is not diagonally dominant and hence the existence of this type of interpolant cannot be shown by that method. Although a Laplace expansion of the determinant can be used to formulate a sufficient condition for existence, it is, however, difficult to verify *a priori*. Consequently, in INQUA1 (Figs. 3.35 and 3.36), we just try to find a solution numerically with TRIDIU (without pivoting).

The result of INQUA1 for abscissas $(-11, -10, -9, -6, -5, -2.5, 0, 5, 8, 9, 10, 11)$, first derivative values $(.5, .7, .1, 1.5, 1., .6, .1, -.8, -.3, 1., 1.1, 1.2)$, $\alpha's$ $(.1, .2, .3, .4, .4, .5, .6, .7, .8, .9., .8, .7)$, and boundary values $y_0 = y_n = 0$ is shown in Fig. 3.34. A critical evaluation should take into consideration the different scalings of the x- and y-axes. Of theoretical interest is what happens in the special cases when all the α_k are 0 or when they are all 1. In the first case, the y'_k are specified at x_k, $k = 2, \cdots, n$, and in the second case at $x_k, k = 1, \cdots, n - 1$. We restrict ourselves to the second case: $z_k = x_k, k = 1, \cdots, n - 1$. Then (3.75) simplifies to:

$$-A_{k-1} + A_k = \tfrac{1}{2} h_{k-1}(y'_{k-1} + y'_k), \qquad k = 2, \cdots, n - 1. \tag{3.76}$$

With boundary conditions (3.71), these equations have a unique solution. Condition (3.76) is identical to that of (3.46).

4

Cubic Spline Interpolants

4.1. First Derivatives as Unknowns

Suppose that we are again given abscissas

$$x_1 < x_2 \cdots < x_n, \qquad n \geq 3, \tag{4.1}$$

and corresponding ordinates y_k, $k = 1, \cdots, n$. We now wish to join cubic polynomials,

$$s_k(x) = A_k + B_k(x - x_k) + C_k(x - x_k)^2 + D_k(x - x_k)^3, \tag{4.2}$$

so that the resulting cubic spline interpolant, s, is *twice* continuously differentiable. From the interpolation conditions, $s_k(x_k) = y_k$, $s_k(x_{k+1}) = y_{k+1}$, $k = 1, \cdots, n-1$, it follows immediately that

$$A_k = y_k \tag{4.3}$$

and

$$h_k B_k + h_k^2 C_k + h_k^3 D_k = \Delta y_k, \tag{4.4}$$

$k = 1, \cdots, n-1$. Now

$$s_k'(x) = B_k + 2C_k(x - x_k) + 3D_k(x - x_k)^2$$

and
$$s''_k(x_k) = 2C_k + 6D_k(x - x_k).$$

Introducing $B_k = s'_k(x_k)$, $k = 1, \cdots, n-1$, and B_n as unknowns, the C^1 conditions, $s'_k(x_{k+1}) = s'_{k+1}(x_{k+1}) = B_{k+1}$, yield

$$B_k + 2h_kC_k + 3h_k^2D_k = B_{k+1}, \qquad k = 1, \cdots, n-1. \qquad (4.5)$$

Equations (4.4) and (4.5) may be solved for C_k and D_k as functions of B_k and B_{k+1}:

$$C_k = \frac{1}{h_k}(3d_k - 2B_k - B_{k+1}),$$

$$\qquad (4.6)$$

$$D_K = \frac{1}{h_k^2}(-2d_k + B_k + B_{k+1}),$$

for $k = 1, \cdots, n-1$. If the B_k were prescribed ahead of time, then we would have a *Hermite interpolation problem*, which could be easily solved from these equations. We will return to this problem later. In the present case, however, we have the C^2 conditions,

$$s''_{k-1}(x_k) = s''_k(x_k), \qquad k = 2, \cdots, n-1,$$

to consider. These are easily expressed as

$$2C_{k-1} + 6h_{k-1}D_{k-1} = 2C_k, \qquad k = 2, \cdots, n-1,$$

and we thus obtain a linear system of equations,

$$\frac{1}{h_{k-1}}B_{k-1} + 2(\frac{1}{h_{k-1}} + \frac{1}{h_k})B_k + \frac{1}{h_k}B_{k+1} = \frac{3}{h_{k-1}}d_{k-1} + \frac{3}{h_k}d_k,$$

$$\qquad (4.7)$$

$$k = 2, \cdots, n-1.$$

As this is a system of $n-2$ equations in n unknowns, $B_1, \cdots, B_n$, we need to add two extra conditions. This will be done in such a way so as to reduce the number of unknowns to $n-2$ and to also guarantee the existence of a solution to (4.7). In principle, one could assign values to B_1 and B_2 and then solve (4.7) by a simple recursion, but from experience such *asymmetric end conditions* are strongly reflected in the resulting spline interpolant. It is better to assign suitable values to B_1 and B_n, as then the coefficient matrix of (4.7) is symmetric, tridiagonal, and strictly diagonally dominant. For example, we could set:

1. B_1 and B_n arbitrarily.

2. $B_1 = d_1$, $B_n = d_{n-1}$, as the slopes of the lines through (x_k, y_k), $k = 1, 2$, and $k = n - 1, n$, respectively.

3. B_1 as the slope at x_1 of the quadratic through points 1 and 2 and having slope B_2 at x_2, and similarly, B_n as the slope at x_n of the quadratic passing through points $n - 1$ and n and having slope B_{n-1} at x_{n-1}.

4. B_1 as the slope at x_1 of the cubic passing through the first four points, and B_n as the slope at x_n of the cubic passing through the last four points. ($n \geq 4$ would be necessary.)

5. $B_1 = B_2$, and $B_{n-1} = B_n$.

We will consider other end conditions later.

For end conditions 1 and 2, the first and last equations of (4.7) must simply be modified to

$$2(\frac{1}{h_1} + \frac{1}{h_2})B_2 + \frac{1}{h_2}B_3 = \frac{3}{h_1}d_1 + \frac{3}{h_2}d_2 - \frac{1}{h_1}B_1$$

and   (4.8)

$$\frac{1}{h_{n-2}}B_{n-2} + 2(\frac{1}{h_{n-2}} + \frac{1}{h_{n-1}})B_{n-1} = \frac{3}{h_{n-2}}d_{n-2} + \frac{3}{h_{n-1}}d_{n-1} - \frac{1}{h_{n-1}}B_n,$$

and then the values of B_1 and B_n substituted for on the right. For the third condition, set

$$p(x) = \alpha + \beta(x - x_1) + \gamma(x - x_1)^2.$$

Then we must have

$$\begin{aligned}
p(x_1) &= \alpha = y_1, \\
p(x_2) &= y_1 + h_1\beta + h_1^2\gamma = y_2, \\
p'(x_2) &= \beta + 2h_1\gamma = B_2.
\end{aligned}$$

These yield

$$B_1 = \beta = 2d_1 - B_2,$$

and so the first equation of (4.7) becomes

$$(\frac{1}{h_1} + \frac{2}{h_2})B_2 + \frac{1}{h_2}B_3 = \frac{1}{h_1}d_1 + \frac{3}{h_2}d_2.$$ (4.9)

Similarly, the last equation becomes

$$\frac{1}{h_{n-2}}B_{n-2} + (\frac{2}{h_{n-2}} + \frac{1}{h_{n-1}})B_{n-1} = \frac{3}{h_{n-2}}d_{n-2} + \frac{1}{h_{n-1}}d_{n-1}.$$ (4.10)

For the fourth end condition we make use of the *Newton interpolating polynomial* (1.6), i. e.,

$$p_3(x) = a_0 + a_1(x - x_1) + a_2(x - x_1)(x - x_2) + a_3(x - x_1)(x - x_2)(x - x_3).$$

The coefficients a_j may be explicitly computed from (1.7) and (1.8) as

$$a_0 = y_1,$$
$$a_1 = d_1,$$
$$a_2 = \frac{d_2 - d_1}{h_1 + h_2},$$

and

$$a_3 = \frac{\dfrac{d_3 - d_2}{h_2 + h_3} - \dfrac{d_2 - d_1}{h_1 + h_2}}{h_1 + h_2 + h_3}.$$

Since

$$p_3'(x) = a_1 + a_2(2x - x_1 - x_2)$$
$$+ a_3[(x - x_1)(x - x_2) + (x - x_1)(x - x_3) + (x - x_2)(x - x_3)],$$
$$p_3''(x) = 2a_2 + 2a_3[(x - x_1) + (x - x_2) + (x - x_3)],$$

and

$$p_3'''(x) = 6a_3, \tag{4.11}$$

it follows that

$$p_3'(x_1) = a_1 - h_1 a_2 + h_1(h_1 + h_2)a_3, \tag{4.12}$$
$$p_3''(x_1) = 2a_2 - 2(2h_1 + h_2)a_3. \tag{4.13}$$

Similarly, for the right endpoint, using

$$\tilde{p}_3(x) = \tilde{a}_0 + \tilde{a}_1(x - x_{n-3}) + \tilde{a}_2(x - x_{n-3})(x - x_{n-2})$$
$$+ \tilde{a}_3(x - x_{n-3})(x - x_{n-2})(x - x_{n-1}),$$

we have

$$\tilde{a}_0 = y_{n-3},$$
$$\tilde{a}_1 = d_{n-3},$$
$$\tilde{a}_2 = \frac{d_{n-2} - d_{n-3}}{h_{n-3} + h_{n-2}},$$
$$\tilde{a}_3 = \frac{\dfrac{d_{n-1} - d_{n-2}}{h_{n-1} + h_{n-2}} - \dfrac{d_{n-2} - d_{n-3}}{h_{n-2} + h_{n-3}}}{h_{n-3} + h_{n-2} + h_{n-1}},$$

and

$$\begin{aligned}
\tilde{p}_3'(x_n) &= \tilde{a}_1 + \tilde{a}_2(h_{n-3} + 2h_{n-2} + 2h_{n-1}) + \tilde{a}_3[(h_{n-2} + h_{n-1})h_{n-1} \\
&\quad + (h_{n-3} + h_{n-2} + h_{n-1})(h_{n-2} + 2h_{n-1})], & (4.14) \\
\tilde{p}_3''(x_n) &= 2\tilde{a}_2 + 2\tilde{a}_3(h_{n-3} + 2h_{n-2} + 3h_{n-1}), & (4.15) \\
\tilde{p}_3'''(x_n) &= 6\tilde{a}_3. & (4.16)
\end{aligned}$$

Hence, for end condition 4, (4.8) would have to be modified by setting
Hence, for end condition 4, (4.8) would have to be modified by setting
$B_1 = p_3'(x_1)$ according to (4.12) and $B_n = \tilde{p}_3'(x_n)$ according to (4.14).

Finally, for the fifth type of end condition, the first equation of (4.7)
becomes

$$(\frac{3}{h_1} + \frac{2}{h_2})B_2 + \frac{1}{h_2}B_3 = \frac{3}{h_1}d_1 + \frac{3}{h_2}d_2, \qquad (4.17)$$

and the last,

$$\frac{1}{h_{n-2}}B_{n-2} + (\frac{2}{h_{n-2}} + \frac{3}{h_{n-1}})B_{n-1} = \frac{3}{h_{n-2}}d_{n-2} + \frac{3}{h_{n-1}}d_{n-1}. \qquad (4.18)$$

In all five cases, neither the number of equations was changed nor was the
the inherent *strict diagonal dominance* of the coefficient matrix affected.
Hence, the associated cubic spline interpolants do exist. The corresponding
systems of equations are set up in CUB1R5 (1st derivatives as unknowns, 5
possible end conditions, Figs. 4.1 and 4.2). It solves the systems by means
of TRIDIS and then computes the remaining coefficients of (4.2) by (4.3)
and (4.6). Examples will be given in the next section, where we first discuss
yet other possible end conditions.

4.2.   Second Derivatives as Unknowns

We now take the second derivatives, $C_k = \frac{1}{2}s_k''(x_k)$, $k = 1, \cdots, n-1$, and
C_n, as unknowns. Since $s_k''(x_{k+1}) = 2C_{k+1}$, we must have

$$2C_k + 6h_k D_k = 2C_{k+1}, \qquad k = 1, \cdots, n-1. \qquad (4.19)$$

Then from (4.4) and (4.19), we can express the B_k and D_k as functions of
C_k and C_{k+1}:

$$\begin{aligned}
B_k &= d_k - \frac{1}{3}h_k(2C_k + C_{k+1}), \\
& \qquad\qquad\qquad\qquad\qquad k = 1, \cdots, n-1, \\
D_k &= \frac{1}{3h_k}(C_{k+1} - C_k).
\end{aligned} \qquad (4.20)$$

```
      SUBROUTINE CUB1R5 (N,X,Y,EPS,A,B,C,D,IR,IFLAG)
      DIMENSION X(N),Y(N),A(N),B(N),C(N),D(N)
      IFLAG=0
      IF(N.LT.3) THEN
          IFLAG=1
          RETURN
      END IF
      IF(IR.LT.1.OR.IR.GT.5) THEN
          IFLAG=5
          RETURN
      END IF
      N1=N-1
      N2=N-2
      N3=N-3
      G1=Y(2)-Y(1)
      G2=Y(3)-Y(2)
      GN1=Y(N1)-Y(N2)
      GN=Y(N)-Y(N1)
      H1=X(2)-X(1)
      H2=X(3)-X(2)
      HN1=X(N1)-X(N2)
      HN=X(N)-X(N1)
      IF(N.GT.3) THEN
          G3=Y(4)-Y(3)
          GN2=Y(N2)-Y(N3)
          H3=X(4)-X(3)
          HN2=X(N2)-X(N3)
      END IF
      C1=0.
      D1=0.
      CN2=0.
      DN2=0.
      IF(IR.EQ.1) GOTO 10
      IF(IR.EQ.2) THEN
          B(1)=G1/H1
          B(N)=GN/HN
          GOTO 10
      END IF
      B(1)=0.
      B(N)=0.
      IF(IR.EQ.3) THEN
          C1=-1./H1
          D1=-2.*G1/(H1*H1)
          CN2=-1./HN
          DN2=-2.*GN/(HN*HN)
          GOTO 10
      END IF
      IF(IR.EQ.4) THEN
          IF(N.LT.4) THEN
              IFLAG=6
              RETURN
          END IF
          A1=G1/H1
          A2=(G2/H2-A1)/(H2+H1)
          A3=(((G3/H3-G2/H2)/(H3+H2))-A2)/(H1+H2+H3)
          B(1)=A1-H1*A2+A3*(H1*(H1+H2))
          A1=GN2/HN2
          A2=(GN1/HN1-A1)/(HN1+HN2)
          A3=(((GN/HN-GN1/HN1)/(HN+HN1))-A2)/(HN2+HN1+HN)
          B(N)=A1+A2*(HN2+2*HN1+2*HN)+
     &        A3*((HN2+HN1+HN)*(HN1+2.*HN)+(HN1+HN)*HN)
```

(cont.)

```
            GOTO 10
          END IF
          IF(IR.EQ.5) THEN
              C1=1./H1
              CN2=1./HN
              GOTO 10
          END IF
10        B1=B(1)
          B(1)=1./H2
          IF (N.EQ.3) THEN
              C(1)=2.*(1./H1+1./H2)+C1+CN2
              D(1)=3.*(G1/(H1*H1)+G2/(H2*H2))-B1/H1-B(N)/HN+D1+DN2
          ELSE
              C(1)=2.*(1./H1+1./H2)+C1
              C(N2)=2.*(1./HN1+1./HN)+CN2
              D(1)=3.*(G1/(H1*H1)+G2/(H2*H2))-B1/H1+D1
              D(N2)=3.*(GN1/(HN1*HN1)+GN/(HN*HN))-B(N)/HN+DN2
          END IF
          E1=H2
          F1=G2
          DO 20 K=2,N3
              K1=K+1
              K2=K+2
              E2=X(K2)-X(K1)
              F2=Y(K2)-Y(K1)
              B(K)=1./E2
              C(K)=2.*(1./E1+1./E2)
              D(K)=3.*(F1/(E1*E1)+F2/(E2*E2))
              E1=E2
              F1=F2
20        CONTINUE
          CALL TRIDIS(N2,C,B,D,EPS,IFLAG)
          IF(IFLAG.NE.0) RETURN
          DO 30 K=N2,1,-1
              B(K+1)=D(K)
30        CONTINUE
          B(1)=B1
          IF(IR.EQ.3) THEN
              B(1)=-B(2)+2.*G1/H1
              B(N)=-B(N1)+2.*GN/HN
          END IF
          IF(IR.EQ.5) THEN
              B(1)=B(2)
              B(N)=B(N1)
          END IF
          DO 40 K=1,N1
              K1=K+1
              DX=X(K1)-X(K)
              DY=Y(K1)-Y(K)
              A(K)=Y(K)
              DYX=DY/DX
              C(K)=(3.*DYX-2.*B(K)-B(K1))/DX
              D(K)=(-2.*DYX+B(K)+B(K1))/(DX*DX)
40        CONTINUE
          RETURN
          END
```

Figure 4.1. Program listing of CUB1R5.

Calling sequence:

CALL CUB1R5(N,X,Y,EPS,A,B,C,D,IR,IFLAG)

Purpose:
Determination of the coefficients A_k, B_k, C_k, and D_k of a cubic spline interpolant. Five different end conditions may be used. They must, however, be of the same type at both ends.

Description of the parameters:

N	Number of given points. N$\geq$ 3 is necessary. End condition 4 requires N$\geq$ 4.
X	ARRAY(N): Upon calling must contain the abscissa values x_k, $k = 1, \cdots, n$, with $x_1 < x_2 < \cdots < x_n$.
Y	ARRAY(N): Upon calling must contain the ordinate values y_k, $k = 1, \cdots, n$.
EPS	see TRIDIS.
A,B,C,D	ARRAY(N): Upon execution with IFLAG=0 contain the desired spline coefficients, $k = 1, \cdots, n - 1$.
IR	$1 \leq$ IR ≤ 5. The number of the desired end condition. For IR=1 y_1' and y_n' must set in B(1) and B(N), respectively.
IFLAG	=0: Normal execution.
	=1: N$\geq$ 3 required.
	=2: Error in solving the system of equations.
	=5: IR$<$ 1 or IR$>$ 5 not allowed.
	=6: End condition 4 requires N$\geq$ 4.

Required subroutine: TRIDIS.

Figure 4.2. Description of CUB1R5.

If the C_k were to be prescribed ahead of time, then we would have a modified *Hermite interpolation problem*, which could easily be solved by means of these equations. In the present case, however, we have yet to satisfy the C^1 conditions,

$$s_{k-1}'(x_k) = s_k'(x_k), \qquad k = 2, \cdots, n - 1.$$

These are easily expressed as

$$B_{k-1} + 2h_{k-1}C_{k-1} + 3h_{k-1}^2 D_{k-1} = B_k,$$

and then an application of (4.20) yields the linear system,

$$h_{k-1}C_{k-1} + 2(h_{k-1} + h_k)C_k + h_kC_{k+1} = 3(d_k - d_{k-1}),$$

$$(4.21)$$

$$k = 2, \cdots, n - 1.$$

This is again a system of $n - 2$ equations in n unknowns $C_1, \cdots, C_n$, and just as before, we must add two extra conditions. We will arrange this so that the number of equations is not increased.

We could, for example, choose:

6. $2C_1 = y_1''$, $2C_n = y_n''$, arbitrarily.

7. $y_1''' = y_2'''$, $y_{n-1}''' = y_n'''$ ($n \geq 5$ would be necessary), i. e., $s_1 = s_2$ and $s_{n-2} = s_{n-1}$ ([1]).

8. $y_1'' = (d_2 - d_1)/(h_1 + h_2)$, $\quad y_n'' = (d_{n-1} - d_{n-2})/(h_{n-2} + h_{n-1})$, i. e., y_1'' as the value of the second derivative of the quadratic through the first three points, and y_n'' that of the quadratic through the last three.

9. y_1'' and y_n'' as values of the second derivatives of the cubics through the first four and last four points evaluated at x_1 and x_n, respectively.

10. y_1''', y_n''' arbitrarily (for $y_1''' = y_n''' = 0$, by (4.20), we would have $y_1'' = y_2''$ and $y_{n-1}'' = y_n''$, i. e., the spline is actually a quadratic in the first and last intervals ([151])).

11. y_1''', y_n''' as the third derivative of the cubic through the first four and last four points ($n \geq 4$ would be necessary), respectively ([44]).

12. $y_1'' = y_n'' = 0$, i. e., $C_1 = C_n = 0$.

For end conditions 6,8,9, and 12, we simply write the first and last equations of (4.21) in the form,

$$2(h_1 + h_2)C_2 + h_2C_3 = 3(d_2 - d_1) - h_1C_1,$$

$$(4.22)$$

$$h_{n-2}C_{n-2} + 2(h_{n-2} + h_{n-1})C_{n-1} = 3(d_{n-1} - d_{n-2}) - h_{n-1}C_n,$$

and substitute $C_1 = \frac{1}{2}y_1''$ and $C_n = \frac{1}{2}y_n''$ on the right with the appropriate values of y_1'' and y_n''.

For end condition 7, we make use of the second equation of (4.20) and the fact that $s_k'''(x_k) = 6D_k$. From these, we easily obtain additional equations,

$$\frac{1}{h_1}(C_2 - C_1) = \frac{1}{h_2}(C_3 - C_2),$$

$$\frac{1}{h_{n-2}}(C_{n-1} - C_{n-2}) = \frac{1}{h_{n-1}}(C_n - C_{n-1}),$$

which can be solved for C_1 and C_n and then substituted into the first and last equations of (4.21). The result is

$$[2(h_1 + h_2) + h_1 + \frac{h_1^2}{h_2}]C_2 + (h_2 - \frac{h_1^2}{h_2})C_3 \;=\; 3(d_2 - d_1),$$

$$(h_{n-2} - \frac{h_{n-1}^2}{h_{n-2}})C_{n-2} + [2(h_{n-2} + h_{n-1}) \tag{4.23}$$

$$+h_{n-1} + \frac{h_{n-1}^2}{h_{n-2}}]C_{n-1} \;=\; 3(d_{n-1} - d_{n-2}).$$

Here, $n \geq 5$ is necessary, as the reader can easily verify.

End conditions 9 are fulfilled by substituting $C_1 = \frac{1}{2}p_3''(x_1)$ from (4.13) and $C_n = \frac{1}{2}\tilde{p}_3''(x_n)$ from (4.15) into (4.22).

End conditions 10 are accomplished by using $y_1''' = 6D_1$ and $y_n''' = 6D_{n-1}$. From (4.20), we see that

$$C_1 = C_2 - \frac{1}{2}h_1 y_1'''$$

and

$$C_n = C_{n-1} + \frac{1}{2}h_{n-1}y_n'''.$$

Now substitute these into the first and last equations of (4.20) to obtain

$$(3h_1 + 2h_2)C_2 + h_2 C_3 \;=\; 3(d_2 - d_1) + \frac{1}{2}h_1^2 y_1''',$$

$$\tag{4.24}$$

$$h_{n-2}C_{n-2} + (2h_{n-2} + 3h_{n-1})C_{n-1} \;=\; 3(d_{n-1} - d_{n-2}) - \frac{1}{2}h_{n-1}^2 y_n'''.$$

Finally, for end conditions 11, we just substitute $y_1''' = 6a_3$ from (4.11) and $y_n''' = 6\tilde{a}_3$ from (4.16) into (4.24).

In all seven cases, we have neither changed the number of equations nor altered the fact that the coefficient matrix is strictly diagonally dominant. This is easily seen from (4.22), (4.23), and (4.24). The corresponding cubic spline interpolants are therefore uniquely determined. The associated linear systems are set up and solved in CUB2R7 (2nd derivatives as unknowns, 7 possible end conditions, Figs. 4.3 and 4.4). The remaining coefficients are calculated from (4.3) and (4.20).

It should be noted that the first five end conditions can also be realized with the second derivatives as unknowns and the last seven with the first

```
SUBROUTINE CUB2R7(N,X,Y,EPS,A,B,C,D,IR,IFLAG)
DIMENSION X(N),Y(N),A(N),B(N),C(N),D(N)
IFLAG=0
IF(N.LT.3) THEN
    IFLAG=1
    RETURN
END IF
IF(IR.LT.6.OR.IR.GT.12) THEN
    IFLAG=5
    RETURN
END IF
N1=N-1
N2=N-2
N3=N-3
G1=Y(2)-Y(1)
G2=Y(3)-Y(2)
GN1=Y(N1)-Y(N2)
GN=Y(N)-Y(N1)
H1=X(2)-X(1)
H2=X(3)-X(2)
HN1=X(N1)-X(N2)
HN=X(N)-X(N1)
IF(N.GT.3) THEN
    G3=Y(4)-Y(3)
    GN2=Y(N2)-Y(N3)
    H3=X(4)-X(3)
    HN2=X(N2)-X(N3)
END IF
B1=0.
C1=0.
D1=0.
BN2=0.
DN2=0.
IF(IR.EQ.6) GOTO 20
C(1)=0.
C(N)=0.
IF(IR.EQ.7) THEN
    IF(N.LT.5) THEN
        IFLAG=6
        RETURN
    END IF
    H=H1*H1/H2
    B1=H1+H
    C1=-H
    H=HN*HN/HN1
    BN2=HN+H
    CN2=-H
    GOTO 20
END IF
IF(IR.EQ.8) THEN
    C(1)=(G2/H2-G1/H1)/(H1+H2)
    C(N)=(GN/HN-GN1/HN1)/(HN+HN1)
    GOTO 20
END IF
IF(IR.EQ.9.OR.IR.EQ.11) THEN
    IF(N.LT.4) THEN
        IFLAG=7
        RETURN
```

(cont.)

```
            END IF
            A2=(G2/H2-G1/H1)/(H2+H1)
            A3=(((G3/H3-G2/H2)/(H3+H2))-A2)/(H1+H2+H3)
            IF(IR.EQ.9) C(1)=2.*A2-2.*A3*(2.*H1+H2)
            IF(IR.EQ.11) D(1)=6.*A3
            A2=(GN1/HN1-GN2/HN2)/(HN1+HN2)
            A3=(((GN/HN-GN1/HN1)/(HN+HN1))-A2)/(HN2+HN1+HN)
            IF(IR.EQ.9) C(N)=2.*A2-2.*A3*(HN2+2.*HN1+3.*HN)
            IF(IR.EQ.11) THEN
                D(N1)=6.*A3
                GOTO 10
            END IF
            GOTO 20
        END IF
10      IF(IR.EQ.10.OR.IR.EQ.11) THEN
            DH1=D(1)
            DHN=D(N1)
            B1=H1
            D1=H1*H1*DH1/2.
            BN2=HN
            DN2=-HN*HN*DHN/2.
        END IF
20      CH1=C(1)/2.
        C(N)=C(N)/2.
        C(1)=H2+C1
        IF(N.EQ.3) THEN
            B(1)=2.*(H1+H2)+B1+BN2
            D(1)=3.*(G2/H2-G1/H1)-H1*CH1-HN*C(N)+D1+DN2
        ELSE
            B(1)=2.*(H1+H2)+B1
            B(N2)=2.*(HN1+HN)+BN2
            D(1)=3.*(G2/H2-G1/H1)-H1*CH1+D1
            D(N2)=3.*(GN/HN-GN1/HN1)-HN*C(N)+DN2
        END IF
        E1=H2
        F1=G2
        DO 30 K=2,N3
            K1=K+1
            K2=K+2
            E2=X(K2)-X(K1)
            F2=Y(K2)-Y(K1)
            C(K)=E2
            B(K)=2.*(E1+E2)
            D(K)=3.*(F2/E2-F1/E1)
            E1=E2
            F1=F2
30      CONTINUE
        IF(IR.EQ.7) C(N3)=C(N3)+CN2
        CALL TRIDIS(N2,B,C,D,EPS,IFLAG)
        IF(IFLAG.NE.0) RETURN
        DO 40 K=N2,1,-1
            C(K+1)=D(K)
40      CONTINUE
        C(1)=CH1
        IF(IR.EQ.7) THEN
            H=H1*H1/H2
            C(1)=((H1+H)*C(2)-H*C(3))/H1
```

(*cont.*)
```
      H=HN*HN/HN1
      C(N)=((H+HN)*C(N1)-H*C(N2))/HN
   END IF
   IF (IR.EQ.10.OR.IR.EQ.11) THEN
      C(1)=C(2)-H1*DH1/2.
      C(N)=C(N1)+HN*DHN/2.
   END IF
   DO 50 K=1,N1
      K1=K+1
      H=X(K1)-X(K)
      A(K)=Y(K)
      B(K)=(Y(K1)-Y(K))/H-1./3.*H*(2*C(K)+C(K1))
      D(K)=1./(3.*H)*(C(K1)-C(K))
50 CONTINUE
   RETURN
   END
```

Figure 4.3. Program listing of CUB2R7.

Calling sequence:

CALL CUB2R7(N,X,Y,EPS,A,B,C,D,IR,IFLAG)

Purpose:
Determination of the coefficients A_k, B_k, C_k, and D_k of a cubic spline interpolant. Seven different end conditions may be used. They must, however, be of the same type at both ends. $N \geq 3$ is required. End condition 7 requires $N \geq 5$. End conditions 9 and 11 require $N \geq 4$.

Description of the parameters:

N,X,Y,EPS,A,B,C,D as in CUB1R5.

IR $6 \leq IR \leq 12$. Number of the desired end condition. For IR=6, y_1'' and y_n'' must be given in C(1) and C(N), respectively. For IR=10, y_1''' and y_n''' must be given in D(1) and D(N$-$1), respectively.

IFLAG   =0:   Normal execution.

           =1: $N \geq 3$ is necessary.

           =2:   Error in solving the system of equations.

           =5:   IR< 6 or IR> 12 not allowed.

           =6: $N \geq 5$ required for end condition 7.

           =7: $N \geq 4$ required for end condition 9 or 11.

Required subroutine: TRIDIS.

Figure 4.4. Description of CUB2R7.

derivatives as unknowns, albeit at the price of having an $n \times n$ instead of an $(n-2) \times (n-2)$ coefficient matrix. The new matrix can, however, be arranged to still be tridiagonal. In principle, we could consider yet other end conditions [1,151,154]. In particular, we could prescribe derivative values at points other than x_1 and x_n ([38]), but this would have led to a program rather difficult to read.

All these splines can be evaluated using CUBVAL (Figs. 4.5 and 4.6). Figures 4.7a–4.10c show the results of the 12 end conditions applied to the same data set. As these plots are all somewhat different, we can see the influence of these various end conditions. The following specific values were used: for IR=1, $y_1' = -2$ and $y_n' = -.2$; for IR=6, $y_1'' = 2$ and $y_n'' = 3$; for IR=10 and IR=12, $y_1'' = y_n'' = y_1''' = y_n''' = 0$. A number of cubic spline examples with IR=8 are given in Figs. 4.11a–4.16b. Later, we shall use these same examples to show how other methods can often be used to obtain even better interpolants. In most cases, only very small differences with the results of the other values of IR could be observed. Hence, in the coming chapters, we will generally only consider one or two types of end conditions. Usually, we will still be able to specify values for the first or second derivative at the endpoints. As such, one can naturally also set these to the values corresponding to one of the 12 cubic spline end conditions.

4.3.   Periodic Cubic Spline Interpolants

A cubic spline interpolant s is said to be periodic if

$$s_1^{(j)}(x_1) = s_{n-1}^{(j)}(x_n), \qquad j = 0, 1, 2. \tag{4.25}$$

As we shall see later, such *periodic spline interpolants* are especially useful for interpolating closed curves in the plane.

Obviously, the case $j = 0$ gives $y_n = y_1$. The rest of the development depends on whether we decide to use the B_k or the C_k as unknowns. Here, we choose the C_k, since the coefficients in (4.21) are more easily computed than those in (4.7). By (4.25), with $j = 2$, we must have

$$C_n = C_1. \tag{4.26}$$

Hence, the last equation of (4.21) becomes

$$h_{n-1}C_1 + h_{n-2}C_{n-2} + 2(h_{n-2} + h_{n-1})C_{n-1} = 3(d_{n-1} - d_{n-2}). \tag{4.27}$$

Here, we must use the fact that $y_n = y_1$ in the calculation of d_{n-1}. The remaining condition of (4.25), for $j = 1$, yields

$$B_1 = B_{n-1} + 2h_{n-1}C_{n-1} + 3h_{n-1}^2 D_{n-1},$$

```
FUNCTION CUBVAL(N,X,A,B,C,D,V,IFLAG)
DIMENSION X(N),A(N),B(N),C(N),D(N)
DATA I/1/
IFLAG=0
IF(N.LT.2) THEN
    IFLAG=1
    RETURN
END IF
CALL INTONE(X,N,V,I,IFLAG)
IF(IFLAG.NE.0) RETURN
DX=V-X(I)
CUBVAL=A(I)+DX*(B(I)+DX*(C(I)+D(I)*DX))
RETURN
END
```

Figure 4.5. Program listing of CUBVAL.

Calling sequence:

CALL CUBVAL(N,X,A,B,C,D,V,IFLAG)

Purpose:
CUBVAL is a FUNCTION subprogram for the calculation of a function value of a cubic spline interpolant at a point $V \in [X(1), X(N)]$.

Description of the parameters:

N	Number of given points. $N \geq 2$ is required.
X	ARRAY(N): x-values.
A,B,C,D	ARRAY(N): Vectors containing the spline coefficients.
V	Value of the point at which the spline interpolant is to be evaluated.
IFLAG	=0: Normal execution.
	=1: $N \geq 2$ is required.
	=3: Error in the determination of the interval (INTONE).

Required subroutine: INTONE.

Remark: The statement 'DATA I/1/' has the effect that I is set to 1 at the first call to CUBVAL.

Figure 4.6. Description of CUBVAL.

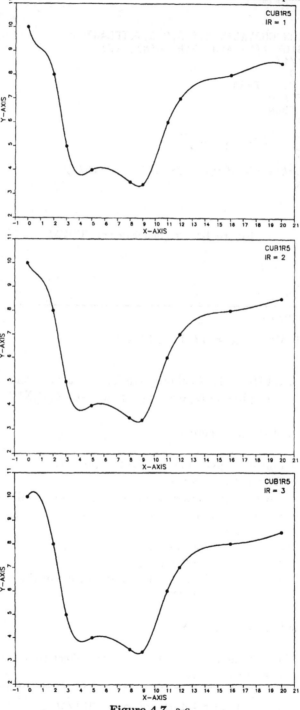

Figure 4.7. a-c.

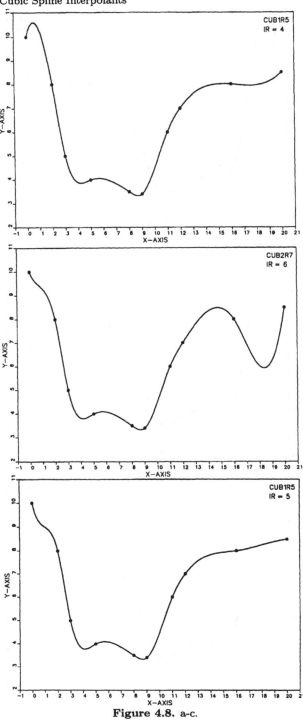

Figure 4.8. a-c.

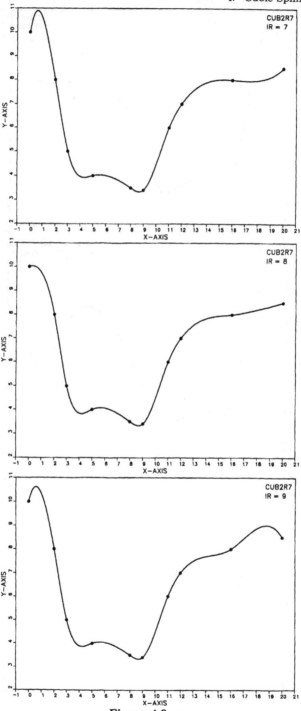

Figure 4.9. a-c.

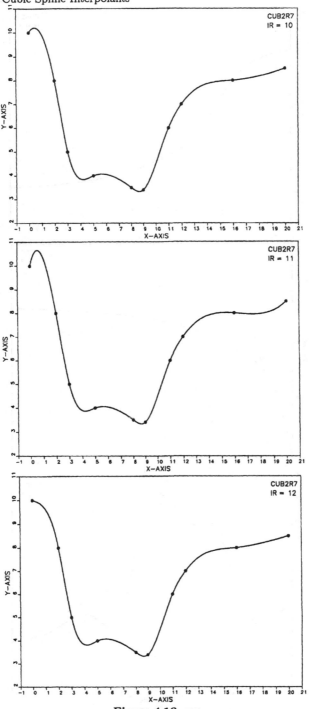

Figure 4.10. a-c.

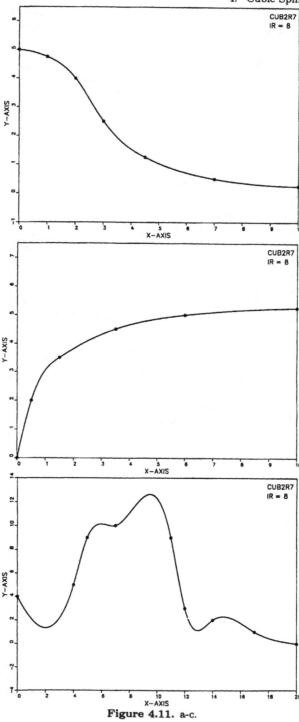

Figure 4.11. a-c.

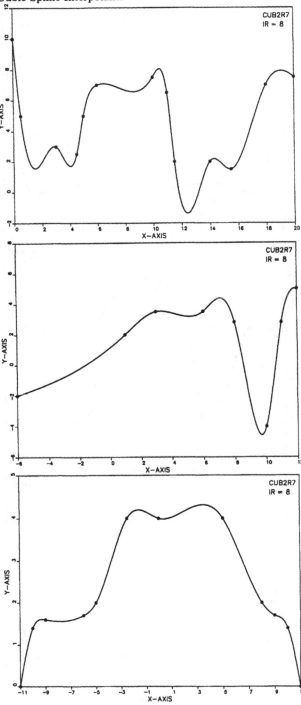

Figure 4.12. a-c.

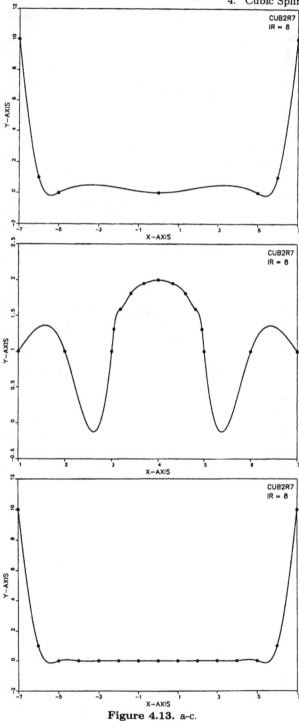

Figure 4.13. a-c.

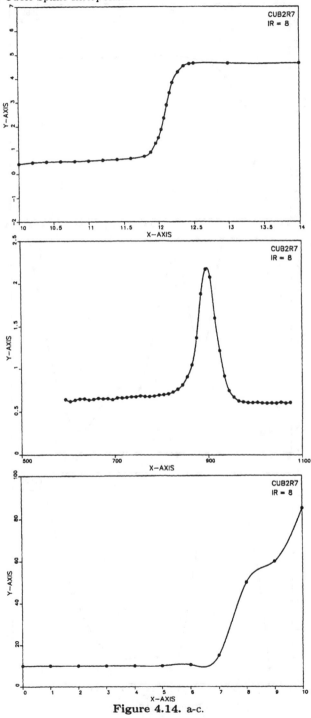

Figure 4.14. a-c.

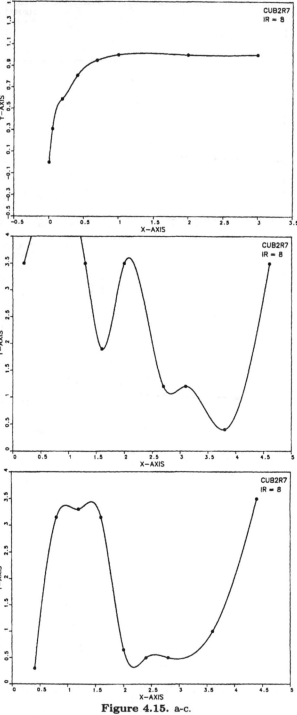

Figure 4.15. a-c.

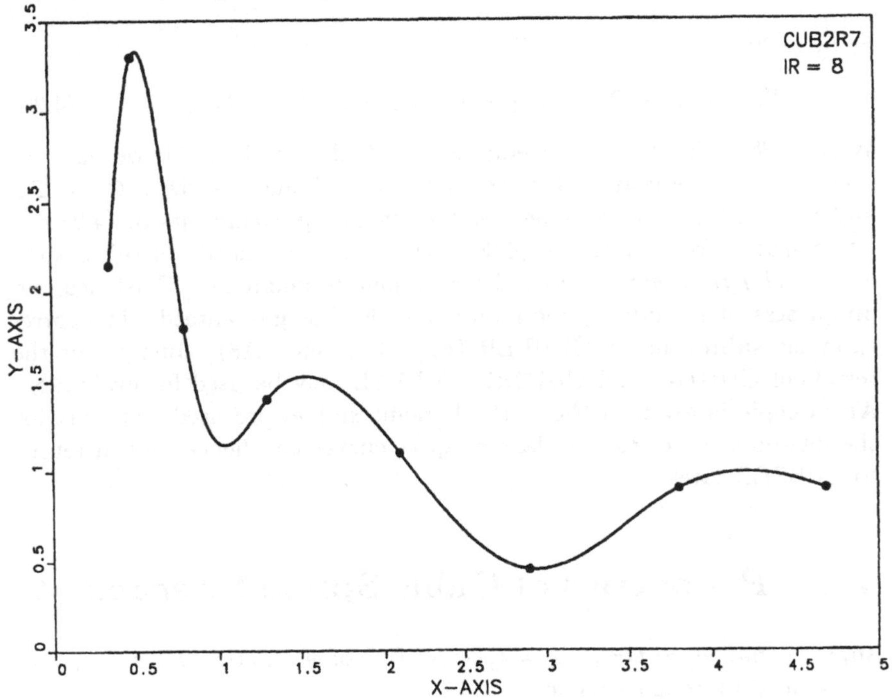

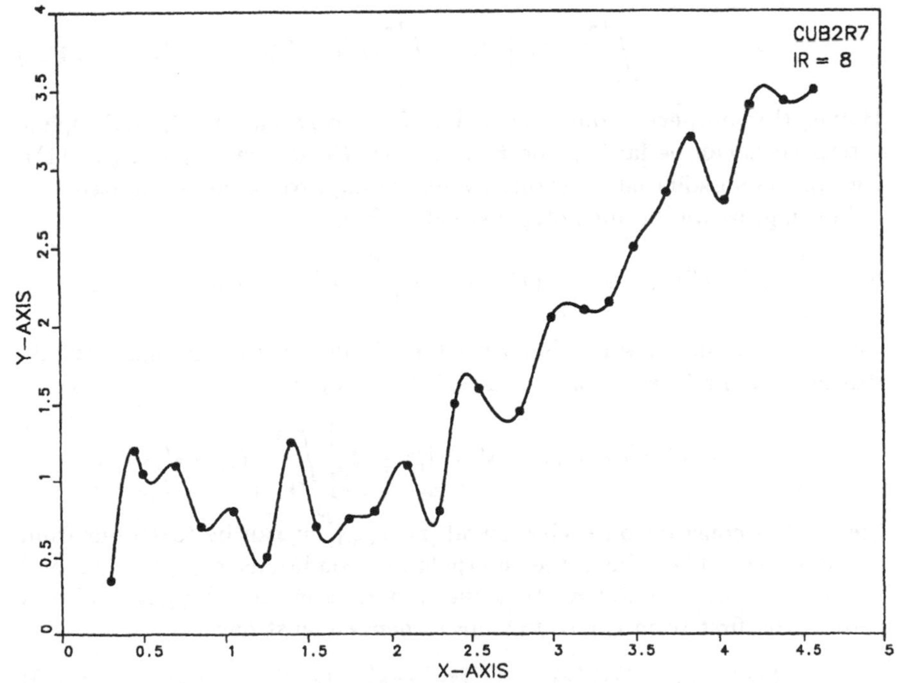

Figure 4.16. a, b.

or by (4.20),

$$2(h_1 + h_{n-1})C_1 + h_1 C_2 + h_{n-1} C_{n-1} = 3(d_1 - d_{n-1}). \qquad (4.28)$$

With (4.28), the first $n - 3$ equations of (4.21), and (4.27), we have altogether $n - 1$ equations for the calculation of the unknowns $C_1 = C_n$, and $C_2, \cdots, C_{n-1}$. Just as before, the remaining coefficients of (4.2) can be computed from (4.3) and (4.20). The coefficient matrix of this system is *cyclically tridiagonal* and strictly diagonally dominant. Existence and uniqueness of periodic spline interpolants is thus guaranteed. The corresponding subroutine is CUBPER (Figs. 4.17 and 4.18). Just as for the results of CUB1R5 and CUB2R7, CUBVAL may be used for evaluation. An example is given in Fig. 4.19. Periodic splines are used primarily for the interpolation of closed plane or space curves and hence we will return to CUBPER later.

4.4.   Properties of Cubic Spline Interpolants

Suppose that $y_k = f(x_k)$, $k = 1, \cdots, n$, for some function $f \in C^2[x_1, x_n]$. We wish to investigate when

$$\int_{x_1}^{x_n} [s''(x)]^2 dx \le \int_{x_1}^{x_n} [f''(x)]^2 dx. \qquad (4.29)$$

Having this property would mean that, for slowly varying s', cubic spline interpolants can be fairly smooth, since then $\tilde{F}_4$ (defined following (3.11)), i. e., the mean squared curvature, would be approximately minimized.

Starting, as we did for polygonal paths, from

$$0 \le \int_{x_1}^{x_n} (f'' - s'')^2 dx = \int_{x_1}^{x_n} (f'')^2 dx - 2\int_{x_1}^{x_n} (f'' - s'')s'' dx + \int_{x_1}^{x_n} (s'')^2 dx,$$

we investigate under what conditions the middle term on the right actually vanishes. Using integration by parts, it follows that

$$\int_{x_1}^{x_n} (f'' - s'')s'' dx = (f' - s')s'' \mid_{x_1}^{x_n} - \sum_{k=1}^{n-1} \int_{x_k}^{x_{k+1}} (f' - s')s''' dx.$$

Since s''' is constant on each interval $[x_k, x_{k+1}]$, it can be taken out from the integral. Then from the interpolation conditions $s_k(x_k) = y_k$ and $s_k(x_{k+1}) = y_{k+1}$, it follows that the second term on the right is always zero. If the first term is also to vanish, then we must have

$$(f'(x_1) - s_1'(x_1))s_1''(x_1) = (f'(x_n) - s_{n-1}'(x_n))s_{n-1}''(x_n). \qquad (4.30)$$

This is certainly the case ([57]) if

$$y_1'' = s_1''(x_1) = 0 = s_{n-1}''(x_n) = y_n''. \tag{4.31}$$

Cubic splines with these end conditions are said to be *natural*, although at times they are not this at all ([151]). On the contrary, they often produce a curve with an unnatural appearance near the endpoints. In fact, this was already evident in some of the examples that we have given. We remark that equations (4.21) with end conditions (4.31) can be derived directly

```
      SUBROUTINE CUBPER(N,X,Y,EPS,A,B,C,D,IFLAG)
      DIMENSION X(N),Y(N),A(N),B(N),C(N),D(N)
      IFLAG=0
      IF(N.LT.4) THEN
          IFLAG=1
          RETURN
      END IF
      N1=N-1
      Y(N)=Y(1)
      E1=X(2)-X(1)
      F1=Y(2)-Y(1)
      EN1=X(N)-X(N1)
      B(1)=2.*(E1+EN1)
      D(1)=E1
      C(1)=3.*(F1/E1-(Y(N)-Y(N1))/EN1)
      DO 10 K=2,N1
          K1=K+1
          E2=X(K1)-X(K)
          F2=Y(K1)-Y(K)
          D(K)=E2
          B(K)=2.*(E1+E2)
          C(K)=3.*(F2/E2-F1/E1)
          E1=E2
          F1=F2
10    CONTINUE
      CALL TRIPES(N1,B,D,C,EPS,IFLAG)
      IF(IFLAG.NE.0) RETURN
      C(N)=C(1)
      DO 20 K=1,N1
          K1=K+1
          H=X(K1)-X(K)
          A(K)=Y(K)
          B(K)=(Y(K1)-Y(K))/H-H*(2.*C(K)+C(K1))/3.
          D(K)=(C(K1)-C(K))/(3.*H)
20    CONTINUE
      RETURN
      END
```

Figure 4.17. Program listing of CUBPER.

Calling sequence:

CALL CUBPER(N,X,Y,EPS,A,B,C,D,IFLAG)

Purpose:
Determination of the coefficients A_k, B_k, C_k, and D_k of a periodic cubic spline interpolant.

Description of the parameters:

N   Number of given points. N$\geq$ 4 is required.
X   ARRAY(N): Upon calling must contain the abscissa values
          x_k, $k = 1, \cdots, n$, with $x_1 < x_2 < \cdots < x_n$.
Y   ARRAY(N): Upon calling must contain the ordinate values
          y_k, $k = 1, \cdots, n$. $y_n = y_1$ is necessary and is enforced
          by the program.
EPS   see TRIDIS.
A,B,C,D   ARRAY(N): Contain upon execution with IFLAG=0 the
                     desired spline coefficients, $k = 1, \cdots, n - 1$.
IFLAG   =0:   Normal execution.
          =1:   N< 4 not allowed.
          =2:   Error in solving the system (TRIPES).

Required subroutine: TRIPES.

Figure 4.18. Description of CUBPER.

from the requirement that

$$\int_{x_1}^{x_n} (s''(x))^2 dx = \sum_{k=1}^{n-1} \int_{x_k}^{x_{k+1}} (s_k''(x))^2 dx$$

be a minimum.

A second possibility for (4.30) to hold is obviously

$$f'(x_1) = s_1'(x_1), f'(x_n) = s_{n-1}'(x_n), \tag{4.32}$$

meaning that only functions with the same first derivative values as s at x_1 and x_n are to be considered as possible competitors to s.

As for periodic splines, with

$$s_1^{(j)}(x_1) = s_{n-1}^{(j)}(x_n), \qquad j = 0, 1, 2,$$

(4.30) holds if

$$f'(x_1) = f'(x_n), \tag{4.33}$$

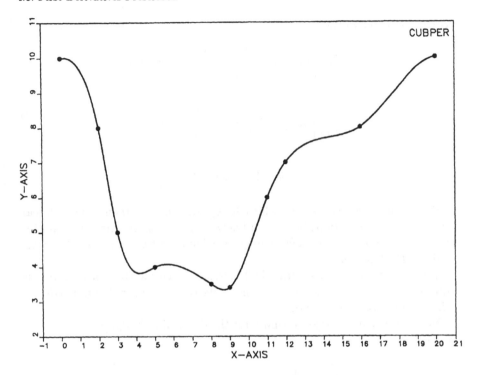

Figure 4.19.

i. e., we minimize only over those functions with the same function values and first derivatives at x_1 and x_n.

4.5. First Derivatives Prescribed

In some physical and technical measurements, it can be the case that values for the first or even second derivatives are collected instead of function values. The problem then is to construct the ordinates y_k of a smooth interpolant that respects this kind of data.

Suppose then that we are given $x_1 < x_2 < \cdots < x_n$ and derivative values $B_k = y'_k$. The corresponding cubic spline interpolant is then characterized by (4.7). By moving the unknowns to the left side and the known values to the right, we obtain

$$-\frac{1}{h_{k-1}^2}y_{k-1} + \left(\frac{1}{h_{k-1}^2} - \frac{1}{h_k^2}\right)y_k + \frac{1}{h_k^2}y_{k+1}$$

$$= \frac{1}{3}\left(\frac{1}{h_{k+1}}y_{k-1}' + 2\left(\frac{1}{h_{k-1}} + \frac{1}{h_k}\right)y_k' + \frac{1}{h_k}y_{k+1}'\right), \qquad (4.34)$$

$$k = 2, \cdots, n-1.$$

If, in addition, we assign values to y_1 and y_n, we then have a linear system of $n-2$ equations in $n-2$ unknowns, $y_2, \cdots, y_{n-2}$. After solving this system, the required coefficients may be calculated from $A_k = y_k$, $B_k = y_k'$, and C_k, D_k from (4.6).

We must of course ask: "In which cases does this system have a unique solution?" For this, the determinant of its tridiagonal, antisymmetric coefficient matrix must be nonzero.

Now there is a recursion formula for the determinant, d_n, of a matrix D_n of the form,

$$D_n = \begin{pmatrix} b_1 & c_1 & & & & & \\ a_1 & b_2 & c_2 & & & & \\ & a_2 & b_3 & c_3 & & & \\ & & & \cdot & \cdot & \cdot & \\ & & & & \cdot & \cdot & \\ & & & & a_{n-2} & b_{n-1} & c_{n-1} \\ & & & & & a_{n-1} & b_n \end{pmatrix}, \qquad (4.35)$$

namely,

$$\begin{aligned} d_0 &= 1, \\ d_1 &= b_1, \\ d_n &= b_k d_{k-1} - a_{k-1}c_{k-1}d_{k-2}, \quad k = 2, \cdots, n. \end{aligned} \qquad (4.36)$$

This is easily derived by expanding the determinant of the kth submatrix, D_k, along the bottom row.

In our case, n is replaced by $n-2$,

$$b_k = \frac{1}{h_k^2} - \frac{1}{h_{k+1}^2},$$

and

$$c_k = -a_k = \frac{1}{h_{k+1}^2}. \qquad (4.37)$$

```
      SUBROUTINE INCUB1(N,X,A,B,C,D,E,EPS,IFLAG)
      DIMENSION X(N),A(N),B(N),C(N),D(N),E(N)
      IFLAG=0
      IF(N.LT.3) THEN
          IFLAG=1
          RETURN
      END IF
      N1=N-1
      N2=N-2
      EX=0.
      VZ=-1.
      DO 10 K=1,N1
          H=X(K+1)-X(K)
          EX=EX+VZ*H*H
          VZ=-VZ
10    CONTINUE
      IF(ABS(EX).LT.EPS) THEN
          IFLAG=8
          RETURN
      END IF
      H1=1./(X(2)-X(1))
      A1=A(1)
      DO 20 K=2,N1
          KP1=K+1
          KM1=K-1
          H2=1./(X(KP1)-X(K))
          C(KM1)=H2*H2
          A(KM1)=-C(KM1)
          E(KM1)=(H1*H1-H2*H2)
          D(KM1)=(H1*B(KM1)+2.*(H1+H2)*B(K)+H2*B(KP1))/3.
          H1=H2
20    CONTINUE
      H=1./(X(2)-X(1))
      D(1)=D(1)+H*H*A1
      D(N2)=D(N2)-H1*H1*A(N)
      CALL TRIDIU(N2,A,E,C,D,EPS,IFLAG)
      IF(IFLAG.NE.0) RETURN
      DO 30 K=N2,1,-1
          A(K+1)=D(K)
30    CONTINUE
      A(1)=A1
      DO 40 K=1,N1
          K1=K+1
          DX=X(K1)-X(K)
          DY=A(K1)-A(K)
          DYX=DY/DX
          C(K)=(3.*DYX-2.*B(K)-B(K1))/DX
          D(K)=(-2.*DYX+B(K)+B(K1))/(DX*DX)
40    CONTINUE
      RETURN
      END
```

Figure 4.20. Program listing of INCUB1.

Calling sequence:

CALL INCUB1(N,X,A,B,C,D,E,EPS,IFLAG)

Purpose:
Determination of the coefficients A_k, B_k, C_k, and D_k of a cubic spline interpolant for given values of the first derivative.

Description of the parameters:

N		Number of given points. $N \geq 3$ is necessary.
X		ARRAY(N): Upon calling must contain the abscissa values x_k, $k = 1, \cdots, n$, with $x_1 < x_2 < \cdots < x_n$.
A		ARRAY(N): Input: Upon calling A(1) must contain the value y_1 and A(N) the value y_n.
B		ARRAY(N): Input: Upon calling must contain the values of the first derivatives y'_k, $k = 1, \cdots, n$.
A,B,C,D		ARRAY(N): Output: Upon completion with IFLAG=0 contain the desired spline coefficients, $k = 1, \cdots, n - 1$.
E		ARRAY(N): Work space.
EPS		see TRIDIU.
IFLAG	=0:	Normal execution.
	=1:	$N < 3$ not allowed.
	=2:	Error in solving the system (TRIDIU).
	=8:	Existence condition violated (see text).

Required subroutine: TRIDIU.

Figure 4.21. Description of INCUB1.

Then (4.36) becomes ([99])

$$d_k = (\frac{1}{h_k^2} - \frac{1}{h_{k+1}^2})d_{k-1} + \frac{1}{h_k^4}d_{k-2},$$

and an easy induction argument then shows that

$$d_k = det(D_k) = \frac{(-1)^{k+1}}{\left[\prod_{j=1}^{k+1} h_j\right]^2} \sum_{j=1}^{k+1}(-1)^j h_j^2. \tag{4.38}$$

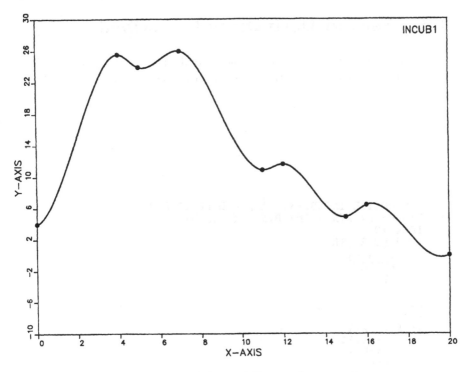

Figure 4.22.

Hence, the system (4.34), for given y_1 and y_n, is uniquely solvable precisely when

$$\sum_{j=1}^{n-1} (-1)^j h_j^2 \neq 0. \tag{4.39}$$

In particular, this shows that for equally spaced points and odd n, the corresponding spline interpolant does not in general exist. (Despite being singular, (4.34) can have several solutions for certain right-hand sides.)

INCUB1 (Figs. 4.20 and 4.21) uses the subroutine TRIDIU (see the appendix) to solve (4.34) numerically. This does not use pivoting and hence it may fail even in cases where (4.39) holds. An example is given in Fig. 4.22. The y_k', $k = 1, \cdots, 9$ were chosen to be $(1, 0, -1, -.5, -.2, 0, .5, 1, 1.5)$ and the end values were set to $y_1 = 4, y_9 = 0$.

4.6.   Second Derivatives Prescribed

If instead of the first derivatives, the second derivatives $y_k'' = 2C_k, k = 1, \cdots, n$ are prescribed, then the problem becomes somewhat simpler. From

```
     SUBROUTINE INCUB2(N,X,A,B,C,D,EPS,IFLAG)
     DIMENSION X(N),A(N),B(N),C(N),D(N)
     IFLAG=0
     IF(N.LT.3) THEN
         IFLAG=1
         RETURN
     END IF
     N1=N-1
     N2=N-2
     A1=A(1)
     H1=X(2)-X(1)
     DO 10 K=2,N1
         KP1=K+1
         KM1=K-1
         H2=X(KP1)-X(K)
         A(KM1)=-1./H2
         B(KM1)=(1./H1+1./H2)
         D(KM1)=-(H1*C(KM1)+2.*(H1+H2)*C(K)+H2*C(KP1))/6.
         H1=H2
10   CONTINUE
     D(1)=D(1)+1./(X(2)-X(1))*A1
     D(N2)=D(N2)+1./(X(N)-X(N1))*A(N)
     CALL TRIDIS(N2,B,A,D,EPS,IFLAG)
     IF(IFLAG.NE.0) RETURN
     DO 20 K=N2,1,-1
         A(K+1)=D(K)
         C(K)=C(K)/2.
20   CONTINUE
     A(1)=A1
     C(N1)=C(N1)/2.
     C(N)=C(N)/2.
     DO 30 K=1,N1
         K1=K+1
         H=X(K1)-X(K)
         B(K)=(A(K1)-A(K))/H-H*(2.*C(K)+C(K1))/3.
         D(K)=(C(K1)-C(K))/(3.*H)
30   CONTINUE
     RETURN
     END
```

Figure 4.23. Program listing of INCUB2.

Calling sequence:

CALL INCUB2(N,X,A,B,C,D,EPS,IFLAG)

Purpose:
Determination of the coefficients A_k, B_k, C_k, and D_k of a cubic spline interpolant for given values of the second derivative.

Description of the parameters:

N,X,EPS as in INCUB1.

A ARRAY(N): Input: Upon calling A(1) must contain the value y_1 and A(N) the value y_n.

C ARRAY(N): Input: Upon calling must contain the values of the second derivative y_k'', $k = 1, \cdots, n$.

A,B,C,D ARRAY(N): Output: Upon completion with IFLAG=0 contain the desired spline coefficients, $k = 1, \cdots, n - 1$.

IFLAG =0: Normal execution.
=1: N< 3 not allowed.
=2: Error in solving the system (TRIDIS).

Required subroutine: TRIDIS.

Figure 4.24. Description of INCUB2.

(4.21), we obtain

$$-\frac{1}{h_{k-1}}y_{k-1} + \left(\frac{1}{h_{k-1}} + \frac{1}{h_k}\right) y_k - \frac{1}{h_k}y_{k+1}$$

$$= -\frac{1}{6}\left[h_{k-1}y_{k-1}'' + 2(h_{k-1} + h_k)y_k'' + h_ky_{k+1}''\right], \tag{4.40}$$

$$k = 2, \cdots, n - 1.$$

If again y_1 and y_n are also given, then the resulting tridiagonal coefficient matrix is always nonsingular. This follows from the fact that it is *irreducible*, diagonally dominant, and strictly diagonally dominant in the first and last rows ([166]). We could also have concluded nonsingularity from a calculation similar to that of the previous section ([100]). Hence, the required cubic spline interpolant always exists and is unique.

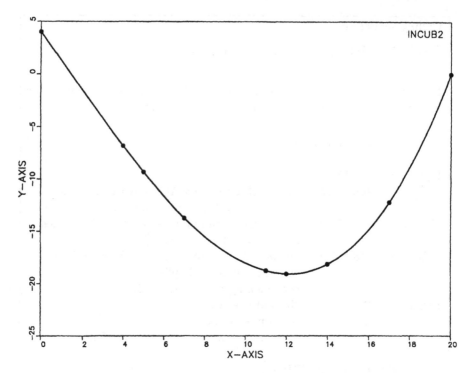

Figure 4.25.

It also has a noteworthy property ([100]). Since

$$s_k''(x) = 2C_k + 6D_k(x - x_k),$$

it follows from (4.20) that

$$s_k''(x) = \frac{2}{h_k}[C_k(x_{k+1} - x) + C_{k+1}(x - x_k)], \qquad k = 1, \cdots, n - 1.$$

$$(4.41)$$

Thus, if all the $y_k'' \geq 0$ or all the $y_k'' \leq 0$, $k = 1, \cdots, n$, then since $C_k = \frac{1}{2}y_k''$, (4.41) shows that $s_k''(x) \geq 0$ or $s_k''(x) \leq 0$ for all $x \in [x_k, x_{k+1}]$ and $k = 1, \cdots, n - 1$, i.e., we obtain a convex or concave cubic spline interpolant. Note that this property is independent of the values of y_1 and y_n.

INCUB2 (Figs. 4.23 and 4.24) may be used for this calculation with arbitrary values of the y_k''. An example is given in Fig. 4.25. Here, the values $y_k'' = (k - 1)/10$, $k = 1, \cdots, 9$, and $y_1 = 4, y_9 = 0$, were chosen. An obvious application of INCUB2 would be to join two points by a concave or convex curve whose shape can be controlled by the y_k''.

4.7. Smoothing with Cubic Spline Functions

In the next two sections, we will turn our attention from the interpolation to the smoothing by means of cubic splines of given data (x_k, y_k), $k = 1, \cdots, n$. We will again require that $x_1 < x_2 < \cdots < x_n$, but we will now also suppose that we are dealing with errors in the y_k. Most of the models that appear in the literature ([17,25,70,71,73,114,115,173]) are based on the minimality property (4.29) of natural cubic splines together with a constraint of the form,

$$\sum_{k=1}^{n} w_k (A_k - y_k)^2 \leq S. \qquad (4.42)$$

This results in a nonlinear optimization problem. Here, the $w_k > 0$ are positive weights and $S \geq 0$ is a global parameter that controls the smoothness of the resulting spline function. (Clearly, if $S = 0$, then we obtain the natural cubic interpolating spline.) A single such parameter S results in a certain inflexibility, as the distances from the curve to each point cannot be controlled individually. Attempting to gain such flexibility results in a relatively expensive iterative method for a *nonlinear* problem. It would be more appropriate to require that

$$w_k (A_k - y_k)^2 \leq S_k, \qquad k = 1, \cdots, n, \qquad (4.43)$$

for given values S_k. However, from a numerical standpoint, this problem is no longer practically solvable ([163]). Hence, we have taken the approach here ([153,154]) of introducing control parameters, p_k, which may be varied as needed and which allow bounds of the type (4.43) to be attained but which do not necessarily guarantee the optimality of a solution. (The precise mathematical context is explained in [163]). In exchange for optimality, we gain the advantage of having only to solve a *linear* system of equations for any particular given values of the p_k.

Our model is similar to (2.19). Specifically, we make the assumption that

$$p_k (y_k - A_k) = 6(D_k - D_{k-1}), \qquad k - 1, \cdots, n, \qquad (4.44)$$
$$D_0 = D_n = 0.$$

Here, the $p_k > 0$ are variable, prescribed control parameters. They have the effect of guaranteeing that the differences between the given and sought-for ordinates, y_k and A_k, in (4.2), are proportional to the jump in the third derivative of the desired cubic spline function, (4.2).

Substituting the values for D_k from (4.20) into (4.44), writing $y_k'' = 2C_k$, and then solving (4.44) for A_k gives

$$A_k = y_k - \frac{1}{p_k} \left[\frac{1}{h_{k-1}} y''_{k-1} - \left(\frac{1}{h_{k-1}} + \frac{1}{h_k} \right) y''_k + \frac{1}{h_k} y''_{k+1} \right],$$

(4.45)

$$k = 1, \cdots, n.$$

Here, as in what follows, we make the convention that undefined terms (e.g., $\frac{1}{h_0} y''_0$ for k=1) are set to zero. Since $d_k = (A_{k+1} - A_k)/h_k$, the substitution of (4.45) into (4.21) then yields:

$$\left[\frac{6}{p_{k-1}} \frac{1}{h_{k-2}} \frac{1}{h_{k-1}} \right] y''_{k-2}$$

$$+ \left[h_{k-1} - \frac{6}{p_{k-1}} \frac{1}{h_{k-1}} \left(\frac{1}{h_{k-2}} + \frac{1}{h_{k-1}} \right) - \frac{6}{p_k} \frac{1}{h_{k-1}} \left(\frac{1}{h_{k-1}} + \frac{1}{h_k} \right) \right] y''_{k-1}$$

$$+ \left[2(h_{k-1} + h_k) + \frac{6}{p_{k-1}} \frac{1}{h^2_{k-1}} + \frac{6}{p_k} \left(\frac{1}{h_{k-1}} + \frac{1}{h_k} \right)^2 + \frac{6}{p_{k+1}} \frac{1}{h^2_k} \right] y''_k$$

$$+ \left[h_k - \frac{6}{p_k} \frac{1}{h_k} \left(\frac{1}{h_{k-1}} + \frac{1}{h_k} \right) - \frac{6}{p_{k+1}} \frac{1}{h_k} \left(\frac{1}{h_k} + \frac{1}{h_{k+1}} \right) \right] y''_{k+1}$$

$$+ \left[\frac{6}{p_{k+1}} \frac{1}{h_k} \frac{1}{h_{k+1}} \right] y''_{k+2} = 6(d_k - d_{k-1}), \qquad k = 2, \cdots, n-1.$$

(4.46)

For simplicity, we take end conditions,

$$y''_1 = y''_n = 0.$$

(4.47)

The coefficient matrix of the linear system (4.46), in the unknowns $y''_2, \cdots, y''_{n-1}$, is symmetric and five-diagonal. It can also be shown to be positive definite ([154, p. 80]) and hence the required smoothing spline function exists and is unique. Its coefficients may be calculated from (4.45), with $C_k = \frac{1}{2} y''_k$, $k = 1, \cdots, n$, and from (4.20). As $p_k \to 0$, we obtain a smoothing line, and as $p_k \to \infty$, the natural cubic interpolating spline.

This calculation is carried out in the subroutine CUBSM1 (Figs. 4.26 and 4.27). In order to avoid some complicated special cases, we require that $n \geq 5$. In practice, this is no real restriction. The five-diagonal system of equations is solved by means of PENTAS (see the appendix).

```
       SUBROUTINE CUBSM1(N,X,Y,P,EPS,Y2,A,B,C,D,IFLAG)
       DIMENSION X(N),Y(N),P(N),Y2(N),A(N),B(N),C(N),D(N)
       IFLAG=0
       IF(N.LT.5) THEN
            IFLAG=1
            RETURN
       END IF
       N1=N-1
       N2=N-2
       DO 10 K=1,N1
            D(K)=1./(X(K+1)-X(K))
10     CONTINUE
       D(N)=0.
       P1=1./P(1)
       P2=1./P(2)
       H1=D(1)
       H2=D(2)
       R1=(Y(2)-Y(1))*H1
       DO 20 K=1,N2
            K1=K+1
            K2=K+2
            H3=D(K2)
            P3=1./P(K2)
            S=H1+H2
            A(K)=2./H1+2./H2+6.*(H1*H1*P1+S*S*P2+H2*H2*P3)
            B(K)=1./H2-6.*H2*(P2*S+P3*(H2+H3))
            C(K)=6.*P3*H2*H3
            R2=(Y(K2)-Y(K1))*H2
            Y2(K)=6.*(R2-R1)
            H1=H2
            H2=H3
            P1=P2
            P2=P3
            R1=R2
20     CONTINUE
       CALL PENTAS(N2,A,B,C,Y2,EPS,IFLAG)
       IF(IFLAG.NE.0) RETURN
       DO 30 K=N2,1,-1
            Y2(K+1)=Y2(K)
30     CONTINUE
       Y2(1)=0.
       Y2(N)=0.
       H1=0.
       DO 40 K=1,N1
            K1=K+1
            B(K)=D(K)
            H2=D(K)*(Y2(K1)-Y2(K))
            D(K)=H2/6.
            A(K)=Y(K)-(H2-H1)/P(K)
            C(K)=Y2(K)/2.
            H1=H2
40     CONTINUE
       A(N)=Y(N)+H1/P(N)
       DO 50 K=1,N1
            K1=K+1
            H=B(K)
            B(K)=(A(K1)-A(K))*H-(Y2(K1)+2.*Y2(K))/(6.*H)
50     CONTINUE
       RETURN
       END
```

Figure 4.26. Program listing of CUBSM1.

Calling sequence:

CALL CUBSM1(N,X,Y,P,EPS,Y2,A,B,C,D,IFLAG)

Purpose:
The coefficients A_k, B_k, C_k, and D_k, $k = 1, \cdots, n-1$, of a cubic smoothing spline are determined.

Description of the parameters:

N		Number of given points. N$\geq$ 5 is required.
X		ARRAY(N): Upon calling must contain the abscissa values
		x_k, $k = 1, \cdots, n$, with $x_1 < x_2 < \cdots < x_n$.
Y		ARRAY(N): Y(1),Y(2),$\cdots$,Y(N) must contain
		the ordinates y_k, $k = 1, \cdots, n$.
P		ARRAY(N): Upon calling must contain the parameters
		p_k, $k = 1, \cdots, n$, with $p_k > 0$.
EPS		Value used in the accuracy test (see PENTAS).
Y2		ARRAY(N): Output: Vector with the values of the
		second derivatives.
A,B,C,D		ARRAY(N): Upon completion with IFLAG=0 contain
		the desired spline coefficients, $k = 1, \cdots, n - 1$.
IFLAG	=0:	Normal execution.
	=1:	N$<$ 5 not allowed.
	=2:	Error in solving the system (PENTAS).

Required subroutine: PENTAS.

Figure 4.27. Description of CUBSM1.

Figures 4.28a–4.29c show a sequence of results from CUBSM1 for $p_k =$ .01, $p_k = .05$, $p_k = .25$, $p_k = 1$, $p_k = 10$ (each for $k = 1, \cdots, n$), and finally for $p = (.1, .5, .5, 5, 5, .1, .5, .1, .1, .1)$. In Fig. 4.29c, we can really see how the shape of the resulting curve can be controlled by adjusting the values of the p_k.

Finally, we mention the works [35,59,105] where in addition to (4.42), other types of side conditions may be imposed.

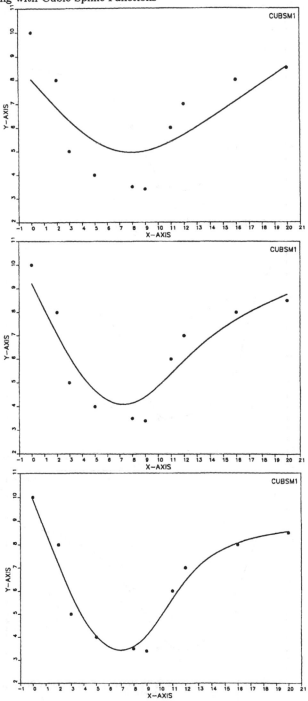

Figure 4.28. a-c.

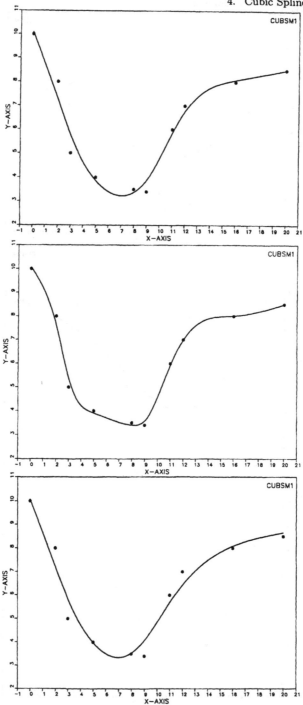

Figure 4.29. a-c.

4.8. Smoothing with Periodic Cubic Spline Functions

In the periodic case, we assume that

$$y_n = y_1. \tag{4.48}$$

From the system of equations for periodic cubic splines, which was derived earlier, with $y_k'' = 2C_k$, $k = 1, \cdots, n - 1$, and $y_n'' = y_1''$, we obtain

$$2(h_1 + h_{n-1})y_1'' + h_1 y_2'' + h_{n-1}y_{n-1}''$$

$$= 6\left[\frac{1}{h_1}(A_2 - A_1) - \frac{1}{h_{n-1}}(A_1 - A_{n-1})\right],$$

$$h_{k-1}y_{k-1}'' + 2(h_{k-1} + h_k)y_k'' + h_k y_{k+1}''$$

$$= 6\left[\frac{1}{h_k}(A_{k+1} - A_k) - \frac{1}{h_{k-1}}(A_k - A_{k-1})\right], \quad k = 2, \cdots, n - 2,$$

$$h_{n-1}y_1'' + h_{n-2}y_{n-2}'' + 2(h_{n-2} + h_{n-1})y_{n-1}''$$

$$= 6\left[\frac{1}{h_{n-1}}(A_1 - A_{n-1}) - \frac{1}{h_{n-2}}(A_{n-1} - A_{n-2})\right]. \tag{4.49}$$

Here, the A_k, $k = 1, \cdots, n - 1$, are the as yet unknown ordinates. In analogy to (4.48), A_n is set to $A_n = A_1$. The model (4.44), appropriately modified, becomes

$$p_k(y_k - A_k) = 6(D_k - D_{k-1}), \qquad k = 1, \cdots, n - 1,$$
$$D_0 = D_{n-1}. \tag{4.50}$$

The last equation of (4.44), for $k = n$, is no longer present because of the assumption of periodicity. We now solve (4.50) for A_k and substitute the expression (4.20) for D_k (with $C_n = C_1$) to obtain formulas for the A_k in terms of the C_k. Substituting these into (4.49) then again yields a linear system of equations. Its coefficient matrix is positive definite ([163]) and *cyclically* five-diagonal, i.e., of the form,

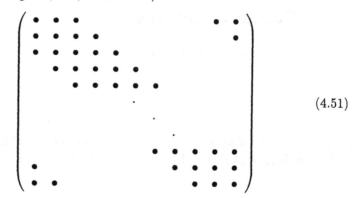

$$\tag{4.51}$$

A detailed description may be found in [46,163]. In this regard, [26] is also of interest.

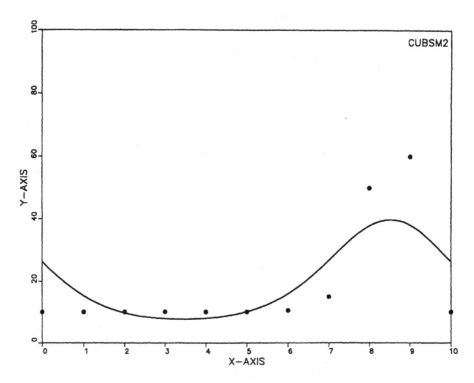

Figure 4.30.

```
      SUBROUTINE CUBSM2(N,X,Y,P,EPS,Y2,A,B,C,D,E,F,IFLAG)
      DIMENSION X(N),Y(N),Y2(N),P(N),A(N),B(N),C(N),D(N),
     &          E(N),F(N)
      IFLAG=0
      IF(N.LT.6) THEN
          IFLAG=1
          RETURN
      END IF
      N1=N-1
      N2=N-2
      P1=1./P(N1)
      P2=1./P(1)
      H1=1./(X(N)-X(N1))
      H2=1./(X(2)-X(1))
      R1=(Y(1)-Y(N1))*H1
      DO 10 K=1,N1
          K1=K+1
          S=H1+H2
          IF(K.LT.N1) THEN
              H3=1./(X(K+2)-X(K1))
              P3=1./P(K1)
              R2=(Y(K1)-Y(K))*H2
          ELSE
              H3=1./(X(2)-X(1))
              P3=1./P(1)
              R2=(Y(1)-Y(N1))*H2
          END IF
          A(K)=2./H1+2./H2+6.*(H1*H1*P1+S*S*P2+H2*H2*P3)
          IF(K.LT.N1) B(K)=1./H2-6.*H2*(P2*S+P3*(H2+H3))
          IF(K.LT.N2) C(K)=6.*P3*H2*H3
          Y2(K)=6.*(R2-R1)
          IF(K.EQ.1) THEN
              H4=1./(X(N1)-X(N2))
              B(N1)=1./H1-6.*H1*(P1*(H1+H4)+P2*(H1+H2))
              C(N2)=6.*P1*H4*H1
          END IF
          IF(K.EQ.2) C(N1)=6.*P1*H1/(X(N)-X(N1))
          H1=H2
          H2=H3
          P1=P2
          P2=P3
          R1=R2
10    CONTINUE
      CALL PENPES(N1,A,B,C,D,E,Y2,F,EPS,IFLAG)
      Y2(N)=Y2(1)
      H1=(Y2(1)-Y2(N1))/(X(N)-X(N1))
      DO 20 K=1,N1
          K1=K+1
          H=1./(X(K1)-X(K))
          B(K)=H
          H2=H*(Y2(K1)-Y2(K))
          D(K)=H2/6.
          A(K)=Y(K)-(H2-H1)/P(K)
          C(K)=Y2(K)/2.
          H1=H2
20    CONTINUE
      A(N)=A(1)
```

(*cont.*)
```
    DO 30 K=1,N1
        K1=K+1
        H=B(K)
        B(K)=(A(K1)-A(K))*H-(Y2(K1)+2.*Y2(K))/(6.*H)
 30 CONTINUE
    RETURN
    END
```

Figure 4.31. Program listing of CUBSM2.

Calling sequence:

CALL CUBSM2(N,X,Y,P,EPS,Y2,A,B,C,D,E,F,IFLAG)

Purpose:
The coefficients A_k, B_k, C_k, and D_k, $k = 1, \cdots, n - 1$, of a periodic cubic smoothing spline are determined.

Description of the parameters:

X,Y2,A,B,C,D as in CUBSM1.

N	Number of given points. N$\geq$ 6 is necessary.
Y	ARRAY(N): Y(1),Y(2),$\cdots$,Y(N) must contain the ordinates y_k, $k = 1, \cdots, n$. $y_1 = y_n$ is required and is enforced by the program.
P	ARRAY(N): Upon calling must contain the parameters p_k, $k = 1 \cdots, n$, with $p_k > 0$. $p_1 = p_n$ is required and is enforced by the program.
EPS	Value used in the accuracy test (see PENPES).
E,F	ARRAY(N): Work space.
IFLAG	=0: Normal execution.
	=1: N< 6 not allowed.
	=2: Error in solving the system (PENPES).

Required subroutine: PENPES.

Figure 4.32. Description of CUBSM2.

The subroutine CUBSM2 (Figs. 4.31 and 4.32) assembles the coefficient matrix (4.51) and solves the corresponding linear system by calling PENPES (see the appendix). It then also calculates the coefficients, A_k, B_k, C_k, D_k, $k = 1, \cdots, n - 1$, of the periodic smoothing spline.

Figure 4.30 shows an example with $p_k = 1$. We shall later return to CUBSM2 when we consider the fitting of planar data by closed curves.

4.9. Cubic X-Spline Interpolants

We now return to the problem of interpolation by cubic splines. Although these do have the same function values and continuous first and second derivatives at the knots x_k, $k = 2, \cdots, n-1$, there can be large jumps in the *third* derivative. The idea ([9]) is now to limit the size of these jumps by allowing some small discontinuities in the second derivative. Specifically, we look for a spline interpolant $s \in C^1[x_1, x_n]$ with $s(x_k) = y_k$, $k = 1, \cdots, n$, and s_k of (4.2), such that

$$s_k''(x_k) - s_{k-1}''(x_k) = \alpha_k \left[s_k'''(x_k) - s_{k-1}'''(x_k) \right],$$

$$(4.52)$$

$$k = 2, \cdots, n-1,$$

where the $\alpha_k \in \mathbb{R}$ are arbitrary. Following [9], when such an s exists and is unique, it is called a cubic X-spline interpolant. The usual cubic spline is recovered by taking $\alpha_k = 0$, $k = 2, \cdots, n-1$.

If we let $B_k = s_k'(x_k)$, $k = 1, \cdots, n-1$, and $B_n = s_{n-1}'(x_n)$ be our unknowns, then the remaining coefficients of (4.2) are given by (4.3) and (4.6). By substituting these in the equations corresponding to (4.52),

$$C_k - C_{k-1} - 3h_{k-1}D_{k-1} = 3\alpha_k(D_k - D_{k-1}), \qquad k = 2, \cdots, n-1,$$

we obtain the $n - 2$ equations in n unknowns, $B_1, \cdots, B_n$,

$$\frac{1}{h_{k-1}} \left(1 - \frac{3\alpha_k}{h_{k-1}} \right) B_{k-1}$$

$$+ \left[\frac{1}{h_{k-1}} \left(2 - \frac{3\alpha_k}{h_{k-1}} \right) + \frac{1}{h_k} \left(2 + \frac{3\alpha_k}{h_k} \right) \right] B_k$$

$$+ \frac{1}{h_k} \left(1 + \frac{3\alpha_k}{h_k} \right) B_{k+1}$$

$$(4.53)$$

$$= \frac{3}{h_{k-1}} \left(1 - \frac{2\alpha_k}{h_{k-1}} \right) d_{k-1} + \frac{3}{h_k} \left(1 + \frac{2\alpha_k}{h_k} \right) d_k,$$

$$k = 2, \cdots, n-1.$$

For given B_1 and B_n (we wish to avoid other end conditions here), the coefficient matrix is not symmetric, but it is again tridiagonal. From (4.53), we see easily that

$$\frac{1}{h_{k-1}}\left|1 - \frac{3\alpha_k}{h_{k-1}}\right| + \frac{1}{h_k}\left|1 + \frac{3\alpha_k}{h_k}\right| < \frac{1}{h_{k-1}}\left|2 - \frac{\alpha_k}{h_{k-1}}\right| + \frac{1}{h_k}\left|2 + \frac{3\alpha_k}{h_k}\right|,$$

$$k = 2, \cdots, n - 1,$$

$$(4.54)$$

is a sufficient condition for strict diagonal dominance and hence for the existence and uniqueness of X-splines. Often, it is more practical to make use of the stronger conditions,

$$1 - \frac{3\alpha_k}{h_{k-1}} \geq 0 \text{ and } 1 + \frac{3\alpha_k}{h_k} \geq 0,$$

or, equivalently,

$$-\frac{h_k}{3} \leq \alpha_k \leq \frac{h_{k-1}}{3}, \qquad k = 2, \cdots, n - 1. \tag{4.55}$$

Typical choices are to take α_k as the midpoint ([9]) of the corresponding interval (4.55), i.e.,

$$\alpha_k = \frac{1}{6}(h_{k-1} - h_k), \qquad k = 2, \cdots, n - 1, \tag{4.56}$$

or as one of its endpoints ([9,95]), i.e.,

$$\alpha_k = -\frac{h_k}{3} \text{ or } \alpha_k = \frac{h_{k-1}}{3}, \qquad k = 2, \cdots, n - 1. \tag{4.57}$$

For the first possibility in (4.57), the tridiagonal coefficient matrix becomes lower *bidiagonal*, and for the second, upper bidiagonal, thus reducing the computational expense ([9,95]). A convex combination of these two kinds of X-splines is considered in [95], and another generalization is given in [5]. If computational expense (very small, anyway) is not the major issue, then (4.56) might be preferable.

A further criterium for choosing the α_k could be the symmetry of the symmetry matrix. This is realized precisely when

$$\frac{1}{h_k}\left(1 - \frac{3\alpha_{k+1}}{h_k}\right) = \frac{1}{h_k}\left(1 + \frac{3\alpha_k}{h_k}\right),$$

i.e., when

$$\alpha_{k+1} = -\alpha_k, \qquad k = 2, \cdots, n - 2. \tag{4.58}$$

```
      SUBROUTINE CUBXSP(N,X,Y,ICASE,ALPHA,EPS,A,B,C,D,IFLAG)
      DIMENSION X(N),Y(N),ALPHA(N),A(N),B(N),C(N),D(N)
      IFLAG=0
      IF(N.LT.3) THEN
          IFLAG=1
          RETURN
      END IF
      IF (ICASE.LT.1.OR.ICASE.GT.3) THEN
          IFLAG=5
          RETURN
      END IF
      N1=N-1
      N2=N-2
      B1=B(1)
      H1=1./(X(2)-X(1))
      H3=(Y(2)-Y(1))*H1
      DO 10 K=2,N1
          S1=0.
          KP1=K+1
          KM1=K-1
          KM2=K-2
          H2=1./(X(KP1)-X(K))
          IF(ICASE.EQ.2) ALPHA(K)=(1./H1-1./H2)/6.
          IF(ICASE.EQ.3) ALPHA(KP1)=-ALPHA(K)
          AK=ALPHA(K)
          H4=(Y(KP1)-Y(K))*H2
          H5=AK*H1
          H6=AK*H2
          H7=3.*H5
          H8=3.*H6
          IF(ICASE.NE.3.AND.K.GT.2) THEN
              A(KM2)=H1*(1.-H7)
              S1=ABS(A(KM2))
          END IF
          IF(K.LT.N1) THEN
              C(KM1)=H2*(1.+H8)
              S1=S1+ABS(C(KM1))
              IF(ICASE.EQ.3.AND.K.GT.2) S1=S1+S1
          END IF
          B(KM1)=H1*(2.-H7)+H2*(2.+H8)
          IF(S1.GE.ABS(B(KM1))) THEN
              IFLAG=9
              RETURN
          END IF
          D(KM1)=3.*H1*(1.-H5-H5)*H3+3.*H2*(1.+H6+H6)*H4
          H1=H2
          H3=H4
10    CONTINUE
      H=1./(X(2)-X(1))
      D(1)=D(1)-H*(1.-3.*ALPHA(2)*H)*B1
      D(N2)=D(N2)-H1*(1.+3.*ALPHA(N1)*H1)*B(N)
      IF(ICASE.LE.2) CALL TRIDIU(N2,A,B,C,D,EPS,IFLAG)
      IF(ICASE.EQ.3) CALL TRIDIS(N2,B,C,D,EPS,IFLAG)
      DO 20 K=N2,1,-1
          B(K+1)=D(K)
20    CONTINUE
      B(1)=B1
```

(cont.)
```
    DO 30 K=1,N1
        K1=K+1
        DX=X(K1)-X(K)
        DY=Y(K1)-Y(K)
        DYX=DY/DX
        A(K)=Y(K)
        C(K)=(3.*DYX-2.*B(K)-B(K1))/DX
        D(K)=(-2.*DYX+B(K)+B(K1))/(DX*DX)
30  CONTINUE
    RETURN
    END
```

Calling sequence:

CALL CUBXSP(N,X,Y,ICASE,ALPHA,EPS,A,B,C,D,IFLAG)

Purpose:
Calculation of a cubic X-spline interpolant.

Description of the parameters:

N,X,Y,EPS,A,B,C,D as in CUB1R5.

ICASE	=1:	All the α_k must be given by the user.
	=2:	The α_k are set to $\alpha_k = (h_{k-1} - h_k)/6$.
	=3:	α_2 must be given by the user.
ALPHA	ARRAY(N):	Upon calling must contain the α_k in ALPHA(K), K=2,3,$\cdots$,N−1 for ICASE=1. For ICASE=3 ALPHA(2) must contain α_2. In any case a test is made to see whether or not the condition, $|(1-3\alpha_k/h_{k-1})|/h_{k-1} + |(1+3\alpha_k/h_k)|/h_k < |(2-3\alpha_k/h_{k-1})/h_{k-1} + (2+3\alpha_k/h_k)/h_k|$ (*), is satisfied.
IFLAG	=0:	Normal execution.
	=1:	N$\geq$ 3 is required.
	=2:	Error in solving the system of equations (TRIDIU, TRIDIS).
	=5:	ICASE< 1 or ICASE> 3 not allowed.
	=9:	The condition (*) for strict diagonal dominance does not hold for at least one k.

Required subroutines: TRIDIU, TRIDIS.

Figure 4.33. CUBXSP and its description.

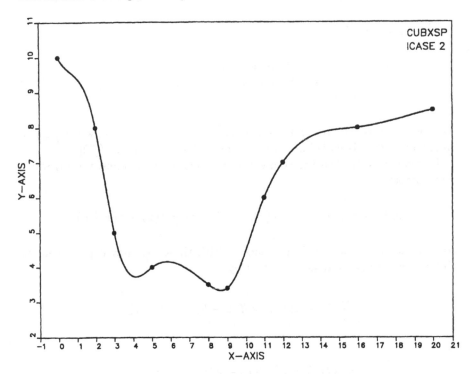

Figure 4.34.

Naturally, (4.54) must also hold. One could, for example, given α_2, compute the remaining α_k's by (4.58) while checking that (4.54) is satisfied. The aforementioned method with these options (except (4.57)) for computing cubic X-spline interpolants is implemented in the subroutine CUBXSP (Fig. 4.33). Figure 4.34 shows an example with the α_k chosen as per (4.56). The user should be cautioned to choose the α_k well away from zero.

4.10. Discrete Cubic Spline Interpolants

Suppose that we are again given abscissas $x_1 < \cdots < x_n$ and ordinates $y_k, \ k = 1, \cdots, n$, as well as a number $h > 0$. Instead of requiring twice continuously differentiability at the knots, as was the case for cubic spline interpolants, we set the following transition conditions for the interior knots of the segments (4.2):

$$
\begin{aligned}
s_k(x_k + h) &= s_{k-1}(x_k + h), \\
s_k(x_k) &= s_{k-1}(x_k), \qquad k = 2, \cdots, n - 1. \\
s_k(x_k - h) &= s_{k-1}(x_k - h).
\end{aligned}
\tag{4.59}
$$

We also require certain end conditions at x_1 and x_n. These will uniquely determine what is called a *discrete cubic spline interpolant* ([72]).

In order to establish the connection to usual cubic splines, we introduce the notation,

$$
\Delta f(x) = f(x + h) - f(x), \qquad \nabla f(x) = f(x) - f(x - h),
$$

for forward and backward differences. With these, we may express central differences of first and second order by

$$
\frac{1}{2}(\nabla + \Delta)f(x) = f(x + h) - f(x - h),
$$

and

$$
\nabla \Delta f(x) = f(x + h) - 2f(x) - f(x - h).
$$

Then the conditions (4.59) may be reexpressed as

$$
\begin{aligned}
s_{k-1}(x_k) &= s_k(x_k), \\
\frac{1}{2}(\nabla + \Delta)s_{k-1}(x_k) &= \frac{1}{2}(\nabla + \Delta)s_k(x_k), \quad k = 2, \cdots, n - 1, \\
\nabla \Delta s_{k-1}(x_k) &= \nabla \Delta s_k(x_k).
\end{aligned}
\tag{4.60}
$$

The second and third equations say that the first and second differences with respect to h are to agree at the knots x_k, $k = 1, \cdots, n - 1$. Hence, the usual C^2 conditions may be obtained by dividing the second equation by h and the third equation by h^2 and letting $h \to 0$.

Since in most applications, usual cubic spline interpolants are more useful than the discrete ones, we avoid discussion of end conditions and systems of equations and refer the reader to [33,34]. A corresponding subroutine, DCSINT, for various end conditions is given in [34]. Further, [34] also gives one called DCSSMO for smoothing with discrete splines. Several other similar splines are discussed in [28,29]. Figures 4.35a,b,c were produced using DCSINT. The h was successively chosen to be $h = .1, 1, 2$. For all three, $y_1' = d_1$ and $y_n' = d_{n-1}$ were used as end conditions. For larger h, it becomes very noticeable that such splines are only C^1.

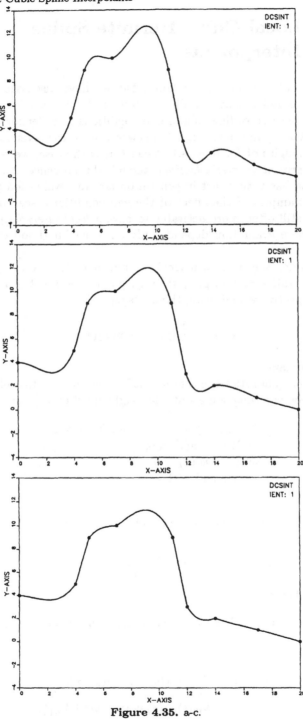

Figure 4.35. a-c.

4.11.   Local Cubic Hermite Spline Interpolants

So far, the cubic spline interpolants that we have discussed all require the formation and solution of a linear system of equations involving *all* of the given abscissas, ordinates and end conditions in order for them to be evaluated at *any* point in $[x_1, x_n]$. (This was also the case even for quadratic splines, although not for linears.) When this is the case, we refer to it as a *global interpolation method*. Even though the influence of a particular piece of data has little effect in general on faraway evaluation points, (for numerical examples of the effect of the end conditions, see, for example, [154]), it is still often more desirable to have a *local interpolation method*, i. e., one where evaluation depends only on a neighborhood of the given point.

For example, one obtains a local C^0 interpolant, s_k on $[x_k, x_{k+1}]$, by putting quadratics p_k and p_{k+1} through points $k-1, k, k+1$ and $k, k+1, k+2$, respectively, and using the average,

$$s_k(x) = \frac{1}{2}\left(p_k(x) + p_{k+1}(x)\right), \tag{4.61}$$

as the interpolant.

These same quadratics can also be used in the construction of a local C^1 interpolant by choosing some suitable weight function $w(t)$ such that

$$
\begin{aligned}
0 &\le w(t) \le 1 && \text{for } 0 \le t \le 1, \\
w(0) &= 1, w(1) = 0, \\
w'(0) &= w'(1) = 0, \\
w(t) &+ w(1-t) = 1,
\end{aligned}
\tag{4.62}
$$

and setting

$$s_k(x) = w(t)p_k(x) + (1 - w(t))p_{k+1}(x), \tag{4.63}$$

where

$$t = \frac{x - x_k}{\Delta x_k}.$$

In fact, we then have

$$s_k(x_k) = p_k(x_k) = y_k, \qquad s_k(x_{k+1}) = p_{k+1}(x_{k+1}) = y_{k+1},$$

and since

$$
\begin{aligned}
s_k'(x) &= \frac{1}{h_k}w'(t)\left[p_k(x) - p_{k+1}(x)\right] \\
&\quad + w(t)p_k'(x) + [1 - w(t)]p_{k+1}'(x),
\end{aligned}
$$

```
SUBROUTINE HECUB(N,X,Y,B,C,D,IFLAG)
DIMENSION X(N),Y(N),B(N),C(N),D(N)
IFLAG=0
IF(N.LT.2) THEN
     IFLAG=1
     RETURN
END IF
DO 10 K=1,N-1
     K1=K+1
     H=X(K1)-X(K)
     DYDX=(Y(K1)-Y(K))/H
     C(K)=(3.*DYDX-2.*B(K)-B(K1))/H
     D(K)=(-2.*DYDX+B(K)+B(K1))/(H*H)
10   CONTINUE
RETURN
END
```

Figure 4.36. Program listing of HECUB.

Calling sequence:

CALL HECUB(N,X,Y,B,C,D,IFLAG)

Purpose:
Determination of a cubic Hermite spline interpolant.

Description of the parameters:

N,X,Y as in CUB1R5.

B	ARRAY(N):	Upon calling must contain the values of the derivatives y'_k, $k = 1, \cdots, n$.
Y,B,C,D	ARRAY(N):	Upon completion with IFLAG=0 contain the desired spline coefficients, $k = 1, 2 \cdots, n - 1$, with A(K)=Y(K).
IFLAG	=0:	Normal execution.
	=1:	N$\geq$ 2 is required.

Figure 4.37. Description of HECUB.

---

also

$$s'_k(x_k) = p'_k(x_k), \qquad s'_k(x_{k+1}) = p'_{k+1}(x_{k+1}).$$

One example ([77]) is

$$w(t) = 1 - 3t^2 + 2t^3.$$

In this case, the interpolating function (4.63) is in fact a polynomial of degree five, which may cause some unwanted oscillations.

A local C^0 method using cubic polynomials is obtained by taking the interpolant on the interval $[x_k, x_{k+1}]$ as that cubic which passed through points $k-1, k, k+1$, and $k+2$.

A local C^1 method using cubic polynomials is obtained by either prescribing the values y_k', $k = 1, \cdots, n$, of the first derivative at x_k or by calculating them from neighboring data, as is done, for example, in the subroutines GRAD1, GRAD2, GRAD3, and GRAD4 of Chapter 3. The resulting *local cubic Hermite spline interpolant* is, for a given $k \in \{1, \cdots, n-1\}$, a polynomial of degree three, p_k, with

$$p_k(x_k) = y_k, \quad p_k(x_{k+1}) = y_{k+1},$$
$$p_k'(x_k) = y_k', \quad p_k'(x_{k+1}) = y_{k+1}'. \tag{4.64}$$

By (4.6), the coefficients of

$$p_k(x) = A_k + B_k(x - x_k) + C_k(x - x_k)^2 + D_k(x - x_k)^3 \tag{4.65}$$

are given by

$$\begin{aligned} A_k &= y_k, \\ B_k &= y_k', \\ C_k &= \frac{1}{h_k}(3d_k - 2y_k' - y_{k+1}'), \\ D_k &= \frac{1}{h_k^2}(-2d_k + y_k' + y_{k+1}'). \end{aligned} \tag{4.66}$$

This method is implemented in the subroutine HECUB (Figs. 4.36 and 4.37). Its quality depends upon the y_k', at least when these have been calculated and not given ahead of time. Hence, we will discuss in the next section some further possibilities for determining the y_k'.

4.12.   Approximation of Values for the First Derivative II

The previously described methods (subroutines GRAD1 through GRAD4) were purely analytical in nature. But in the special case when the concern is shape-preserving interpolation, methods modified or constructed on the basis of geometric considerations are often more successful.

If, for example, monotone data with $y_1 \leq y_2 \leq \cdots \leq y_n$ is to be reproduced, then $y_k' \geq 0$ must be obtained. More generally, it would be necessary that

$$\begin{aligned} y_k' = y_{k+1}' = 0 & \qquad \text{for } d_k = 0, \\ sgn(y_k') = sgn(y_{k+1}') = sgn(d_k) & \qquad \text{for } d_k \neq 0. \end{aligned} \tag{4.67}$$

Thus, in this case, for example, the method (3.57) corresponding to the subroutine GRAD2 is modified to ([49])

$$y_k' = \begin{cases} 0 & \text{if } d_{k-1} = 0 \text{ or } d_k = 0 \\ & \qquad\qquad\qquad k = 2, \cdots, n-1, \\ \dfrac{h_k d_{k-1} + h_{k-1} d_k}{h_k + h_{k-1}} & \text{otherwise} \end{cases}$$

$$y_1' = \begin{cases} 0 & \text{if } d_1 = 0 \text{ or } sgn(\tilde{y}_1') \neq sgn(d_1) \\ \tilde{y}_1' = d_1 + h_1 \dfrac{(d_1 - d_2)}{h_1 + h_2} & \text{otherwise} \end{cases} \quad , \quad (4.68)$$

$$y_n' = \begin{cases} 0 & \text{if } d_{n-1} = 0 \\ & \text{or } sgn(\tilde{y}_n') \neq sgn(d_{n-1}) \\ \\ \tilde{y}_n' = d_{n-1} + h_{n-1} \dfrac{(d_{n-1} - d_{n-2})}{h_{n-2} + h_{n-1}} & \text{otherwise} \end{cases} .$$

This method, which could also be used even if there is no underlying monotonicity, is implemented in GRAD2B (Figs. 4.38 and 4.39).

Now, in formula (4.68), y_k' cannot go to zero as one of d_{k-1} or d_k (but not both) approach zero. Hence, we try to find a method that does not have this disadvantage. One such method is based on the interpolation of three consecutive points $(x_k, y_k), (k = 1, 2, 3$ without loss of generality), not by quadratics but rather by a simple rational function involving three parameters, namely,

$$r(x) = a + \frac{b}{x - c}. \tag{4.69}$$

Since we are looking for

$$r'(x_2) = -\frac{b}{(x_2 - c)^2}, \tag{4.70}$$

and thus require b and c, we first calculate from the interpolation conditions,

$$r(x_k) = y_k = a + \frac{b}{x_k - c}, \quad k = 1, 2, 3,$$

the differences,

$$\Delta y_1 = b \left(\frac{1}{x_2 - c} - \frac{1}{x_1 - c} \right),$$

$$\Delta y_2 = b \left(\frac{1}{x_3 - c} - \frac{1}{x_2 - c} \right).$$

```
      SUBROUTINE GRAD2B(N,X,Y,B,EPS,IFLAG)
      DIMENSION X(N),Y(N),B(N)
      IFLAG=0
      IF(N.LT.3) THEN
          IFLAG=1
          RETURN
      END IF
      N1=N-1
      N2=N-2
      H1=X(2)-X(1)
      H2=X(3)-X(2)
      H3=(Y(2)-Y(1))/H1
      IF(ABS(H3).LT.EPS) THEN
          B(1)=0.
      ELSE
          Z=H3+(H3-(Y(3)-Y(2))/H2)*H1/(X(3)-X(1))
          IF(Z*H3.LT.0.) THEN
              B(1)=0.
          ELSE
              B(1)=Z
          END IF
      END IF
      DO 10 K=2,N1
          KM1=K-1
          KP1=K+1
          H2=X(KP1)-X(K)
          H4=(Y(KP1)-Y(K))/H2
          IF(ABS(H3).LT.EPS.OR.ABS(H4).LT.EPS) THEN
              B(K)=0.
          ELSE
              B(K)=(H3*H2+H4*H1)/(X(KP1)-X(KM1))
          END IF
          H1=H2
          H3=H4
 10   CONTINUE
      H1=X(N1)-X(N2)
      IF(ABS(H4).LT.EPS) THEN
          B(N)=0.
      ELSE
          Z=H4+(H4-(Y(N1)-Y(N2))/H1)*H2/(X(N)-X(N2))
          IF(Z*H4.LT.0) THEN
              B(N)=0.
          ELSE
              B(N)=Z
          END IF
      END IF
      RETURN
      END
```

Figure 4.38. Program listing of GRAD2B.

Calling sequence:

CALL GRAD2B(N,X,Y,B,EPS,IFLAG)

Purpose:
Determination of values for the first derivative by modified quadratic interpolation.

Description of the parameters:

N,X,Y,B as in GRAD1.
EPS   Value for the accuracy test. Recommedation: $EPS=10^{-t}$
          (t: number of available decimal places). If the absolute
          value of a number is smaller than EPS, it is
          interpreted as zero.
IFLAG =0:   Normal execution.
IFLAG =1: $N \geq 3$ is required.

Figure 4.39. Description of GRAD2B.

From these, it follows that

$$- d_1(x_2 - c)(x_1 - c) = b = -d_2(x_3 - c)(x_2 - c), \qquad (4.71)$$

and hence,

$$c = \frac{d_2 x_3 - d_1 x_1}{d_2 - d_1}. \qquad (4.72)$$

From (4.70) and (4.71), we have

$$r'(x_2) = \frac{x_3 - c}{x_2 - c} d_2,$$

and hence an application of (4.72) shows that

$$r'(x_2) = \frac{d_1 d_2}{\left(\dfrac{y_3 - y_1}{x_3 - x_1} \right)}. \qquad (4.73)$$

Here, the condition $d_2 - d_1 \neq 0$, necessary for (4.72), no longer appears. Just as for (4.68), this leads to the setting of ([49])

$$y'_k = \begin{cases} 0 & \text{if } y_{k+1} - y_{k-1} = 0 \\[2ex] \dfrac{d_{k-1} d_k}{\left(\dfrac{y_{k+1} - y_{k-1}}{x_{k+1} - x_{k-1}} \right)} & \text{otherwise} \end{cases} \qquad k = 2, \cdots, n-1,$$

```
      SUBROUTINE GRAD2R(N,X,Y,B,EPS,IFLAG)
      DIMENSION X(N),Y(N),B(N)
      IFLAG=0
      IF(N.LT.3) THEN
          IFLAG=1
          RETURN
      END IF
      N1=N-1
      N2=N-2
      H1=(Y(2)-Y(1))/(X(2)-X(1))
      H3=Y(3)-Y(1)
      IF(ABS(H3).LT.EPS) THEN
          B(1)=0.
      ELSE
          B(1)=H1*H1*(X(3)-X(1))/H3
      END IF
      DO 10 K=2,N1
          KP1=K+1
          KM1=K-1
          H2=(Y(KP1)-Y(K))/(X(KP1)-X(K))
          H3=Y(KP1)-Y(KM1)
          IF(ABS(H3).LT.EPS) THEN
              B(K)=0.
          ELSE
              B(K)=H2*H1*(X(KP1)-X(KM1))/H3
          END IF
          H1=H2
10    CONTINUE
      IF(ABS(H3).LT.EPS) THEN
          B(N)=0.
      ELSE
          B(N)=H2*H2*(X(N)-X(N2))/H3
      END IF
      RETURN
      END
```

Figure 4.40. Program listing of GRAD2R.

Calling sequence:

CALL GRAD2R(N,X,Y,B,EPS,IFLAG)

Purpose:
Determination of values for the first derivative by modified local rational interpolation.

Description of the parameters:

N,X,Y,B,EPS,IFLAG as in GRAD2B.

Figure 4.41. Description of GRAD2R.

$$y_1' = \begin{cases} 0 & \text{if } y_3 - y_1 = 0 \\ \dfrac{d_1^2}{\left(\dfrac{y_3 - y_1}{x_3 - x_1}\right)} & \text{otherwise} \end{cases} , \tag{4.74}$$

$$y_n' = \begin{cases} 0 & \text{if } y_n - y_{n-2} = 0 \\ \dfrac{d_{n-1}^2}{\left(\dfrac{y_n - y_{n-2}}{x_n - x_{n-2}}\right)} & \text{otherwise} \end{cases} .$$

This method is implemented in GRAD2R (Figs. 4.40 and 4.41) and can also be used even if there is no underlying monotonicity. Because of the pole in (4.69), this rational interpolant may be better suited to rapidly decreasing or increasing monotone or convex data than is a quadratic interpolant.

The following well-known method of Akima ([3]) is based purely on geometric considerations. Let the indices $j = k - 2, k - 1, k, k + 1, k + 2$ of interior points (x_j, y_j) be denoted by $1, 2, 3, 4, 5$. Let A be the point of intersection of the lines **12** and **34** and B that of **23** and **45**. Further, let C be the point of intersection of a line through point 3 with **12** and D of the same line with **45** (see Fig. 4.42). The desired slope t at point 3, corresponding to y_k' at x_k, is then defined by imposing the proportionality condition,

$$\left| \frac{\overline{2C}}{\overline{CA}} \right| = \left| \frac{\overline{4D}}{\overline{DB}} \right|. \tag{4.75}$$

Suppose that the coordinates of the points $1, 2, 3, 4, 5, A, B, C$, and D in Fig. 4.42 are denoted by (x, y) with indices $1, 2, 3, 4, 5, a, b, c$, and d. As usual, by d_j, we mean

$$d_j = \frac{\Delta y_j}{\Delta x_j}, \qquad j = 1, 2, 3, 4. \tag{4.76}$$

The desired t is defined by

$$t = \frac{y_d - y_c}{x_d - x_c}. \tag{4.77}$$

From Fig. 4.42 we see that the following equations must hold:

$$d_1 = \frac{y_a - y_2}{x_a - x_2} = \frac{y_c - y_2}{x_c - x_2}, \tag{4.78}$$

$$d_4 = \frac{y_4 - y_b}{x_4 - x_b} = \frac{y_4 - y_a}{x_4 - x_a}, \tag{4.79}$$

$$d_2 = \frac{y_b - y_3}{x_b - x_3}, \tag{4.80}$$

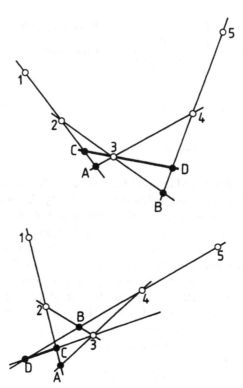

Figure 4.42.

$$d_3 = \frac{y_3 - y_a}{x_3 - x_a}, \tag{4.81}$$

$$t = \frac{y_3 - y_c}{x_3 - x_c} = \frac{y_d - y_3}{x_d - x_3}. \tag{4.82}$$

Since condition (4.75) can also be written in the form,

$$\left| \frac{x_2 - x_c}{x_c - x_a} \right| = \left| \frac{x_4 - x_d}{x_b - x_d} \right|, \tag{4.83}$$

we want to eliminate, in sequence, all the unknown ordinate differences. For $y_3 - y_a$, this is done, for example, using (4.78) and (4.81) in the following manner. On the one hand,

$$d_1 = \frac{y_a - y_2}{x_a - x_2} = \frac{(y_a - y_3) + (y_3 - y_2)}{(x_a - x_3) + (x_3 - x_2)} = \frac{(y_3 - y_a) - \Delta y_2}{(x_3 - x_a) - \Delta x_2},$$

and on the other hand,

$$y_3 - y_a = (x_3 - x_a)d_3.$$

This yields

$$\frac{x_3 - x_a}{\Delta x_3} = \frac{\Delta x_1 \Delta y_2 - \Delta x_2 \Delta y_1}{\Delta x_1 \Delta y_3 - \Delta x_3 \Delta y_1} = \frac{S_{12}}{S_{13}}.$$

In a similar manner, we successively find that

$$\frac{x_b - x_3}{\Delta x_2} = \frac{\Delta x_3 \Delta y_4 - \Delta x_4 \Delta y_3}{\Delta x_2 \Delta y_4 - \Delta x_4 \Delta y_2} = \frac{S_{34}}{S_{24}},$$

$$x_3 - x_c = \frac{\Delta x_1 \Delta y_2 - \Delta x_2 \Delta y_1}{\Delta x_1 t - \Delta y_1},$$

$$x_d - x_3 = \frac{\Delta x_3 \Delta y_4 - \Delta x_4 \Delta y_3}{\Delta y_4 - \Delta x_4 t}.$$

We now write (4.83) in the form,

$$\left| \frac{(x_3 - x_c) - \Delta x_2}{(x_3 - x_a) - (x_3 - x_c)} \right| = \left| \frac{\Delta x_3 - (x_d - x_3)}{(x_b - x_3) - (x_d - x_3)} \right|,$$

and substitute in the preceding four expressions to finally obtain, after a lengthy calculation,

$$|S_{12}S_{24}|(\Delta y_3 - \Delta x_3 t)^2 = |S_{13}S_{34}|(\Delta y_2 - \Delta x_2 t)^2. \qquad (4.84)$$

If in addition to (4.75), we require that points 2 and 4 lie in the same half-space defined by the line **CD**, then we must have

$$(t - d_2)(t - d_3) \le 0.$$

Then (4.84) reduces to

$$\sqrt{|S_{12}S_{24}|}(\Delta y_3 - \Delta x_3 t) = \sqrt{|S_{13}S_{34}|}(\Delta x_2 t - y_2),$$

and so finally we obtain

$$t = \frac{w_2 \Delta y_2 + w_3 \Delta y_3}{w_2 \Delta x_2 + w_3 \Delta x_3}, \qquad (4.85)$$

where

$$w_2 = \sqrt{|S_{13}S_{34}|}, \qquad w_3 = \sqrt{|S_{12}S_{24}|}. \qquad (4.86)$$

Alternatively, we may write

$$t = \frac{\sqrt{|(d_3 - d_1)(d_4 - d_3)|}\, d_2 + \sqrt{|(d_2 - d_1)(d_4 - d_2)|}\, d_3}{\sqrt{|(d_3 - d_1)(d_4 - d_3)|} + \sqrt{|(d_2 - d_1)(d_4 - d_2)|}}. \qquad (4.87)$$

From (4.87), it follows that

$$d_1 = d_2, d_3 \neq d_1, d_4 \neq d_3 \quad \Rightarrow \quad t = d_2,$$

(4.88)

$$d_3 = d_4, d_1 \neq d_2, d_4 \neq d_2 \quad \Rightarrow \quad t = d_3.$$

These are consistent with the geometry.

In the special cases of $d_1 = d_2 = d_3$ and $d_2 = d_3 = d_4$, (4.87) is not defined and we then set $t = d_2 = d_3$. An unpleasant consequence of (4.87)

```
      SUBROUTINE GRAD5(N,X,Y,ICASE,EPS,B,IFLAG)
      DIMENSION X(N),Y(N),B(N),T(4)
      IFLAG=0
      IF(N.LT.2) THEN
          IFLAG=1
          RETURN
      END IF
      B(1)=(Y(2)-Y(1))/(X(2)-X(1))
      IF(N.GT.2) B(2)=(Y(3)-Y(1))/(X(3)-X(1))
      DO 20 K=3,N-2
          DO 10 J=1,4
              KJ=K+J-2
              KJ1=KJ-1
              T(J)=(Y(KJ)-Y(KJ1))/(X(KJ)-X(KJ1))
10        CONTINUE
          IF(ICASE.EQ.1) THEN
              W2=SQRT(ABS((T(3)-T(1))*(T(4)-T(3))))
              W3=SQRT(ABS((T(2)-T(1))*(T(4)-T(2))))
          ELSE
              W2=ABS(T(4)-T(3))
              W3=ABS(T(2)-T(1))
          END IF
          W23=W2+W3
          IF(W23.LT.EPS) THEN
              B(K)=(T(2)+T(3))/2.
              GOTO 20
          END IF
          B(K)=(W2*T(2)+W3*T(3))/W23
20    CONTINUE
      IF(N.GT.3) B(N-1)=(Y(N)-Y(N-2))/(X(N)-X(N-2))
      B(N)=(Y(N)-Y(N-1))/(X(N)-X(N-1))
      RETURN
      END
```

Figure 4.43. Program listing of GRAD5.

Calling sequence:

CALL GRAD5(N,X,Y,ICASE,EPS,B,IFLAG)

Purpose:
Determination of values for the first derivative by Akima's method (two variants).

Description of the parameters:

N,X,Y,B as in GRAD1.

ICASE =1: Option 1.
=2: Option 2.

EPS If the sum of W2 and W3 (see the text) is smaller than EPS, then these are interpreted as exactly zero. Recommendation: EPS=10^{-t+2}, where t is the number of available decimal digits.

IFLAG =0: Normal execution.
=1: N$\geq$ 2 is required.

Figure 4.44. Description of GRAD5.

is that

$$d_2 = d_4, d_3 \neq d_1, d_4 \neq d_3 \quad \Rightarrow \quad t = d_2,$$

$$(4.89)$$

$$d_3 = d_1, d_2 \neq d_1, d_4 \neq d_2 \quad \Rightarrow \quad t = d_3,$$

as these are not geometrically meaningful. Hence, the modification,

$$t = \frac{|d_4 - d_3|d_2 + |d_2 - d_1|d_3}{|d_4 - d_3| + |d_2 - d_1|},$$

$$(4.90)$$

has been proposed ([3]). For this, (4.88) still holds but (4.89) does not.

In the subroutine GRAD5 (Figs. 4.43 and 4.44), ICASE=1 implements (4.87) and ICASE=2, (4.90). If $d_3 = d_4$ and $d_2 = d_1$ (in the numerical sense), t is set to

$$t = \frac{1}{2}(d_2 + d_3).$$

$$(4.91)$$

At the boundary points with indices $1, 2, n - 1, n$, we have chosen

$$y_1' = d_1, \qquad y_n' = d_{n-1}$$

$$y_2' = \frac{y_3 - y_1}{x_3 - x_1}, \qquad y_{n-1}' = \frac{y_n - y_{n-2}}{x_n - x_{n-2}}.$$

$$(4.92)$$

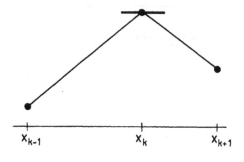

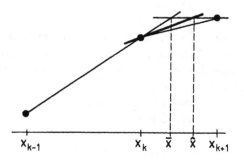

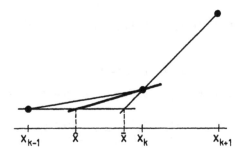

Figure 4.45.

Appropriate values could also have been obtained by interpolation with a quadratic at points $1, 2, 3$ and $n - 2, n - 1, n$, respectively.

Yet another method is that of McAllister and Roulier ([79,80]). Here, for $k = 2, \cdots, n - 1$, the y'_k (see Fig. 4.45) are chosen as follows:

$$
y'_k = \begin{cases}
0 \text{ if } d_{k-1}d_k \leq 0 \\[2ex]
\dfrac{y_{k+1} - y_k}{\hat{x} - x_k}, \text{ where } \hat{x} = \dfrac{1}{2}(\bar{x} + x_{k+1}) \text{ if } |d_{k-1}| > |d_k| > 0 \\[2ex]
(\bar{x} \text{ is the abscissa of the point of intersection of the} \\
\text{line through } (x_k, y_k) \text{ with slope } d_{k-1}, \text{ and the line} \\
\text{through } (x_{k+1}, y_{k+1}) \text{ with slope zero}) \\[2ex]
\dfrac{y_k - y_{k-1}}{x_k - \hat{x}}, \text{ where } \hat{x} = \dfrac{1}{2}(x_{k-1} + \bar{x}) \text{ if } 0 < |d_k| \leq |d_{k+1}| \\[2ex]
(\text{here, } \bar{x} \text{ is the abscissa of the point of intersection of the} \\
\text{line through } (x_k, y_k) \text{ with slope } d_k \text{ and the line} \\
\text{through } (x_{k-1}, y_{k-1}) \text{ with slope zero})
\end{cases}
\tag{4.93}
$$

For $k = n$, we set

$$
y'_n = \begin{cases}
2d_{n-1} & \text{if } d_{n-2}d_{n-1} < 0 \\
0 & \text{if } dd_{n-1} \leq 0 \\
d & \text{if } dd_{n-1} > 0
\end{cases}, \tag{4.94}
$$

where

$$
d = \frac{y_n - \bar{y}}{x_n - \bar{x}}, \quad \bar{x} = \frac{x_{n-1} + x_n}{2}, \quad \bar{y} = y_{n-1} + y'_{n-1}(\bar{x} - x_{n-1}).
$$

For $k = 1$, there is an analogous definition. As can be seen from Fig. 4.45, and which can also be proven formally ([80]), this choice of the y'_k, $k = 1, \cdots, n$, is consistent with the monotonicity and convexity of the data, just as a polygonal path would reproduce these properties. Moreover, if $sgn(d_{k-1}) = sgn(d_k)$, the y'_k depend continuously on d_{k-1} and d_k. For numerical purposes, possible instabilities ([79]) should be intercepted. This method is implemented in the subroutine GRAD6 (Figs. 4.46 and 4.47), which is closely based on the subroutine SLOPES of [80].

In [143], the question is posed how, in principle, the y'_k might be chosen so that the resulting cubic Hermite spline interpolant has no local extremum in $[x_k, x_{k+1}]$, i.e., so that the well-known possible over- and under-shoots of cubic splines (see for example Figs. 4.7a, 4.11c–4.13c, 4.14c, 4.15c, and

```
      SUBROUTINE GRAD6(N,X,Y,EPS,B)
      REAL X(N),Y(N),B(N)
      N1=N-1
      IH1=1
      I=2
      I1=3
      YDIF1=Y(2)-Y(1)
      YDIF2=Y(3)-Y(2)
      H1=YDIF1/(X(2)-X(1))
      H11=H1
      H2=YDIF2/(X(3)-X(2))
      H22=H2
10    IF(ABS(H1).LE.EPS.OR.ABS(H2).LE.EPS.OR.(H1*H1).LE.EPS)
&         THEN B(I)=0.
      ELSE
          IF(ABS(H1).GT.ABS(H2)) THEN
              XBAR=(YDIF2/H1)+X(I)
              XHAT=(XBAR+X(I1))/2.
              B(I)=YDIF2/(XHAT-X(I))
          ELSE
              XBAR=(-YDIF1/H2)+X(I)
              XHAT=(X(IH1)+XBAR)/2.
              B(I)=YDIF1/(X(I)-XHAT)
          END IF
      END IF
      IH1=I
      I=I1
      I1=I1+1
      IF(I.GT.N1) THEN
          IF((H1*H2).LT.0.) THEN
              B(N)=2.*H2
              GOTO 20
          END IF
          XMID=(X(N1)+X(N))/2.
          YXMID=B(N1)*(XMID-X(N1))+Y(N1)
          B(N)=(Y(N)-YXMID)/(X(N)-XMID)
          IF((B(N)*H2).LT.0.) B(N)=0.
          GOTO 20
      ELSE
          YDIF1=Y(I)-Y(IH1)
          YDIF2=Y(I1)-Y(I)
          H1=YDIF1/(X(I)-X(IH1))
          H2=YDIF2/(X(I1)-X(I))
          GOTO 10
      END IF
20    IF((H11*H22).LT.0.) THEN
          B(1)=2.*H11
          RETURN
      END IF
      XMID=(X(1)+X(2))/2.
      YXMID=B(2)*(XMID-X(2))+Y(2)
      B(1)=(YXMID-Y(1))/(XMID-X(1))
      IF((B(1)*H11).LT.0.) B(1)=0.
      RETURN
      END
```

Figure 4.46. Program listing of GRAD6.

Calling sequence:

CALL GRAD6(N,X,Y,EPS,B)

Purpose:
Calculation of values for first derivatives by the method of McAllister and Roulier.

Description of the parameters:

X,Y,B,N as in GRAD1.
EPS   Value for the accuracy test. Recommendation: $EPS=10^{-t}$ (t : number of available decimal digits). If the absolute value of a number is less than EPS, then it is interpreted as zero.

Figure 4.47. Description of GRAD6.

4.16a) can be avoided here. To this end, we write the desired polynomial in the form,

$$p_k(x) = y_k u + y_{k+1} t + tu \left[(t - u) \Delta y_k - \Delta x_k (t y'_{k+1} - u y'_k) \right],$$

$$\text{(4.95)}$$

$$t = \frac{x - x_k}{\Delta x_k}, \quad u = 1 - t.$$

Since

$$p'_k(x) = 2 \left(3 d_k - (y'_k + y'_{k+1}) \right) tu + y'_k u^2 + y'_{k+1} t^2, \qquad \text{(4.96)}$$

the requirements (4.64) do hold. If under- and over-shoots are to be eliminated, then for the chosen y'_k, it is necessary to require ([143]) that

$$y'_k = 0 \qquad \text{if } d_{k-1} d_k < 0, \qquad \text{(4.97)}$$

$$y'_k = y'_{k+1} = 0 \qquad \text{if } d_k = 0, \qquad \text{(4.98)}$$

and that

$$sgn(y'_k) = sgn(y'_{k+1}) = sgn(d_k). \qquad \text{(4.99)}$$

If $d_k = 0$, then we right away set $y'_k = 0$. Suppose then that $d_k \neq 0$. Then (4.96) may be written in the form,

$$\frac{1}{3d_k} p'_k(x) = \left(\sqrt{\frac{y'_k}{3d_k}}\, u - \sqrt{\frac{y'_{k+1}}{3d_k}}\, t \right)^2$$

$$+ 2 \left(1 + \sqrt{\frac{y'_k}{3d_k} \frac{y'_{k+1}}{3d_k}} - \frac{y'_k + y'_{k+1}}{3d_k} \right) tu. \quad (4.100)$$

By (4.99), the square roots are well-defined. Since $0 < t < 1$ and $0 < u < 1$ in the interior of $[x_k, x_{k+1}]$, (4.100) has no zeros there when

$$q_k := \frac{y'_k + y'_{k+1}}{3d_k} - \sqrt{\frac{y'_k}{3d_k} \frac{y'_{k+1}}{3d_k}} \leq 1. \quad (4.101)$$

Since every interior knot x_{k+1} belongs to two intervals, $[x_k, x_{k+1}]$ and $[x_{k+1}, x_{k+2}]$, (4.101) must also hold when k is replaced by $k + 1$. This defines a relationship between the first derivatives at adjoining nodes.

Now we must choose the y'_k appropriately. A method of selection is given in [143], but tests have shown that it does not always work satisfactorily. Our own method, incorporating a modification suggested by the author of [143], works as follows. First, the desired y'_k are initialized according to

$$y'_1 = d_1, \quad y'_k = \frac{d_{k-1} + d_k}{2}, \quad k = 2, \cdots, n-1, \quad y'_n = d_{n-1}. \quad (4.102)$$

Then those y'_k are set to zero as required by (4.97) and (4.98). Further, those y'_k for which (4.99) does not hold are also set to zero.

With these values for the y'_k, so obtained, we proceed as follows. First, we set the factors $p_k = q_k = 0, k = 1, \cdots, n - 1$. If $y'_k \neq 0$ and $y'_{k+1} \neq 0$, then we define the quantities (if $d_k \neq 0$),

$$r_k = \frac{2y'_k + y'_{k+1}}{3d_k}, \quad t_k = \frac{y'_k + 2y'_{k+1}}{3d_k}, \quad k = 1, \cdots, n-1. \quad (4.103)$$

Since

$$s''_k(x_k) = \frac{2}{h_k}(3d_k - 2y'_k - y'_{k+1}) \quad (4.104)$$

and

$$s''_k(x_{k+1}) = \frac{2}{h_k}(-3d_k + y'_k + 2y'_{k+1}), \quad (4.105)$$

there will be no inflection point in the interior of $[x_k, x_{k+1}]$ if

$$sgn\,(s''_k(x_k)) = sgn\,(s''_k(x_{k+1})),$$

corresponding to the condition,

$$sgn(1 - r_k) = sgn(t_k - 1). \tag{4.106}$$

If there is one, i. e., if

$$(1 - r_k)(t_k - 1) < 0, \tag{4.107}$$

then $p_k = 0$ is adjusted to be

$$p_k = \min(r_k, t_k). \tag{4.108}$$

Subsequently, we calculate, for $d_k \neq 0$, the factors q_k of (4.101) and finally set

$$\alpha_k = \max(p_k, q_k, 1), \qquad k = 1, \cdots, n - 1. \tag{4.109}$$

Next, the y'_k are reduced according to

$$
\begin{aligned}
y'_1 &= & y'_1/\alpha_1, & \\
y'_k &= & y'_k/\max(\alpha_{k-1}, \alpha_k), & \quad k = 2, \cdots, n - 1, \\
y'_n &= & y'_n/\alpha_{n-1}. &
\end{aligned}
\tag{4.110}
$$

If $\alpha_k = 1$, there is no change at all in y'_k, and since $q_k \leq 1$, there is no critical point in the interior of $[x_k, x_{k+1}]$. If $\alpha_k = q_k$, then the values of y'_k and y'_{k+1} are so reduced that q_k becomes ≤ 1 and hence there is no longer a critical point. Finally, if $\alpha_k = p_k$, then the reduction of y'_k is such that if $p_k = r_k$, then the new r_k becomes one, i. e., the inflection point has been moved to x_k, and if $p_k = t_k$, then the new t_k becomes one, i. e., the inflection point has been moved to x_{k+1}. The factor $\max(\alpha_{k-1}, \alpha_k)$ comes about because, as we have already mentioned, (4.101) must hold for both k and $k - 1$ at each interior node $x_k, k = 2, \cdots, n - 1$. This method is implemented in GRAD7 (Figs. 4.48 and 4.49).

For the following method, suppose that the given data set is convex, i. e.,

$$d_1 \leq d_2 \leq \cdots \leq d_{n-1}. \tag{4.111}$$

If the local cubic Hermite spline interpolant (4.65) is also to be convex, then since $p''_k(x)$ is linear, we must have

$$p''_k(x_k) = 2C_k \geq 0$$

and

$$p''_k(x_{k+1}) = 2C_k + 6h_k D_k \geq 0.$$

Using (4.66), this becomes the system of inequalities,

$$
\begin{aligned}
2y'_k + y'_{k+1} &\leq 3d_k \\
& \qquad\qquad\qquad\qquad k = 1, \cdots, n - 1. \\
-y'_k - 2y'_{k+1} &\leq -3d_k
\end{aligned}
\tag{4.112}
$$

```fortran
      SUBROUTINE GRAD7(N,X,Y,EPS,B,IFLAG,A)
      DIMENSION X(N),Y(N),B(N),A(N)
      IFLAG=0
      IF(N.LT.3) THEN
          IFLAG=1
          RETURN
      END IF
      N1=N-1
      N2=N-2
      DO 10 K=1,N1
          K1=K+1
          A(K)=(Y(K1)-Y(K))/(X(K1)-X(K))
10    CONTINUE
      DO 20 K=2,N1
          B(K)=(A(K)+A(K-1))/2.
20    CONTINUE
      B(1)=A(1)
      B(N)=A(N1)
      DO 30 K=1,N2
          K1=K+1
          IF(ABS(A(K)).LE.EPS) THEN
              B(K)=0.
              B(K1)=0.
              GOTO 30
          END IF
          IF((A(K)*A(K1)).LT.0.) B(K1)=0.
30    CONTINUE
      IF(ABS(A(N1)).LE.EPS) THEN
          B(N1)=0.
          B(N)=0.
      END IF
      DO 40 K=1,N1
          K1=K+1
          IF((B(K)*A(K)).LT.0.) B(K)=0.
          IF((B(K1)*A(K)).LT.0.) B(K1)=0.
40    CONTINUE
      IF((B(N)*A(N1)).LT.0.) B(N)=0.
      DO 50 K=1,N1
          PK=0.
          QK=0.
          K1=K+1
          IF((ABS(B(K))+ABS(B(K1))).GT.EPS) THEN
              H=3.*A(K)
              RK=(2.*B(K)+B(K1))/H
              SK=(2.*B(K1)+B(K))/H
              IF((1.-RK)*(SK-1.).LT.0.) PK=AMIN1(RK,SK)
          END IF
          IF(ABS(A(K)).GT.EPS) THEN
              H=3.*A(K)
              QK=(B(K)+B(K1))/H-SQRT(B(K)*B(K1))/ABS(H)
          END IF
          A(K)=AMAX1(QK,PK,1.)
50    CONTINUE
      B(1)=B(1)/A(1)
      B(N)=B(N)/A(N1)
      DO 60 K=1,N2
          K1=K+1
          B(K1)=B(K1)/AMAX1(A(K),A(K1))
60    CONTINUE
      RETURN
      END
```

Figure 4.48. Program listing of GRAD7.

Calling sequence:

CALL GRAD7(N,X,Y,EPS,B,IFLAG,A)

Purpose:
Determination of values for first derivatives by the method of R. Schmidt (with modifications).

Description of the parameters:

N,X,Y,B as in GRAD1.

EPS Value for the accuracy test. Recommendation: EPS=10^{-t} (t : the number of available decimal digits). If the absolute value of a number is less than EPS, then it is interpreted as zero.

IFLAG =0: Normal execution.

   =1: N$\geq$ 3 is required.

A ARRAY(N): Work space.

Figure 4.49. Description of GRAD7.

If the y'_k satisfy this system, then the p_k, and hence the resulting global function, is convex. Now, (4.112) may be rewritten in the form,

$$\frac{1}{2}(3d_k - y'_k) \leq y'_{k+1} \leq 3d_k - 2y'_k. \qquad (4.113)$$

In particular,

$$\frac{1}{2}(3d_k - y'_k) \leq 3d_k - 2y'_k,$$

and hence,

$$y'_k \leq d_k, \qquad k = 1, \cdots n - 1. \qquad (4.114)$$

Since (4.114) must also hold for $k + 1$, it follows from (4.113) that

$$y'_k \geq 3d_k - 2d_{k+1}, \qquad k = 1, \cdots, n - 2, \qquad (4.115)$$

and thus, putting this together with (4.114),

$$3d_k - 2d_{k+1} \leq y'_k \leq d_k. \qquad (4.116)$$

A simple procedure to choose the y'_k so that they satisfy (4.113) and (4.116) goes as follows ([88,89,90]):

Choose $y'_k \in [u_k, v_k]$ with

$$
\begin{aligned}
u_1 &= 3d_1 - 2d_2, \\
v_1 &= d_1, \\
u_k &= \max\left\{ \frac{1}{2}(3d_{k-1} - y'_{k-1}), 3d_k - 2d_{k+1} \right\}, \\
v_k &= \min\left\{ 3d_{k-1} - 2y'_{k-1}, d_k \right\}, \\
&\quad k = 2, \cdots, n-2, \\
u_{n-1} &= \frac{1}{2}(3d_{n-2} - y'_{n-2}), \\
v_{n-1} &= \min\left\{ 3d_{n-2} - 2y'_{n-2}, d_{n-1} \right\}, \\
u_n &= \frac{1}{2}(3d_{n-1} - y'_{n-1}), \\
v_n &= 3d_{n-1} - 2y'_{n-1}.
\end{aligned}
\tag{4.117}
$$

For example, one can set

$$
y'_k = \begin{cases} 0 & \text{if } 0 \in [u_k, v_k] \\ \frac{1}{2}(u_k + v_k) & \text{otherwise} \end{cases},
\tag{4.118}
$$

and succesively test whether the algorithm can proceed i. e., whether $u_k \leq v_k$; if $u_k > v_k$, it has failed. The subroutine GRAD8 (Figs. 4.50 and 4.51) works in that sense. It, in turn, makes use of COMPB (Figs. 4.52 and 4.53).

The examples in Figs. 4.54a–4.68b correspond (except for 4.61a,b,c) to those from 4.11a–4.16b, where the results for twice continuously differentiable cubic spline interpolation with end conditions IR=8 were given. They are the results of HECUB with the determination of the y'_k according to the methods corresponding to subroutines GRAD1, GRAD2, GRAD2B, GRAD2R, GRAD3, GRAD4, GRAD5 (ICASE=1), GRAD5 (ICASE=2), GRAD6, GRAD7, and, for convex data, GRAD8. From the almost 200 plots that were produced, some were chosen because they gave particularly pleasing results, but also others were chosen because they illustrate the deficiencies of the methods. Of course, it is not possible to quantify such decisions.

Generally speaking, one can say that GRAD1, GRAD2, GRAD3, and GRAD4 do not work well, even though, as was almost always the case, CUB2R7 with the corresponding end conditions also did not produce any good results. GRAD1, however, seems to perform the best of these. GRAD2B works better than GRAD2. GRAD2R, on the one hand, is sometimes substantially better than all the other subroutines with the exceptions of GRAD6 and GRAD7, but on the other hand, sometimes is substantially

```
      SUBROUTINE GRAD8(N,X,Y,EPS,B,IFLAG,BH,BL,C,H)
      DIMENSION X(N),Y(N),B(N),BH(N),BL(N),C(N),H(N)
      IFLAG=0
      IF(N.LT.3) THEN
          IFLAG=1
          RETURN
      END IF
      N1=N-1
      N2=N-2
      DO 10 K=1,N1
          H1=X(K+1)-X(K)
          H(K)=H1
          C(K)=(Y(K+1)-Y(K))/H1
10    CONTINUE
      DO 20 K=1,N2
          IF(C(K+1).LT.C(K)) THEN
              IFLAG=15
              GOTO 40
          END IF
20    CONTINUE
      BL(1)=3.*C(1)-2.*C(2)
      BH(1)=C(1)
      CALL COMPB(BL(1),BH(1),BB,IFLAG)
      IF(IFLAG.EQ.0) THEN
          B(1)=BB
      ELSE
          GOTO 40
      END IF
      DO 30 K=2,N2
          K1=K-1
          B1=(3.*C(K1)-B(K1))*0.5
          B2=3.*C(K)-2.*C(K+1)
          IF(B1.GE.B2) THEN
              BL(K)=B1
          ELSE
              BL(K)=B2
          END IF
          B1=3.*C(K1)-2.*B(K1)
          B2=C(K)
          IF(B1.LE.B2) THEN
              BH(K)=B1
          ELSE
              BH(K)=B2
          END IF
          CALL COMPB(BL(K),BH(K),BB,IFLAG)
          IF(IFLAG.EQ.0) THEN
              B(K)=BB
          ELSE
              GOTO 40
          END IF
30    CONTINUE
      BL(N1)=(3.*C(N2)-B(N2))*0.5
      B1=3.*C(N2)-2.*B(N2)
      B2=C(N1)
      IF(B1.LE.B2) THEN
          BH(N1)=B1
```

(*cont.*)

```
      ELSE
         BH(N1)=B2
      END IF
      CALL COMPB(BL(N1),BH(N1),BB,IFLAG)
      IF(IFLAG.EQ.0) THEN
         B(N1)=BB
      ELSE
         GOTO 40
      END IF
      BL(N)=(3.*C(N1)-B(N1))*0.5
      BH(N)=3.*C(N1)-2.*B(N1)
      CALL COMPB(BL(N),BH(N),BB,IFLAG)
      IF(IFLAG.EQ.0) B(N)=BB
40    RETURN
      END
```

Figure 4.50. Program listing of GRAD8.

Calling sequence:

CALL GRAD8(N,X,Y,EPS,B,IFLAG,BH,BL,C,H)

Purpose:
Attempt the selection of shape-preserving values for first derivatives for convex data by the method of Mettke and Lingner.

Description of the parameters:

N,X,Y,B as in GRAD1.

EPS   Value for the accuracy test. Recommendation: $EPS=10^{-t}$ (t : the number of available decimal digits). If the absolute value of a number is less than EPS, then it is interpreted as zero.

IFLAG   =0:   Normal execution.
        =15:   Given points are not convex.
        =16:   Failure of the selection principle (COMPB).

BH,BL,C,H ARRAY(N): Work space.

Required subroutine: COMPB.

Figure 4.51. Description of GRAD8.

```
SUBROUTINE COMPB(BL,BH,B,IFLAG)
IFLAG=0
IF (BL.GT.BH) THEN
    IFLAG=16
ELSE
    IF (BL.LE.0..AND.BH.GE.0.) THEN
        B=0.
    ELSE
        B=(BH+BL)*0.5
    END IF
END IF
RETURN
END
```

Figure 4.52. Program listing of COMPB.

Calling sequence:

CALL COMPB(BL,BH,B,IFLAG)

Purpose:
For a given interval (BL,BH) with BL≤BH, COMPB calculates a value
B for the derivative according to the following rule:

$$B = \begin{cases} 0 & \text{if } BL \leq 0 \leq BH \\ (BL + BH)/2 & \text{otherwise} \end{cases}.$$

Description of the parameters:

BL	Input:	Left-interval endpoint.
BH	Input:	Right-interval endpoint.
B	Output:	Value for the derivative.
IFLAG	=0:	Normal execution.
	=16:	BL>BH not allowed.

Figure 4.53. Description of COMPB.

---

worse. The variants ICASE=1 and ICASE=2 of GRAD5 work equally
well, but almost always GRAD6 and GRAD7 perform better. In most
cases GRAD7, is the best in the sense that the resulting curve is the most
pleasing. GRAD8 fulfills its purpose with convex data, although it can be
shown ([90]) that there exists a convex data set of five elements for which

no convex cubic Hermite spline interpolant exists. Necessary and sufficient conditions for GRAD8 (without the special choice of (4.118)) to run to completion are not known ([88,89]).

In Figs. 4.69a–4.73b, most of the examples given previously are again computed with HEQUA, mostly with knots selected by QUAFZ. Comparing these results, we see that quadratic Hermite spline interpolants with suitable choice of knots (usually the midpoints of the interpolation points suffice) perform just as well as do cubic Hermite spline interpolants. Both methods are superior to C^2 cubic interpolants.

There are numerous other works on shape-preserving cubic spline interpolation ([7,15,16,30,40,131,132,133,134,135,136,137,138,139]). We explicitly point out the subroutine PCHIM ([41,42,43]), which interpolates preserving monotonicity of the data. Although we do not discuss all these publications any further, it does not mean they are not mathematically interesting, but that in our opinion and experience, special rational spline interpolants are better suited to reproduce positivity, monotonicity, and convexity than are polynomial (especially cubic) splines.

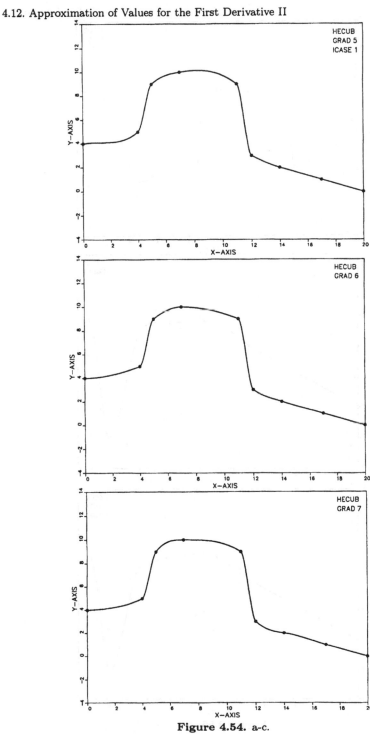

Figure 4.54. a-c.

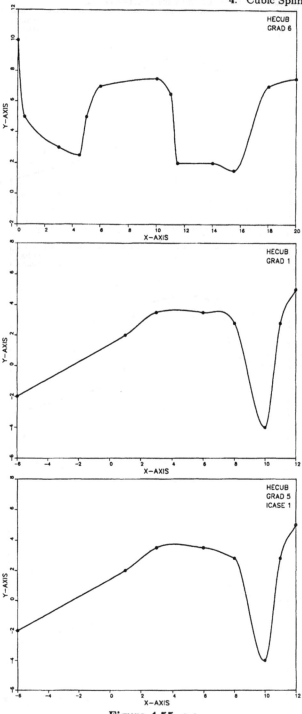

Figure 4.55. a-c.

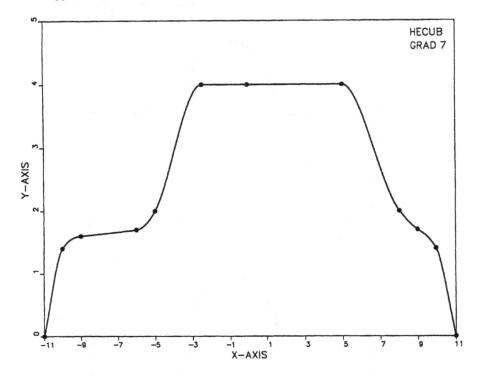

Figure 4.56.

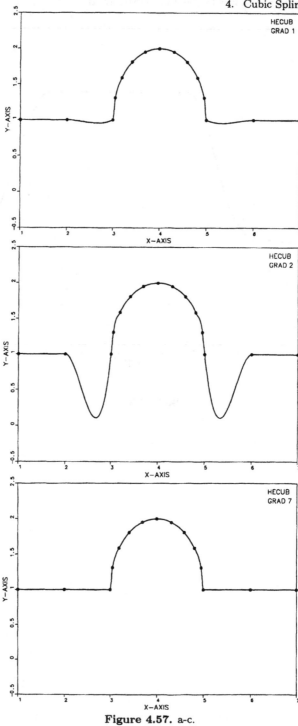

Figure 4.57. a-c.

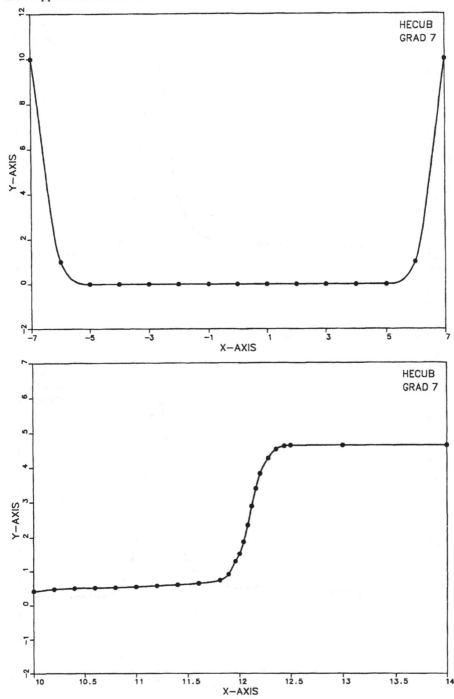

Figure 4.58. a, b.

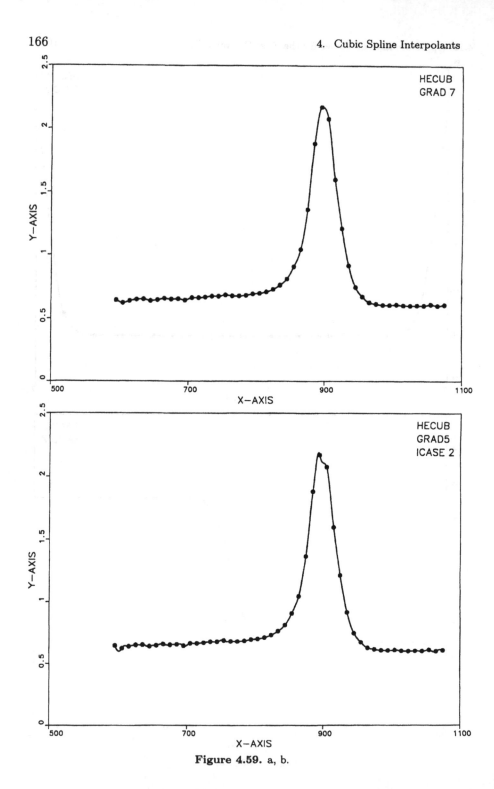

Figure 4.59. a, b.

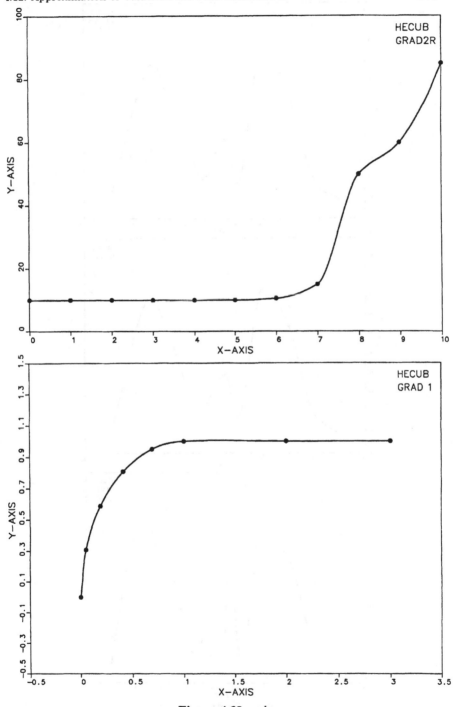

Figure 4.60. a, b.

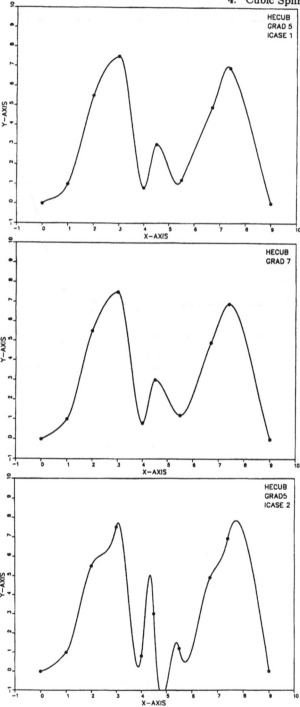

Figure 4.61. a-c.

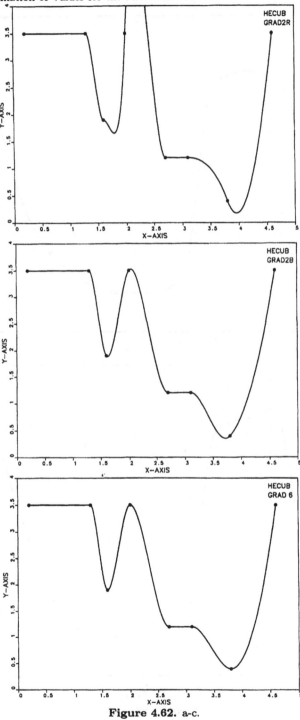

Figure 4.62. a-c.

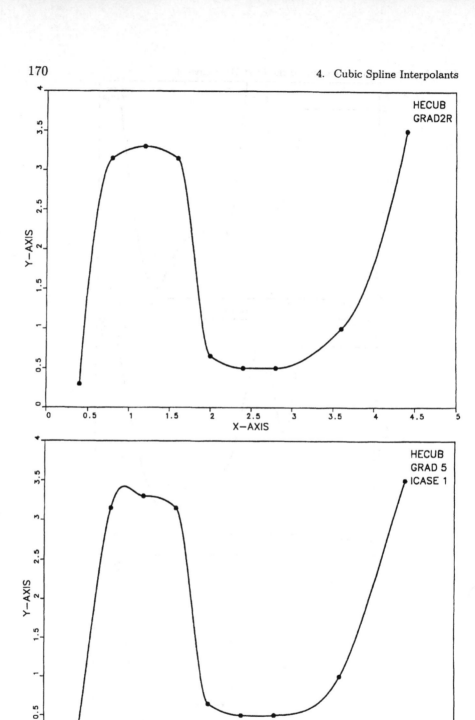

Figure 4.63. a, b.

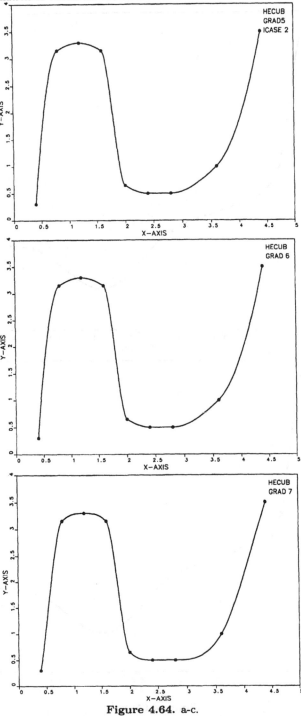

Figure 4.64. a-c.

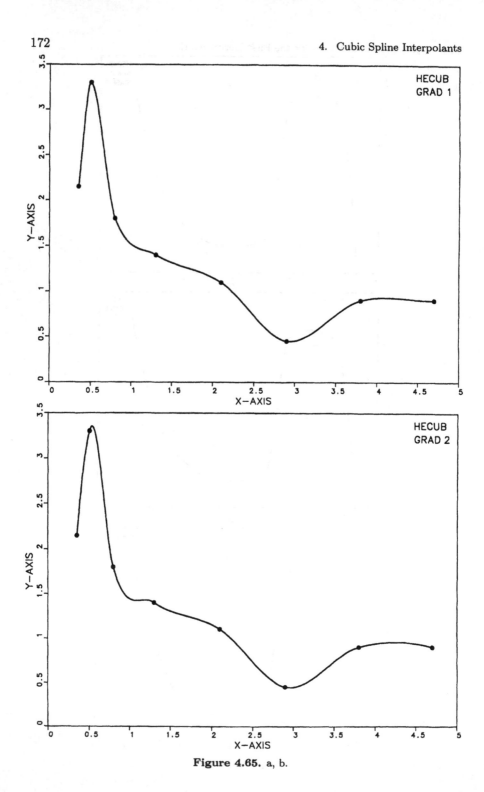

Figure 4.65. a, b.

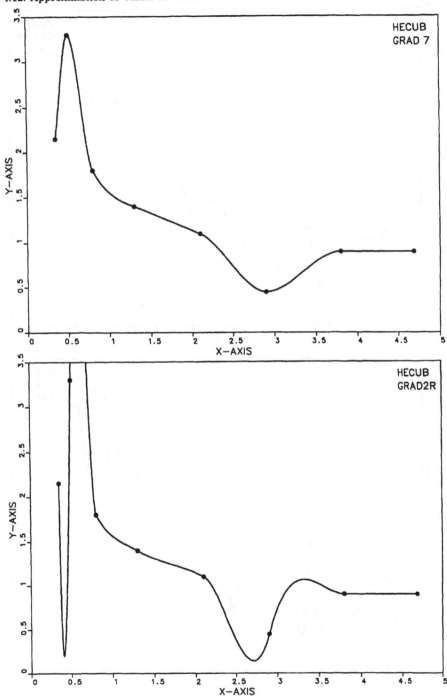

Figure 4.66. a, b.

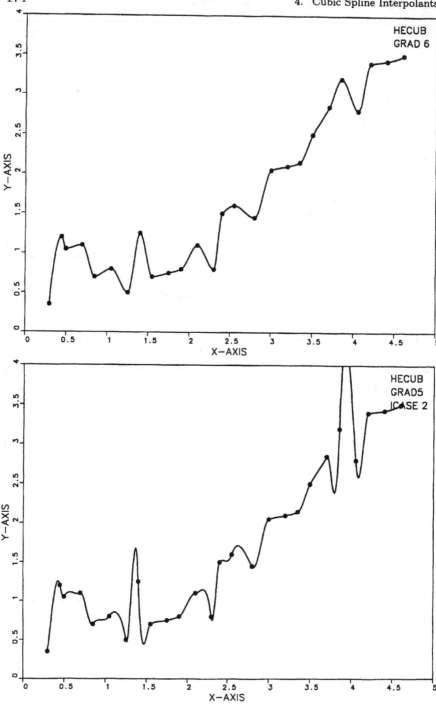

Figure 4.67. a, b.

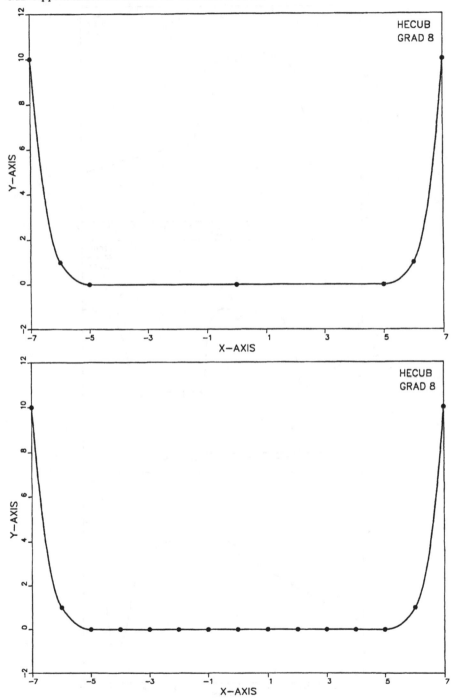

Figure 4.68. a, b.

4. Cubic Spline Interpolants

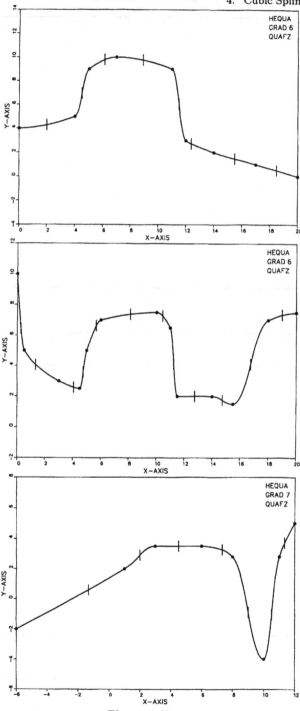

Figure 4.69. a-c.

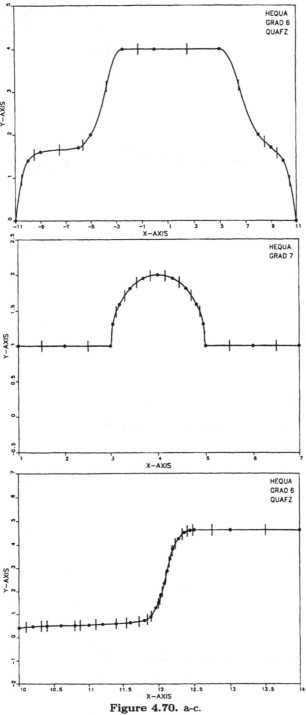

Figure 4.70. a-c.

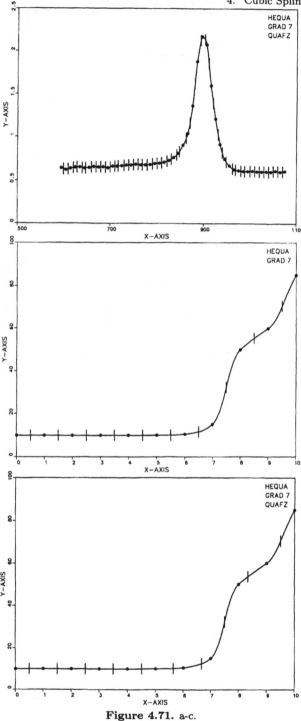

Figure 4.71. a-c.

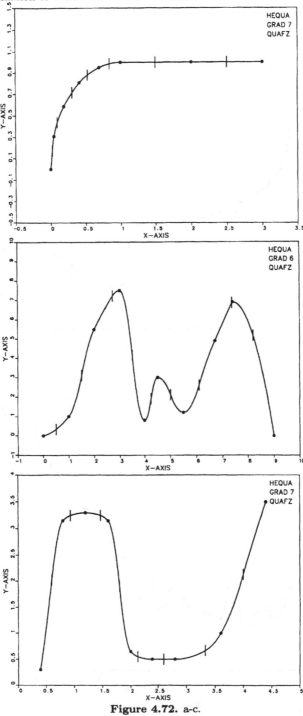

Figure 4.72. a-c.

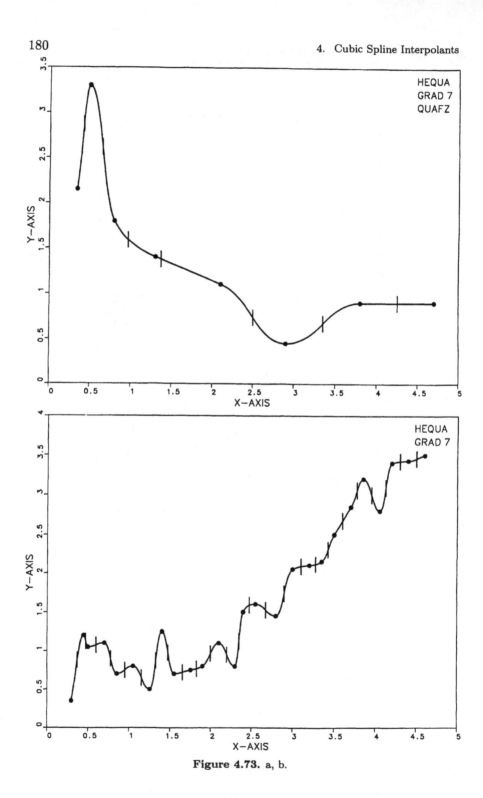

Figure 4.73. a, b.

5

Polynomial Spline Interpolants of Degree Five and Higher

5.1. Spline Interpolants of Degree Five

We deliberately skip spline interpolants of degree four. In the same way as for quadratic splines, the cases of knots the same as, and knots differing from, interpolation points can be characterized by means of systems of equations. It is also possible to consider three times continuously differentiable quartic Hermite spline interpolation, where first and second derivative values are given at the interpolation points. The addition of some extra knots would be required in this case. Much more of quadratic splines also goes over to the quartic case, although with some difficulties similar in nature to those arising when one passes from cubic to quintic splines. Since we feel that polynomial spline interpolants of degree higher than three have little practical application, we go directly to quintic splines, although we will not treat them in such detail and only then because the associated mathematical problems are especially interesting or because special cases may be of use in subsequent sections.

Assume again that (2.1) holds. On each interval $[x_k, x_{k+1}]$, $k = 1, \cdots, n - 1$, we set a quintic polynomial, s_k, of the form,

$$s_k(x) = A_k + B_k(x - x_k) + C_k(x - x_k)^2 + D_k(x - x_k)^3$$
$$+ E_k(x - x_k)^4 + F_k(x - x_k)^5, \tag{5.1}$$

and try to join these together so as to form a four times continuously differentiable function. The interpolation condition $s_k(x_k) = y_k$ gives

$$A_k = y_k, \qquad k = 1, \cdots, n - 1, \tag{5.2}$$

while $s_k(x_{k+1}) = y_{k+1}$ gives

$$B_k + h_k C_k + h_k^2 D_k + h_k^3 E_k + h_k^4 F_k = d_k, \qquad k = 1, \cdots, n - 1. \tag{5.3}$$

The continuity conditions $s_k^{(j)}(x_{k+1}) = s_{k+1}^{(j)}(x_{k+1})$, $k = 1, \cdots, n - 2$, yield

$$B_k + 2h_k C_k + 3h_k^2 D_k + 4h_k^3 E_k + 5h_k^4 F_k = B_{k+1}, \tag{5.4}$$
$$2C_k + 6h_k D_k + 12h_k^2 E_k + 20h_k^3 F_k = 2C_{k+1}, \tag{5.5}$$
$$6D_k + 24h_k E_k + 60h_k^2 F_k = 6D_{k+1}, \tag{5.6}$$
$$24E_k + 120h_k F_k = 24E_{k+1}, \tag{5.7}$$

for $j = 1, 2, 3, 4$, respectively.

Now, there are several possibilities for which parameters to use as unknowns. The pairs $(B_k, C_k), (B_k, D_k)$ ([2]), (B_k, E_k) ([1]), (C_k, D_k) and (C_k, E_k) ([157,158]), $k = 1, \cdots, n$, are all possible choices. For reasons that will become clear in what follows, we choose B_k and C_k, $k = 1, \cdots, n$, i.e., the values of the first and, up to a factor of $1/2$, second derivatives at each of the given abscissas. Then, not only (5.3) holds for $k = n - 1$, but also (5.4) and (5.5). If we write these three conditions in the form,

$$\begin{aligned} h_k^2 D_k + h_k^3 E_k + h_k^4 F_k &= d_k - B_k - h_k C_k, \\ 3h_k^2 D_k + 4h_k^3 E_k + 5h_k^4 F_k &= B_{k+1} - B_k - 2h_k C_k, \\ 6h_k^2 D_k + 12h_k^3 E_k + 20h_k^4 F_k &= 2h_k(C_{k+1} - C_k), \end{aligned} \tag{5.8}$$

we obtain for each $k = 1, \cdots, n - 1$, a system of three linear equations in the three unknowns D_k, E_k, F_k. Solving these by Gaussian elimination, we get

$$D_k = \frac{1}{h_k^2} \{10d_k - 6B_k - 4B_{k+1} - 3h_k C_k + h_k C_{k+1}\}, \tag{5.9}$$

$$E_k = \frac{1}{h_k^3} \{7B_{k+1} + 8B_k - 15d_k - 2h_k C_{k+1} + 3h_k C_k\}, \tag{5.10}$$

$$F_k = \frac{1}{h_k^4} \{6d_k - 3B_k - 3B_{k+1} + h_k C_{k+1} - h_k C_k\}. \tag{5.11}$$

Now replace k by $k-1$ in (5.6) and substitute in (5.9), (5.10), and (5.11). This gives

$$-\frac{1}{2h_{k-1}}C_{k-1} + \frac{3}{2}\left(\frac{1}{h_{k-1}} + \frac{1}{h_k}\right)C_k - \frac{1}{2h_k}C_{k+1}$$

$$-\frac{2}{h_{k-1}^2}B_{k-1} - 3\left(\frac{1}{h_{k-1}^2} - \frac{1}{h_k^2}\right)B_k + \frac{2}{h_k^2}B_{k+1} \qquad (5.12)$$

$$= 5\left(\frac{d_k}{h_k^2} - \frac{d_{k-1}}{h_{k-1}^2}\right), \qquad k = 2, \cdots, n-1.$$

Similarly, replace k by $k-1$ in (5.7) and substitute in (5.10) and (5.11). This results in

$$\frac{2}{h_{k-1}^2}C_{k-1} - 3\left(\frac{1}{h_{k-1}^2} - \frac{1}{h_k^2}\right)C_k - \frac{2}{h_k^2}C_{k+1}$$

$$+\frac{7}{h_{k-1}^3}B_{k-1} + 8\left(\frac{1}{h_{k-1}^3} + \frac{1}{h_k^3}\right)B_k + \frac{7}{h_k^3}B_{k+1} \qquad (5.13)$$

$$= 15\left(\frac{d_k}{h_k^3} - \frac{d_{k-1}}{h_{k-1}^3}\right), \qquad k = 2, \cdots, n-1.$$

Together, the systems (5.12) and (5.13) yield $2(n-2)$ equations in the $2n$ unknowns $B_k, C_k, k = 1, \cdots, n$. If we prescribe values for B_1, B_n, C_1, and C_n, i. e., the first and second (up to a factor of $1/2$) derivatives at the end-points, we are left with $2(n-2)$ unknowns $B_2, \cdots, B_{n-1}$ and $C_2, \cdots, C_{n-1}$. The coefficient matrix then has the form,

$$\mathbf{T} = \begin{pmatrix} \mathbf{U} & \mathbf{D}+\mathbf{V} \\ \mathbf{D}-\mathbf{V} & \mathbf{W} \end{pmatrix}, \qquad (5.14)$$

where $\mathbf{U}, \mathbf{V}$ and $\mathbf{W}$ are tridiagonal matrices and $\mathbf{D}$ is diagonal.

Gaussian elimination would usually be used to numerically solve the linear system (5.12) and (5.13); however, in contrast to the strictly diagonally dominant tridiagonal systems that almost always arise, it is not clear whether or not it can safely be solved without any pivoting. Only in that case could the band structure be exploited to make a more efficient algorithm.

It is always possible to use a *block under relaxation method* ([154,159]), as this requires only the positive definiteness of $\mathbf{U}$ and $\mathbf{W}$. However, because of the simple structure of the given system, this may be computationally more expensive than direct methods due to the large number of iterations that it requires. A corresponding Fortran subroutine, QUINTP, for *natural quintic spline interpolants* $(C_1 = D_1 = C_n = D_n = 0)$ is given in [154]. The Algol procedure "quintic" of [159] handles alternatively the end conditions that we described $(B_1, C_1, B_n, C_n$ prescribed), *natural* end conditions, or prescribed values for C_1, E_1, C_n and E_n.

If we use the ordering $C_2, B_2, C_3, B_3, \cdots, C_{n-1}, B_{n-1}$ of the unknowns, interchange the columns of $\mathbf{T}$ appropriately, and rewrite the equations in the sequence $1, n-1, 2, n, 3, n+1, \cdots, n-2, 2(n-2)$, then $\mathbf{T}$ is transformed into a new coefficient matrix,

$$
\tilde{\mathbf{T}} = \begin{pmatrix}
P_1 & Q_2 & & & & \\
R_2 & P_2 & Q_3 & & & \\
 & R_3 & P_3 & Q_4 & & \\
 & & \cdot & \cdot & \cdot & \\
 & & & \cdot & \cdot & \cdot \\
 & & & R_{n-3} & P_{n-3} & Q_{n-2} \\
 & & & & R_{n-2} & P_{n-2}
\end{pmatrix}, \tag{5.15}
$$

where P_k, Q_k, and R_k are the 2×2 matrices given by

$$
P_k = \begin{pmatrix}
\dfrac{3}{2}\left(\dfrac{1}{h_k} + \dfrac{1}{h_{k+1}}\right) & -3\left(\dfrac{1}{h_k^2} - \dfrac{1}{h_{k+1}^2}\right) \\[4mm]
-3\left(\dfrac{1}{h_k^2} - \dfrac{1}{h_{k+1}^2}\right) & 8\left(\dfrac{1}{h_k^3} + \dfrac{1}{h_{k+1}^3}\right)
\end{pmatrix}, \tag{5.16}
$$

$$
Q_k = \begin{pmatrix}
-\dfrac{1}{2h_k} & \dfrac{2}{h_k^2} \\[4mm]
-\dfrac{2}{h_k^2} & \dfrac{7}{h_k^3}
\end{pmatrix}, \tag{5.17}
$$

and

$$
R_k = \begin{pmatrix}
-\dfrac{1}{2h_k} & -\dfrac{2}{h_k^2} \\[4mm]
\dfrac{2}{h_k^2} & \dfrac{7}{h_k^3}
\end{pmatrix}. \tag{5.18}
$$

The block tridiagonal matrix $\mathbf{T}$ of two diagonal blocks thus becomes a tridiagonal matrix of 2×2 blocks. For such matrices, a variant of Gaussian elimination may be used, where 2×2 matrices are zeroed out instead of numbers ([96]). Unfortunately, this procedure may break down. The Fortran subroutine INSPL5 ([96]) also suffers from this defect, although one could of course always check whether or not the 2×2 matrices on the diagonal become numerically singular during the course of the elimination. (P_1 at least is invertible, but already for $P_2 - R_2 P_1^{-1} Q_2$ this is uncertain.)

The most elegant but also the most laborious method consists of simultaneously eliminating either the B_k or the C_k by writing out (5.12) and (5.13) for $k-1, k$, and $k+1$, eliminating either $B_{k-2}, B_{k-1}, B_k, B_{k+1}, B_{k+2}$ or $C_{k-2}, C_{k-1}, C_k, C_{k+1}, C_{k+2}$ from five of the six equations and then substituting these into the sixth. This gives a system of linear equations with a somewhat complicated five-diagonal coefficient matrix. The details are carried out in [1, p. 122], using B_k and E_k as the initial unknowns.

Such a five-diagonal system can also be derived by writing the third derivative of the desired quintic spline interpolant as a linear combination of *second-order B-splines* and using those coefficients as the unknowns ([54]). We give here the subroutine ([55]) QUINAT (Figs. 5.1 and 5.2), which is based on this procedure. It has the additional advantage of allowing the specification of first or first and second derivative values at selectively one or several or even all of the knots. However, specifying the first derivatives drops the continuity at that knot to three times continuously differentiable, and specifying both first and second derivatives drops it to only twice continuously differentiable. Technically, this is done by introducing double knots, $x_k = x_{k+1}$, or triple knots, $x_k = x_{k+1} = x_{k+2}$, and interpreting the ordinate values corresponding to x_{k+1} and x_{k+2} (for triple knots) as values for the first and second derivatives. The total number of abscissas must then be correspondingly increased. The desired coefficients for the interval $[x_k, x_{k+2}]$, or $[x_k, x_{k+3}]$ for triple knots, are then found in $Y(K+1), B(K+1), \cdots, F(K+1)$, and $Y(K+2), B(K+2), \cdots, F(K+2)$ respectively.

```
SUBROUTINE QUINAT(N,X,Y,B,C,D,E,F,EPS,IFLAG)
REAL X(N),Y(N),B(N),C(N),D(N),E(N),F(N)
IFLAG=0
IF(N.LE.2) THEN
    IFLAG=1
    RETURN
END IF
M=N-2
Q=X(2)-X(1)
R=X(3)-X(2)
Q2=Q*Q
R2=R*R
QR=Q+R
```

(cont.)

```
      D(1)=0.
      E(1)=0.
      D(2)=0.
      IF(ABS(Q).GT.EPS) D(2)=6.*Q*Q2/(QR*QR)
      IF(M.LT.2) GOTO 40
      DO 30 I=2,M
        P=Q
        Q=R
        R=X(I+2)-X(I+1)
        P2=Q2
        Q2=R2
        R2=R*R
        PQ=QR
        QR=Q+R
        IF(Q) 20, 10, 20
10      D(I+1)=0.
        E(I)=0.
        F(I-1)=0.
        GOTO 30
20      Q3=Q2*Q
        PR=P*R
        PQQR=PQ*QR
        D(I+1)=6.*Q3/(QR*QR)
        D(I)=D(I)+(Q+Q)*(15.*PR*PR+(P+R)*Q*(20.*PR+7.*Q2)
     &  +Q2*(8.*(P2+R2)+21.*PR+Q2+Q2)))/(PQQR*PQQR)
        D(I-1)=D(I-1)+6.*Q3/(PQ*PQ)
        E(I)=Q2*(P*QR+3.*PQ*(QR+R+R))/(PQQR*QR)
        E(I-1)=E(I-1)+Q2*(R*PQ+3.*QR*(PQ+P+P))/(PQQR*PQ)
        F(I-1)=Q3/PQQR
30    CONTINUE
40    IF(ABS(R).GT.EPS) D(M)=D(M)+6.*R*R2/(QR*QR)
      DO 60 I=2,N
        IF(X(I).NE.X(I-1)) GOTO 50
        B(I)=Y(I)
        Y(I)=Y(I-1)
        GOTO 60
50      B(I)=(Y(I)-Y(I-1))/(X(I)-X(I-1))
60    CONTINUE
      DO 80 I=3,N
        IF (X(I).NE.X(I-2)) GOTO 70
        C(I)=B(I)*0.5
        B(I)=B(I-1)
        GOTO 80
70      C(I)=(B(I)-B(I-1))/(X(I)-X(I-2))
80    CONTINUE
      IF(M.LT.2) GOTO 100
      P=0.
      C(1)=0.
      E(M)=0.
      F(1)=0.
      F(M-1)=0.
      F(M)=0.
      C(2)=C(4)-C(3)
      D(2)=1./D(2)
      IF(M.LT.3) GOTO 100
      DO 90 I=3,M
        Q=D(I-1)*E(I-1)
```

(cont.)

```
        D(I)=1./(D(I)-P*F(I-2)-Q*E(I-1))
        E(I)=E(I)-Q*F(I-1)
        C(I)=C(I+2)-C(I+1)-P*C(I-2)-Q*C(I-1)
        P=D(I-1)*F(I-1)
 90   CONTINUE
100   I=N-1
        C(N-1)=0.
        C(N)=0.
        IF(N.LT.4) GOTO 120
        DO 110 M=4,N
          I=I-1
          C(I)=(C(I)-E(I)*C(I+1)-F(I)*C(I+2))*D(I)
110   CONTINUE
120   M=N-1
        Q=X(2)-X(1)
        R=X(3)-X(2)
        B1=B(2)
        Q3=Q*Q*Q
        QR=Q+R
        IF(QR) 140, 130, 140
130   V=0.
        T=0.
        GOTO 150
140   V=C(2)/QR
        T=V
150   F(1)=0.
        IF(ABS(Q).GT.EPS) F(1)=V/Q
        DO 180 I=2,M
          P=Q
          Q=R
          R=0.
          IF(I.NE.M) R=X(I+2)-X(I+1)
          P3=Q3
          Q3=Q*Q*Q
          PQ=QR
          QR=Q+R
          S=T
          T=0.
          IF(ABS(QR).GT.EPS) T=(C(I+1)-C(I))/QR
          U=V
          V=T-S
          IF(PQ) 170, 160, 170
160       C(I)=C(I-1)
          D(I)=0.
          E(I)=0.
          F(I)=0.
          GOTO 180
170       F(I)=F(I-1)
          IF(ABS(Q).GT.EPS) F(I)=V/Q
          E(I)=5.*S
          D(I)=10.*(C(I)-Q*S)
          C(I)=D(I)*(P-Q)+(B(I+1)-B(I)+(U-E(I))*P3
     &          -(V+E(I))*Q3)/PQ
          B(I)=(P*(B(I+1)-V*Q3)+Q*(B(I)-U*P3))/PQ-
     &    P*Q*(D(I)+E(I)*(Q-P))
180   CONTINUE
        P=X(2)-X(1)
```

(cont.)

```
      S=F(1)*P*P*P
      E(1)=0.
      D(1)=0.
      C(1)=C(2)-10.*S
      B(1)=B1-(C(1)+S)*P
      Q=X(N)-X(N-1)
      T=F(N-1)*Q*Q*Q
      E(N)=0.
      D(N)=0.
      C(N)=C(N-1)+10.*T
      B(N)=B(N)+(C(N)-T)*Q
      RETURN
      END
```

Figure 5.1. Program listing of QUINAT.

Calling sequence:

CALL QUINAT(N,X,Y,B,C,D,E,F,EPS,IFLAG)

Purpose:
Calculation of a quintic spline interpolant. Double and triple knots are allowed.

Description of the parameters:

N	Number of given points. $N \geq 3$ is required.
X	ARRAY(N): Vector of abscissas.
Y	ARRAY(N): Vector of ordinates.
Y,B,C,	ARRAY(N): Output: Contain upon execution with
D,E,F	IFLAG=0 the desired spline coefficients,
	$K=1,2,\cdots,N-1$. In the case of multiple knots,
	the corresponding coefficients are to be skipped
	(see the text).
EPS	Value used for accuracy test. Recommendation: $EPS=10^{-t+2}$
	(t : the number of available decimal digits). If the absolute
	value of a quantity is less than EPS, it is interpreted as zero.
IFLAG	=0: Normal execution.
	=1: $N \geq 3$ is required.

Figure 5.2. Description of QUINAT.

The boundary conditions consist of setting the values of the third and fourth derivatives at x_1 and x_n to zero. The quintic spline can be evaluated at a point $V \in [x, x_n]$ by the subroutine QUIVAL (Figs. 5.3 and 5.4). We will also later make use of this for another purpose. Should it be the case that the abscissas are equally spaced, then the subroutine QUINEQ([55]) would be more efficient.

As the examples selected in Figs. 5.5a–5.8c show, *natural quintic spline interpolation* is without exception more unusable than natural cubic spline interpolation or even cubic Hermite or quadratic spline interpolation. In this regard, other end conditions are not much different.

Figures 5.9a,b contrast single and multiple knots. The knots $x_k, k = 6, 7, 8$, were chosen in Fig. 5.9b to form a triple knot with $y_k^{(j)} = 0, j = 1, 2$. From these examples, we see that, provided we drop the requirement of four times continuous differentiability, QUINAT can even be used for shape preservation. (In the example, knots 3 and 4 should also have had their multiplicity increased.)

5.2.   Quintic Hermite Spline Interpolants where First Derivative Values Are Specified

Besides giving the (x_k, y_k), suppose now that in (5.1), the values $y_k' = B_k, k = 1, \cdots, n$, of the first derivatives are also specified ([154, 156]). Then the coefficient equations (5.9), (5.10), and (5.11), as well as the C^3 condition (5.12), still hold, except that now the B_k are already known.

```
FUNCTION QUIVAL(N,X,A,B,C,D,E,F,V,IFLAG)
DIMENSION X(N),A(N),B(N),C(N),D(N),E(N),F(N)
DATA I/1/
IFLAG=0
IF(N.LT.2) THEN
   IFLAG=1
   RETURN
END IF
CALL INTONE(X,N,V,I,IFLAG)
IF(IFLAG.NE.0) RETURN
DX=V-X(I)
QUIVAL=A(I)+DX*(B(I)+DX*(C(I)+DX*(D(I)+DX*(E(I)+F(I)*DX))))
RETURN
END
```

Figure 5.3. Program listing of QUIVAL.

FUNCTION QUIVAL(N,X,A,B,C,D,E,F,V,IFLAG)

Purpose:
QUIVAL is a FUNCTION subroutine for the calculation of a function value of a quintic spline interpolant at a point $V \in [X(1),X(N)]$.

Description of the parameters:

N	Number of given points. $N \geq 2$ is required.
X	ARRAY(N): x−values.
A,B,C,	ARRAY(N): Vectors of the spline coefficients.
D,E,F	
V	Value of the point at which the spline function is to be evaluated.
IFLAG	=0:　　　Normal execution.
	=1:　　　$N \geq 2$ is required.
	=3:　　　Error in interval determination (INTONE).

Required subroutine: INTONE.

Remark: The statement 'DATA I/1/' has the effect that I is set to 1 at the first call to QUIVAL.

Figure 5.4. Description of QUIVAL.

If the values $y_1'' = 2C_1$ and $y_n'' = 2C_n$ for the second derivative at the endpoints are also given values, then (5.12) is a linear system with symmetric, tridiagonal, and strictly diagonally dominant coefficient matrix. It can thus be solved for $C_2, \cdots, C_{n-1}$, and then the remaining coefficients, D_k, E_k, and F_k, $k = 1, \cdots, n - 1$, can be calculated from (5.9), (5.10), and (5.11). (The C^4 condition (5.13) has simply been dropped.) In this way, we obtain a three times continuously differentiable quintic spline interpolant. It is calculated by the subroutine HEQUI1 (Figs. 5.10 and 5.11). If values for the y_k' are not available, then they could be approximated by one of the previously given routines such as GRAD1 or GRAD7. This results in a C^3 interpolant, whereas cubic spline interpolants are only C^2; of course, undesirable oscillations must be reckoned with.

Some typical examples are shown in Figs. 5.12a–5.13c. For these, the y_k' were computed by GRAD7. The results of several GRAD subroutines applied to the same dataset are shown in Figs. 5.14a–c. Evidently, HEQUI1 is not suitable for smooth Hermite interpolation. We remark that HEQUI1 is merely the special case of all double knots in QUINAT.

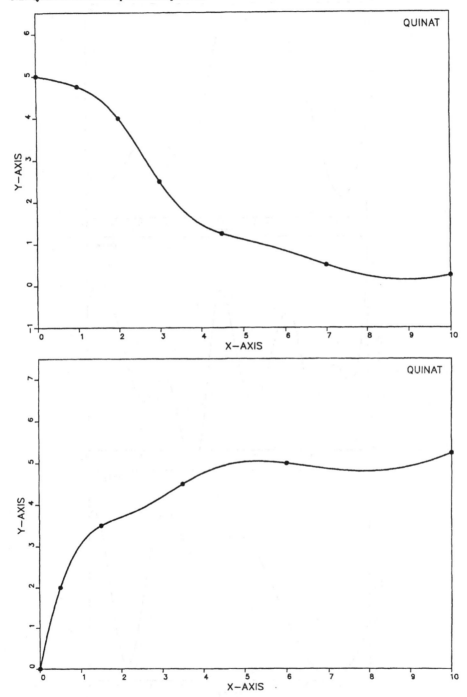

Figure 5.5. a, b.

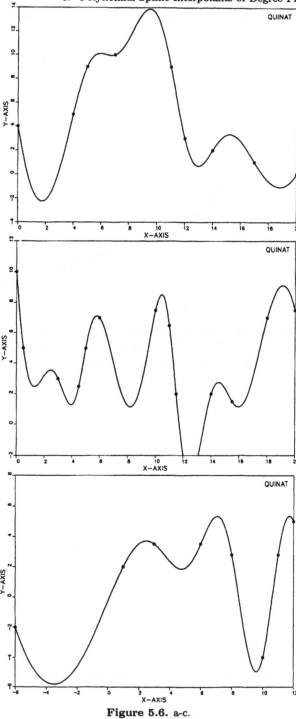

Figure 5.6. a-c.

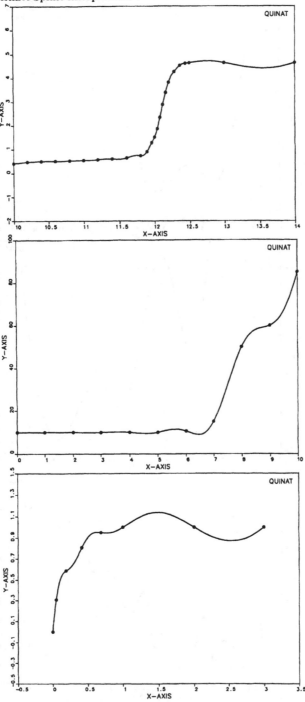

Figure 5.7. a-c.

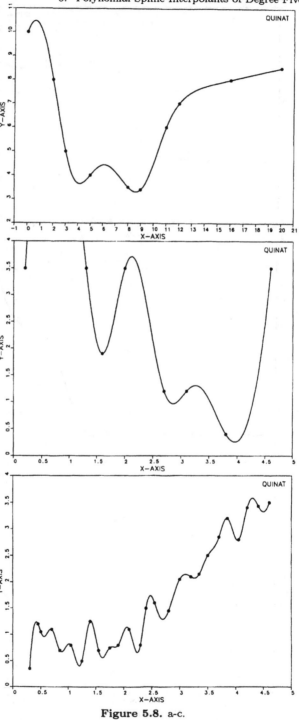

Figure 5.8. a-c.

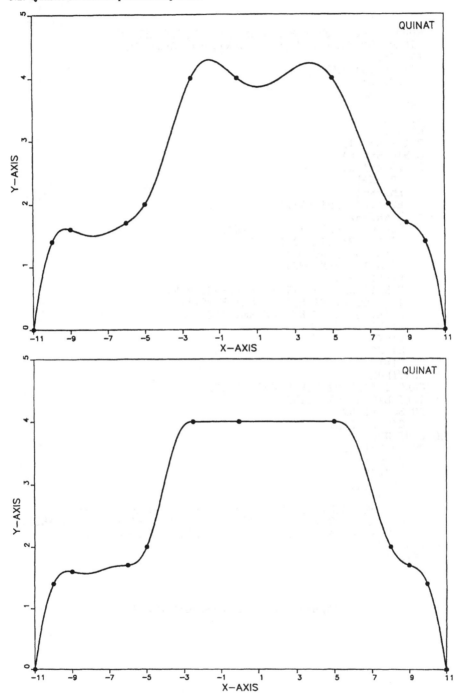

Figure 5.9. a, b.

```
      SUBROUTINE HEQUI1(N,X,Y,B,EPS,C,D,E,F,IFLAG)
      DIMENSION X(N),Y(N),B(N),C(N),D(N),E(N),F(N)
      IFLAG=0
      IF(N.LT.3) THEN
         IFLAG=1
         RETURN
      END IF
      N1=N-1
      N2=N-2
      C(1)=.5*C(1)
      C(N)=.5*C(N)
      DO 20 K=1,N1
         KP1=K+1
         KM1=K-1
         H2=1./(X(KP1)-X(K))
         V2=H2*H2
         S2=10.*V2*H2*(Y(KP1)-Y(K))
         T2=4.*V2*(B(KP1)+B(K))
         IF(K.EQ.1) GOTO 10
         D(KM1)=3.*(H1+H2)
         E(KM1)=-H2
         F(KM1)=S2-S1-T2+T1+2.*(V1-V2)*B(K)
         IF(K.EQ.2) F(KM1)=F(KM1)+H1*C(1)
         IF(K.EQ.N1) F(KM1)=F(KM1)+H2*C(N)
   10    H1=H2
         V1=V2
         S1=S2
         T1=T2
   20 CONTINUE
      CALL TRIDIS(N2,D,E,F,EPS,IFLAG)
      IF(IFLAG.NE.0) RETURN
      DO 30 K=1,N2
         C(K+1)=F(K)
   30 CONTINUE
      DO 40 K=1,N1
         K1=K+1
         G=Y(K1)-Y(K)
         H=1./(X(K1)-X(K))
         HG=H*G
         D(K)=H*(C(K1)-3.*C(K)-H*(6.*B(K)+4.*B(K1)-10.*HG))
         E(K)=H*H*(3.*C(K)-2.*C(K1)+H*(8.*B(K)+7.*B(K1)-15.*HG))
         F(K)=H*H*H*(C(K1)-C(K)-3.*H*(B(K)+B(K1)-2.*HG))
   40 CONTINUE
      RETURN
      END
```

Figure 5.10. Program listing of HEQUI1.

Calling sequence:

CALL HEQUI1(N,X,Y,B,EPS,C,D,E,F,IFLAG)

Purpose:
Calculation of a quintic Hermite spline interpolant for given values of the first derivative. Values for y_1'' and y_n'' must be given in C(1) and C(N), respectively.

Description of the parameters:

N	Number of given points. $N \geq 3$ is required.	
X	ARRAY(N):	Upon calling must contain the abscissa values x_k, $k = 1, \cdots, n$ with $x_1 < x_2 < \cdots < x_n$.
Y	ARRAY(N):	Upon calling must contain the ordinate values y_k, $k = 1, \cdots, n$.
B	ARRAY(N):	Upon calling must contain the values of the derivatives y_k', $k = 1, \cdots, n$.
EPS	see TRIDIS.	
Y,B,C,	ARRAY(N):	Upon execution with IFLAG=0 contain
D,E,F		the desired spline coefficients.
IFLAG	=0:	Normal execution.
	=1:	$N \geq 3$ is required.
	=2:	Error in solving the linear system (TRIDIS).

Required subroutine: TRIDIS.

Figure 5.11. Description of HEQUI1.

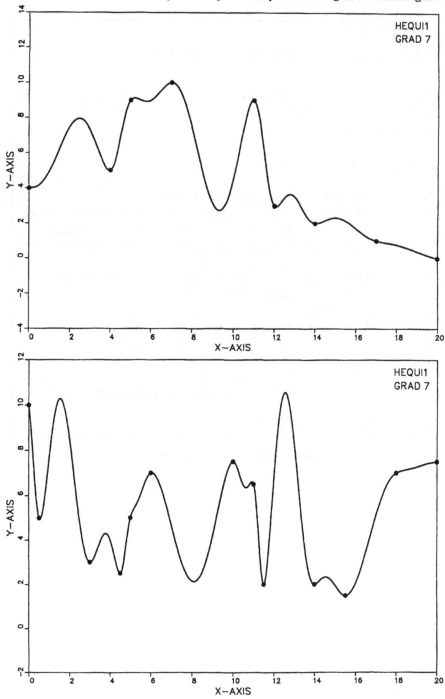

Figure 5.12. a, b.

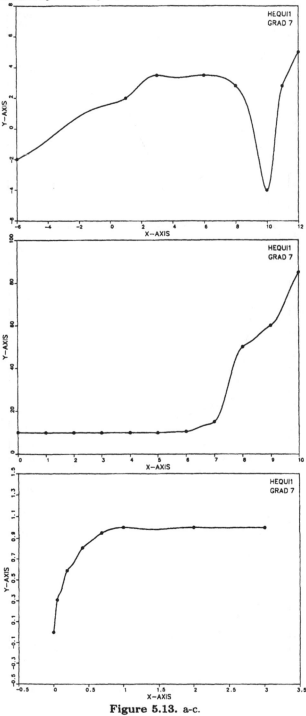

Figure 5.13. a-c.

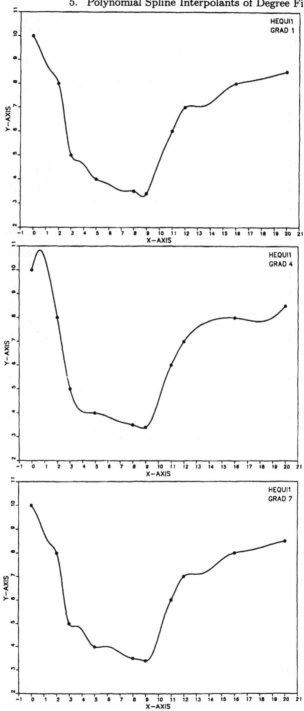

Figure 5.14. a-c.

5.3. Periodic Quintic Hermite Spline Interpolants

If $y_n = y_1$ and $y_1' = B_1 = B_n = y_n'$, then a *periodic quintic Hermite spline interpolant* may be constructed ([38]). For $k = n - 1$, C_n should be set to C_1 in (5.12). An additional equation comes from the $k = n - 1$ case of the C^3 condition (5.6), specifically from

$$6D_{n-1} + 24h_{n-1}E_{n-1} + 60h_{n-1}^2 F_{n-1} = D_1,$$

which after the substitution of (5.9), (5.10), and (5.11) becomes

$$\frac{1}{2}\left(\frac{1}{h_1} + \frac{1}{h_{n-1}}\right)C_1 - \frac{1}{2h_1}C_2 - \frac{1}{2h_{n-1}}C_{n-1}$$

$$\tag{5.19}$$

$$= 5\left(\frac{d_1}{h_1^2} - \frac{d_{n-1}}{h_{n-1}^2}\right) - 3\left(\frac{1}{h_1^2} - \frac{1}{h_{n-1}^2}\right)B_1 - \frac{2}{h_1^2}B_2 + \frac{2}{h_{n-1}^2}B_{n-1}.$$

Such splines may be computed using HEQUIP (Figs. 5.16 and 5.17), and an example using GRAD2 to calculate the y_k' is given in Fig. 5.15.

5.4. Quintic Hermite Spline Interpolants where Second Derivative also Are Specified

If the values $y_k'' = 2C_k$, $k = 1, \cdots, n$, for the second derivative are given, then in (5.12), the C_k are known and the B_k, $k = 2, \cdots, n - 1$, are to be found. We again assume that B_1 and B_n are also given. We mention that the special case for this problem of equally spaced abscissas was considered in [86, 101].

The resulting linear system with skew-symmetric tridiagonal coefficient matrix is then not always solvable. In the corresponding subroutine HEQUI2 (Figs. 5.18 and 5.19), TRIDIU decides whether or not it can be solved numerically. We will not bother with a theoretical analysis such as was done for the case of cubic spline interpolation with specified first derivatives (INCUB1). It should be noted that both these problems may not have a solution. In Fig. 5.20, we simply took $B_1 = d_1$ and $B_n = d_{n-1}$ together with $y_k'' = C_k = 0$, $k = 1, \cdots, 10$, i. e., we created inflection points at each of the knots. This catastrophic result shows that in HEQUI2 the y_k'' must be chosen very carefully.

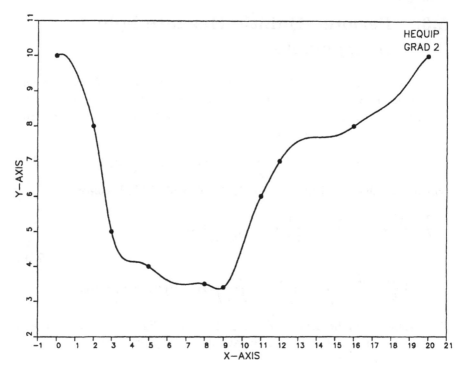

Figure 5.15.

If it is desired to find a quintic spline with specified values $y'_k = B_k$ and $y''_k = 2C_k$, $k = 1, \cdots, n$, for both first *and* second derivatives, then one can simply use formulas (5.9), (5.10), and (5.11) for the remaining coefficients. In contrast to the preceding, one then has a local interpolation method similar to HECUB. Methods for approximating the y''_k would also be needed. In any case, QUIVAL can be used for evaluation of the spline. There are a multitude of possible quintic splines for various interpolations of the jth derivative, $j = 0, 1, 2, \cdots$, at points other than the knots. Such are considered in [167], where values are specified for the function value and second derivative value at certain additional fixed points between the knots. In [168], the fourth derivative is given instead of the second. In [108], the function value and third derivatives are specified at knot midpoints. All of these result in C^2 interpolants. Other special cases may be found in [50, 92]. These and further possibilities seem more likely to be useful for the approximation of functions, rather than the interpolation of given data.

5.5.   Quartic Histosplines

If a spline function $s \in C^m[x_1, x_n]$ with segments,

$$s_k(x) = A_k + B_k(x - x_k) + C_k(x - x_k)^2 + D_k(x - x_k)^3 + \cdots, \quad (5.20)$$

```
SUBROUTINE HEQUIP(N,X,Y,B,EPS,C,D,E,F,IFLAG)
DIMENSION X(N),Y(N),B(N),C(N),D(N),E(N),F(N)
IFLAG=0
IF(N.LT.4) THEN
    IFLAG=1
    RETURN
END IF
N1=N-1
Y(N)=Y(1)
B(N)=B(1)
H1=1./(X(N)-X(N1))
V1=H1*H1
S1=10.*V1*H1*(Y(N)-Y(N1))
T1=4.*V1*(B(N)+B(N1))
DO 10 K=1,N1
    K1=K+1
    H2=1./(X(K1)-X(K))
    V2=H2*H2
    S2=10.*V2*H2*(Y(K1)-Y(K))
    T2=4.*V2*(B(K1)+B(K))
    D(K)=3.*(H1+H2)
    E(K)=-H2
    C(K)=S2-S1-T2+T1+2.*(V1-V2)*B(K)
    H1=H2
    V1=V2
    S1=S2
    T1=T2
10 CONTINUE
CALL TRIPES(N1,D,E,C,EPS,IFLAG)
IF(IFLAG.NE.0) RETURN
C(N)=C(1)
DO 20 K=1,N1
K1=K+1
G=Y(K1)-Y(K)
H=1./(X(K1)-X(K))
HG=H*G
D(K)=H*(C(K1)-3.*C(K)-H*(6.*B(K)+4.*B(K1)-10.*HG))
E(K)=H*H*(3.*C(K)-2.*C(K1)+H*(8.*B(K)+7.*B(K1)-15.*HG))
F(K)=H*H*H*(C(K1)-C(K)-3.*H*(B(K)+B(K1)-2.*HG))
20 CONTINUE
RETURN
END
```

Figure 5.16. Program listing of HEQUIP.

Calling sequence:

CALL HEQUIP(N,X,Y,B,EPS,C,D,E,F,IFLAG)

Purpose:
Determination of a periodic quintic Hermite spline interpolant for given values of the first derivatives. $y_1 = y_n$ and $y'_1 = y'_n$ are required and are enforced by the program.

Description of the parameters:

N,X,Y,B,C,D,E,F as in HEQUI1 (N$\geq$ 4 is required).
EPS   see TRIPES.
IFLAG   =0:   Normal execution.
            =1:   N$<$ 4 not allowed.
            =2:   Error in solving the linear system (TRIPES).

Required subroutine: TRIPES.

Figure 5.17. Description of HEQUIP.

could be constructed so as to satisfy the interpolation and end conditions,

$$\begin{aligned}
s(x_k) &= y_k, & k &= 1, \cdots, n, \\
s'(x_1) &= B_1, & s'(x_n) &= B_n, \\
s''(x_1) &= 2C_1, & s''(x_n) &= 2C_n, & \cdots & \quad \text{etc.},
\end{aligned}$$
(5.21)

then by a simple procedure ([144, 159]), we could also construct a spline $\tilde{s} \in C^{m-1}[x_1, x_n]$ with segments,

$$\tilde{s}_k(x) = \tilde{A}_k + \tilde{B}_k(x - x_k) + \tilde{C}_k(x - x_k)^2 + \tilde{D}_k(x - x_k)^3 + \cdots, \quad (5.22)$$

which satisfies the *area conditions*,

$$\int_{x_k}^{x_{k+1}} \tilde{s}_k(x)dx = h_k \tilde{y}_k, \qquad k = 1, \cdots, n-1, \quad (5.23)$$

for prescribed altitudes $\tilde{y}_k$ as well as the end conditions,

$$\tilde{s}(x_1) = \tilde{y}_0, \quad \tilde{s}(x_n) = \tilde{y}_n, \quad \tilde{s}'(x_1) = \tilde{y}'_0, \quad \tilde{s}'(x_n) = \tilde{y}'_n, \quad \cdots. \quad (5.24)$$

This is done by setting

$$y_1 = 0, \quad y_{k+1} = y_k + h_k \tilde{y}_k, \qquad k = 1, \cdots, n-1, \quad (5.25)$$

$$B_1 = \tilde{y}_0, \quad B_n = \tilde{y}_n, \quad C_1 = \tfrac{1}{2}\tilde{y}'_0, \quad C_n = \tfrac{1}{2}\tilde{y}'_n, \quad \cdots \quad \text{etc.},$$

```
      SUBROUTINE HEQUI2(N,X,Y,B,EPS,C,D,E,F,IFLAG)
      DIMENSION X(N),Y(N),B(N),C(N),D(N),E(N),F(N)
      IFLAG=0
      IF(N.LT.3) THEN
         IFLAG=1
         RETURN
      END IF
      N1=N-1
      N2=N-2
      B1=B(1)
      DO 20 K=1,N1
         KP1=K+1
         KM1=K-1
         H2=1./(X(KP1)-X(K))
         V2=H2*H2
         S2=2.5*V2*H2*(Y(KP1)-Y(K))
         T2=H2*(C(KP1)-C(K))/8.
         IF(K.EQ.1) GOTO 10
         B(KM1)=V2
         D(KM1)=-1.5*(V2-V1)
         E(KM1)=-V2
         F(KM1)=S1-S2+T1-T2+(H1+H2)*C(K)/4.
         IF(K.EQ.2) F(KM1)=F(KM1)-V1*B1
         IF(K.EQ.N1) F(KM1)=F(KM1)+V2*B(N)
   10    H1=H2
         V1=V2
         S1=S2
         T1=T2
   20 CONTINUE
      CALL TRIDIU(N2,B,D,E,F,EPS,IFLAG)
      IF(IFLAG.NE.0) RETURN
      DO 30 K=1,N2
         B(K+1)=F(K)
         C(K)=C(K)/2.
   30 CONTINUE
      B(1)=B1
      C(N1)=C(N1)/2.
      C(N)=C(N)/2.
      DO 40 K=1,N1
         K1=K+1
         G=Y(K1)-Y(K)
         H=1./(X(K1)-X(K))
         HG=H*G
         D(K)=H*(C(K1)-3.*C(K)-H*(6.*B(K)+4.*B(K1)-10.*HG))
         E(K)=H*H*(3.*C(K)-2.*C(K1)+H*(8.*B(K)+7.*B(K1)-15.*HG))
         F(K)=H*H*H*(C(K1)-C(K)-3.*H*(B(K)+B(K1)-2.*HG))
   40 CONTINUE
      RETURN
      END
```

Figure 5.18. Program listing of HEQUI2.

Calling sequence:

CALL HEQUI2(N,X,Y,B,EPS,C,D,E,F,IFLAG)

Purpose:
Calculation of a quintic Hermite spline interpolant for given values of the second derivatives. Values for y_1' and y_n' must be given in B(1) and B(N), respectively.

Description of the parameters:

N,X,Y as in HEQUI1.		
C	ARRAY(N):	Input: Upon calling must contain the values of the second derivatives y_k'', $k = 1, \cdots, n$.
EPS	see TRIDIU.	
Y,B,C, D,E,F	ARRAY(N):	Output: Upon execution with IFLAG=0 contain the desired spline coefficients, K=1,2,$\cdots$,N−1.
IFLAG	=0:	Normal execution.
	=1:	N≥ 3 is required.
	=2:	Error in solving the linear system (TRIDIU).

Required subroutine: TRIDIU.

Figure 5.19. Description of HEQUI2.

calculating s, and then taking $\tilde{s} = s'$. In the case of polynomial splines, then

$$\tilde{A}_k = B_k, \quad \tilde{B}_k = 2C_k, \quad \tilde{C}_k = 3D_k, \quad \tilde{D}_k = 4E_k, \quad \cdots. \qquad (5.26)$$

As we shall later see for rational spline interpolation, this method of construction works equally well if s is a *non-polynomial* spline function. This follows from the facts that

$$
\begin{aligned}
h_k\tilde{y}_k &= \int_{x_k}^{x_{k+1}} \tilde{s}_k(x)dx = \int_{x_k}^{x_{k+1}} s'(x)dx \\
&= s(x_{k+1}) - s(x_k) = y_{k+1} - y_k, \qquad (5.27) \\
&\quad k = 1, \cdots, n - 1,
\end{aligned}
$$

and that both s and $s + y_1$ are antiderivatives of $\tilde{s} = s'$, so that without loss of generality we may take $y_1 = 0$.

We will first of all carry out this procedure for *quadratic histosplines*. Previously, for pedagogical reasons, we gave a direct derivation of this

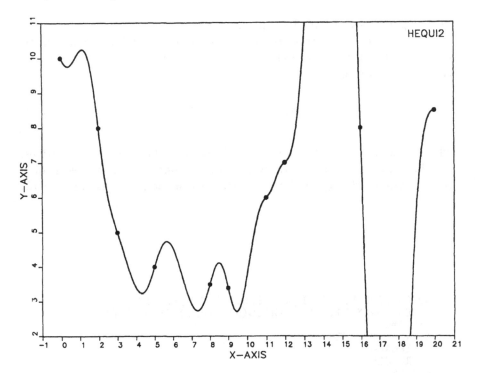

Figure 5.20.

case. If we equip the coefficients of the quadratic with a tilde, then to the system (3.43), i.e.,

$$\frac{1}{h_{k-1}}\tilde{A}_{k-1} + 2\left(\frac{1}{h_{k-1}} + \frac{1}{h_k}\right)\tilde{A}_k + \frac{1}{h_k}\tilde{A}_{k+1} = 3\left(\frac{\tilde{y}_{k-1}}{h_{k-1}} + \frac{\tilde{y}_k}{h_k}\right),$$

corresponds directly that of (4.7), namely,

$$\frac{1}{h_{k-1}}B_{k-1} + 2\left(\frac{1}{h_{k-1}} + \frac{1}{h_k}\right)B_k + \frac{1}{h_k}B_{k+1} = 3\left(\frac{d_{k-1}}{h_{k-1}} + \frac{d_k}{h_k}\right),$$

seeing that

$$d_k = \frac{y_{k+1} - y_k}{h_k} = \tilde{y}_k.$$

The end conditions also correspond upon setting

$$\tilde{A}_1 = B_1 = \tilde{y}_0, \qquad \tilde{A}_n = B_n = \tilde{y}_n.$$

By (4.6), we have

$$\tilde{B}_n = 2C_k, \quad \tilde{C}_k = 3D_k, \quad k = 1, \cdots, n-1,$$

for the previously obtained coefficients (3.42).

Now, in order to construct three times continuously differentiable quartic histosplines, it is necessary to solve the linear systems (5.12) and (5.13) with the end conditions,

$$B_1 = \tilde{y}_0, \quad B_n = \tilde{y}_n, \quad C_1 = \tilde{y}_1' \text{ and } C_n = \tilde{y}_n', \tag{5.28}$$

where $\tilde{y}_1'$ and $\tilde{y}_n'$ are additionally specified values for the first derivatives. As this, however, is not entirely simple, we content ourselves with only *twice continuously differentiable quartic histosplines*, i.e., use only (5.12)

```
      SUBROUTINE HIST42(N,X,Y,EPS,A,B,C,D,E,F,IFLAG)
      DIMENSION X(N),Y(N),A(N),B(N),C(N),D(N),E(N),F(N)
      IFLAG=0
      IF(N.LT.3) THEN
         IFLAG=1
         RETURN
      END IF
      N1=N-1
      A(1)=E(1)*Y(1)
      DO 10 K=2,N1
         A(K)=E(K)*Y(K-1)+(1.-E(K))*Y(K)
   10 CONTINUE
      A(N)=E(N)*Y(N1)
      C(1)=B(1)
      C(N)=B(N)
      B(1)=0.
      DO 20 K=1,N1
         K1=K+1
         B(K1)=B(K)+(X(K1)-X(K))*Y(K)
   20 CONTINUE
      CALL HEQUI1(N,X,B,A,EPS,C,D,E,F,IFLAG)
      IF(IFLAG.NE.0) RETURN
      DO 30 K=1,N1
         B(K)=2.*C(K)
         C(K)=3.*D(K)
         D(K)=4.*E(K)
         E(K)=5.*F(K)
   30 CONTINUE
      B(N)=2.*C(N)
      RETURN
      END
```

Figure 5.21. Program listing of HIST42.

Calling sequence:

CALL HIST42(N,X,Y,EPS,A,B,C,D,E,F,IFLAG)

Purpose:
Determination of flexible quartic histosplines. N$\geq$ 3 is required.

Description of the parameters:

N,X,EPS as in QUHIST.

Y	ARRAY(N):	Input: Y(1),Y(2),$\cdots$,Y(N$-$1) must contain the rectangle heights y_k, $k = 1, \cdots, n - 1$.
B	ARRAY(N):	Input: Upon calling y_1' must be given in B(1) and y_n' in B(N).
E	ARRAY(N):	Input: E(1),E(2),$\cdots$,E(N) must contain the factors w_k, $k = 1, \cdots, n$.
A,B,C, D,E	ARRAY(N):	Upon execution contain the desired spline coefficients.
F	ARRAY(N):	Work space.
IFLAG	=0:	Normal execution.
	=1:	N$\geq$ 3 is required.
	=2:	Error in solving the linear system (TRIDIS).

Required subroutines: HEQUI1, TRIDIS.

Figure 5.22. Description of HIST42.

and not (5.13), which means that in the corresponding system,

$$-\frac{1}{h_{k-1}}\tilde{B}_{k-1} + 3\left(\frac{1}{h_{k-1}} + \frac{1}{h_k}\right)\tilde{B}_k - \frac{1}{h_{k+1}}\tilde{B}_{k+1}$$

$$= 20\left[\frac{\tilde{y}_k}{h_k^2} - \frac{\tilde{y}_{k-1}}{h_{k-1}^2}\right] + \frac{8}{h_{k-1}^2}\tilde{A}_{k-1} + 12\left(\frac{1}{h_{k-1}^2} - \frac{1}{h_k^2}\right)\tilde{A}_k - \frac{8}{h_k^2}\tilde{A}_{k+1}, \tag{5.29}$$

$$k = 2, \cdots, n - 1,$$

besides $\tilde{B}_1 = \tilde{y}_1'$ and $\tilde{B}_n = \tilde{y}_n'$, the $\tilde{A}_k$, $k = 1, \cdots, n$, must also be given. As the $\tilde{A}_k$ are the function values at x_k of the spline segments,

$$\tilde{s}_k(x) = \tilde{A}_k + \tilde{B}_k(x - x_k) + \tilde{C}_k(x - x_k)^2 + \tilde{D}_k(x - x_k)^3 + \tilde{E}_k(x - x_k)^4, \tag{5.30}$$

```
FUNCTION QATVAL(N,X,A,B,C,D,E,V,IFLAG)
DIMENSION X(N),A(N),B(N),C(N),D(N),E(N)
DATA I/1/
IFLAG=0
IF(N.LT.2) THEN
    IFLAG=1
    RETURN
END IF
CALL INTONE(X,N,V,I,IFLAG)
IF(IFLAG.NE.0) RETURN
DX=V-X(I)
QATVAL=A(I)+DX*(B(I)+DX*(C(I)+DX*(D(I)+E(I)*DX)))
RETURN
END
```

Figure 5.23. Program listing of QATVAL.

FUNCTION QATVAL(N,X,A,B,C,D,E,V,IFLAG)

Purpose:
QATVAL is a FUNCTION subroutine for the calculation of a function value of a quartic spline function at a point $V \in [X(1),X(N)]$.

Description of the parameters:

N	Number of given points.
X	ARRAY(N): x-values.
A,B,C,	ARRAY(N): Vectors of the spline coefficients.
D,E	
V	Value of the point at which the spline function is to be evaluated.
IFLAG	=0: Normal execution.
	=1: $N \geq 2$ is required.
	=3: Error in interval determination (INTONE).

Required subroutine: INTONE.

Remark: The statement 'DATA I/1/' has the effect that I is set to 1 at the first call to QATVAL.

Figure 5.24. Description of QATVAL.

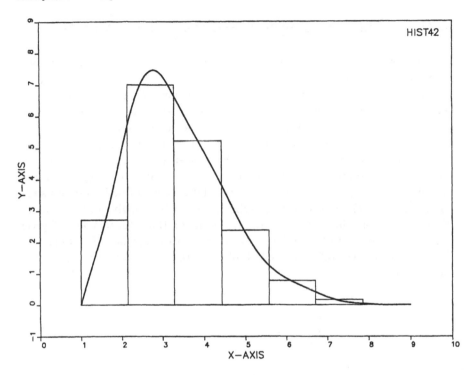

Figure 5.25.

which we are trying to compute, they allow for a certain flexibility in the shape of the resulting curve. By introducing factors, w_k, $k = 1, \cdots, n$, for example, and setting,

$$\tilde{A}_1 = w_1\tilde{y}_1,$$
$$\tilde{A}_k = w_k\tilde{y}_{k-1} + (1 - w_k)\tilde{y}_k, \quad k = 2, \cdots, n-1, \qquad (5.31)$$
$$\tilde{A}_n = w_n\tilde{y}_{n-1},$$

we obtain a useful set of geometrically significant control parameters. This method ([154, 155]) is implemented in the subroutine HIST42 (Figs. 5.21 and 5.22). (We have dropped the tildes from the parameters in the program description.) It calculates the $\tilde{A}_k$ by (5.31) and then calls HEQUI1 with $\tilde{Y}$ instead of Y and $\tilde{A}$ instead of B. Finally, the coefficients of (5.30) are calculated according to (5.26). The given values for $\tilde{y}'_1$ and $\tilde{y}'_n$ are assumed to be in B_1 ($= \tilde{B}_1$) and B_n ($= \tilde{B}_n$). QATVAL (Figs. 5.23 and 5.24) may be used for evaluation. Figure 5.25 shows the result of an example with $w = (0, .3, .8, .5, .3, .4, .2, 0)^t$. The desired and so-attained effects of varying the w_k are clearly visible. In general, however, the right choice of the w_k does not appear to be entirely simple.

5.6.   Higher-Degree Spline Interpolants

Linear systems for *even-degree spline interpolants* can in general be derived in a manner similar to that for quadratic and quartic (which we did not do) splines. The problem of unbalanced end conditions remains. For *odd-degree spline interpolants*, linear systems can be obtained both by the introduction of the even-order derivatives ([157, 158, 4]) as well as odd-order derivatives ([2]). Iterative methods for the solution of such systems for odd degree splines and equally spaced abscissas have been proposed. At present, we are of the opinion that for polynomial spline interpolation of degree four and higher, and especially for equally spaced data, it would be better to use *B-splines* ([18,19,27]) of the corresponding order. We strongly doubt, however, that such interpolants will be at all useful for the interpolation of given data. For the sake of completeness, we refer to a number of publications [12, 13, 14, 31, 81, 102, 103, 104, 106] on shape preservation using higher-degree splines.

6

Rational Spline Interpolants

6.1. Rational Splines with One Freely Varying Pole

The most simply constructed rational function, defined on $[x_k, x_{k+1}]$, involving (just as for cubic splines) four parameters is probably ([128,161])

$$s_k(x) = A_k + B_k(x - x_k) + \frac{C_k}{1 + D_k(x - x_k)}. \qquad (6.1)$$

If the s_k and the subsequent rational spline interpolant that they form are to be twice continuously differentiable on $[x_1, x_n]$ and in particular, continous there, then the pole of s_k,

$$x = x_k - \frac{1}{D_k}, \qquad (6.2)$$

must lie outside of $[x_k, x_{k+1}]$, i. e., either

$$D_k > 0 \quad \text{or} \quad -\frac{1}{D_k} > h_k = x_{k+1} - x_k. \qquad (6.3)$$

In order to see how to handle the difficulties caused by this restriction, we will first of all consider the simplest possible rational function having one pole and two parameters (just as for polygonal paths (2.4)), i. e.,

$$s_k(x) = \frac{C_k}{1 + D_k(x - x_k)}. \tag{6.4}$$

Here, the D_k, because of the stipulated continuity, must also satisfy (6.3). The interpolation conditions (2.2), in this case, become

$$C_k = y_k, \tag{6.5}$$

$$\frac{y_k}{1 + h_k D_k} = y_{k+1}, \quad i. e., \quad D_k = -\frac{d_k}{y_{k+1}}. \tag{6.6}$$

If $d_k = 0$, i. e., $y_{k+1} = y_k$, and $y_{k+1} \neq 0$, then $D_k = 0$ and $s_k \equiv y_k$. If $y_k = y_{k+1} = 0$, then D_k is arbitrary and $s_k \equiv 0$. If $d_k \neq 0$ and $y_k = 0$, then $C_k = 0$, i. e., $s_k \equiv 0$, and hence the interpolation problem has no solution. Finally, if $d_k \neq 0$ and $y_{k+1} = 0$, then the constant D_k is undefined and there is also no solution. There remains the case of $y_k \neq y_{k+1}$, $y_k \neq 0$, $y_{k+1} \neq 0$, for which there arises the restriction that $sgn(y_k) = sgn(y_{k+1})$, as s_k cannot have any zeros. Altogether, we see that the interpolation problem with nonconstant denominator in (6.4) is only solvable in the cases:

$$y_{k+1} > y_k > 0,$$

$$y_k > y_{k+1} > 0,$$

$$y_{k+1} < y_k < 0, \tag{6.7}$$

$$y_k < y_{k+1} < 0.$$

By substituting (6.6) into (6.3), we see that no pole arises when (6.7) holds. If the segments (6.4) are, as for polygonal paths, to be joined continuously, then we must necessarily have $y_k > 0$, $k = 1, \cdots, n$ or $y_k < 0$, $k = 1, \cdots, n$ as well as the corresponding conditions of (6.7). The resulting spline interpolants would then be composed of hyperbolic segments and would certainly not be as useful as are the corresponding polygonal paths. An exception would be if one of the alternatives of (6.7) holds at the *same time* for *all* $k = 1, \cdots, n - 1$, i. e., either

$$y_1 > y_2 > \cdots > y_n > 0,$$

$$y_n > y_{n-1} > \cdots > y_1 > 0, \tag{6.8}$$

$$y_1 < y_2 < \cdots < y_n < 0,$$

or

$$y_n < y_{n-1} < \cdots < y_1 < 0.$$

In that case, since

$$s_k'(x) \;=\; \frac{-C_k d_k}{[1 + D_k(x - x_k)]^2}, \tag{6.9}$$

$$s_k''(x) \;=\; \frac{2C_k D_k^2}{[1 + D_k(x - x_k)]^3}, \tag{6.10}$$

and

$$C_k D_k \;=\; -\frac{y_k}{y_{k+1}} \frac{y_{k+1} - y_k}{h_k},$$

the global spline interpolant, s, is strictly decreasing or increasing and either convex or concave. This follows from the facts that the denominators of (6.9) and (6.10) have no zeros in $[x_k, x_{k+1}]$, and consequently are of constant sign there, and that the numerators are constant. This property goes over naturally to rationals of the form (6.1) as the linear part, $A_k + B_k(x - x_k)$, becomes constant after taking one derivative and zero after taking the second derivative.

We will not discuss the case of rational spline interpolants with one free pole and three parameters, corresponding in number of parameters to quadratic splines, and instead return now to those of the form (6.1). For these, we could also have chosen

$$s_k(x) \;=\; \frac{A_k + B_k(x - x_k) + C_k(x - x_k)^2}{1 + D_k(x - x_k)}, \tag{6.11}$$

$$s_k(x) \;=\; A_k + B_k(x - x_k) + \frac{C_k(x - x_k)^2}{1 + D_k(x - x_k)}, \tag{6.12}$$

or

$$s_k(x) \;=\; A_k + B_k(x - x_k) + \frac{C_k(x - x_k)(x - x_{k+1})}{1 + D_k(x - x_k)}. \tag{6.13}$$

Forms (6.11) and (6.12) are treated in [128, 130]. Form (6.13), taken from [116], has the advantage that the calculations we are about to carry out are somewhat simpler. We stick to (6.1), since this form of rational spline is the least expensive to evaluate.

Now, for (6.1), we have

$$s_k'(x) = B_k - \frac{C_k D_k}{[1 + D_k(x - x_k)]^2} \tag{6.14}$$

and

$$s_k''(x) = \frac{2C_k D_k^2}{[1 + D_k(x - x_k)]^3}.$$ (6.15)

The interpolation conditions (2.2) yield

$$y_k = A_k + C_k,$$ (6.16)

$$y_{k+1} = A_k + h_k B_k + \frac{C_k}{1 + h_k D_k}.$$ (6.17)

Introducing first the values y_k' of the first derivatives at x_k as unknowns, we have

$$y_k' = B_k - C_k D_k,$$ (6.18)

$$y_{k+1}' = B_k - \frac{C_k D_k}{[1 + h_k D_k]^2}.$$ (6.19)

From (6.16), it follows that

$$A_k = y_k - C_k,$$ (6.20)

and from (6.18) that

$$B_k = y_k' + C_k D_k.$$ (6.21)

Now, substituting (6.20) and (6.21) into (6.17), an easy calculation shows that

$$C_k D_k = \frac{1 + h_k D_k}{h_k D_k}(d_k - y_k').$$ (6.22)

Then substituting this expression together with (6.21) into the as yet unused condition (6.19) and solving for D_k results in

$$D_k = \frac{2d_k - y_k' - y_{k+1}'}{h_k(y_{k+1}' - d_k)},$$ (6.23)

so that by (6.22),

$$C_k = \frac{h_k(d_k - y_k')^2(y_{k+1}' - d_k)}{(2d_k - y_k' - y_{k+1}')^2}.$$ (6.24)

Of course, we must impose the restrictions that the denominators of (6.23) and (6.24) are nonzero. Further, we must have $y_k' \neq d_k$ so that $C_k \neq 0$. If appropriate values for the y_k' are given, then the *Hermite interpolation problem* has already been solved. However, if we want s to be twice continuously differentiable, then the extra continuity conditions,

$$s_{k-1}''(x_k) = s_k''(x_k), \quad k = 2, \cdots, n - 1,$$ (6.25)

must be satisfied. Using (6.15) and the expressions,

$$\frac{C_k D_k^2}{1 + h_k D_k} = \frac{1}{h_k}(d_k - y_k'),$$

$$1 + h_k D_k = \frac{d_k - y_k'}{y_{k+1}' - d_k},$$

we see from (6.18) that

$$\frac{2}{h_{k-1}} \frac{(y_k' - d_{k-1})^2}{d_{k-1} - y_{k-1}'} = \frac{2}{h_k} \frac{(d_k - y_k')^2}{y_{k-1}' - d_k},$$

or, in other words,

$$h_{k-1} \frac{d_{k-1} - y_{k-1}'}{(d_{k-1} - y_k')^2} + h_k \frac{d_k - y_{k+1}'}{(d_k - y_k')^2} = 0, \quad k = 2, \cdots, n-1. \qquad (6.26)$$

If, for example, y_1' and y_n' are given and if this *nonlinear system of equations* has a solution $y_2', \cdots, y_{n-1}'$, then the coefficients of the C^2 rational spline interpolant being sought can easily be calculated from (6.23), (6.24), (6.20), and (6.21). We will return to a method of solving such a system ([47]) in a subsequent section of this chapter.

A somewhat more convenient system is obtained by using the values y_k'' of the second derivatives as unknowns. In this case, in addition to the interpolation conditions (6.16) and (6.17), we have, because of (6.15), the equations,

$$y_k'' = 2C_k D_k^2 \qquad (6.27)$$

and

$$y_{k+1}'' = \frac{2C_k D_k^2}{[1 + h_k D_k]^3}, \qquad (6.28)$$

in place of (6.18) and (6.19). In contrast to those equations, which also involve B_k, (6.27) and (6.28) can be solved directly for C_k and D_k as functions of the y_k'', or better, in terms of the quantities g_k defined by

$$g_k^3 = y_k'', \qquad k = 1, \cdots, n. \qquad (6.29)$$

In fact, we obtain

$$C_k = \frac{g_k^3}{2D_k^2}, \qquad (6.30)$$

$$D_k = \frac{g_k - g_{k+1}}{h_k g_{k+1}}, \qquad k = 1, \cdots, n-1. \qquad (6.31)$$

The interpolation conditions thus reduce to

$$A_k = y_k - C_k, \qquad (6.32)$$

and

$$B_k = d_k - \frac{1}{h_k}\frac{C_k}{1 + h_k D_k} = d_k - \frac{1}{2}h_k\frac{g_k^2 g_{k+1}^2}{g_{k+1} - g_k}. \qquad (6.33)$$

Because of (6.30) and (6.31), we require that

$$g_k \neq 0, \quad k = 1, \cdots, n, \quad \text{and} \quad g_k \neq g_{k+1}, \quad k = 1, \cdots, n-1, \qquad (6.34)$$

for otherwise either s_k reduces to a linear or D_k is not defined.

The C^1 continuity conditions,

$$s'_{k-1}(x_k) = s'_k(x_k), \qquad k = 2, \cdots, n-1, \qquad (6.35)$$

using (6.14) and the expressions,

$$C_k D_k = \frac{1}{2}h_k\frac{g_{k+1}g_k^3}{g_k - g_{k+1}}$$

and

$$1 + h_k D_k = \frac{g_k}{g_{k+1}},$$

yield the nonlinear system of equations ([128]),

$$h_{k-1}g_{k-1}g_k^2 + h_k g_k^2 g_{k+1} = 2(d_k - d_{k-1}), \quad k = 2, \cdots, n-1. \qquad (6.36)$$

If $y_1'' = g_1^3$ and $y_h'' = g_n^3$ are given, then (6.36) is a system with the same number of unknowns as equations. However, in what follows, we will only want to consider the case when values y_1' and y_n' are given for the first derivative at x_1 and x_n and so will do without the specification of y_1'' and y_n''. Now from

$$\begin{aligned} y_1' &= s_1'(x_1) &= B_1 - C_1 D_1, \\ y_n' &= s_{n-1}'(x_n) &= B_{n-1} - \frac{C_{n-1}D_{n-1}}{[1 + h_{n-1}D_{n-1}]^2}, \end{aligned}$$

and (6.30), (6.31), and (6.32), we obtain the two additional equations,

$$h_1 g_1^2 g_2 = 2(d_1 - y_1') \qquad (6.37)$$

and

$$h_{n-1}g_{n-1}g_n^2 = 2(y_n' - d_{n-1}). \qquad (6.38)$$

Together, (6.37), (6.36), and (6.38) form a system of n nonlinear equations in n unknowns $g_1, \cdots, g_n$. The first equation involves only g_1 and g_2, the kth only g_{k-1}, g_k, g_{k+1}, $k = 2, \cdots, n - 1$, and the nth equation involves only g_{n-1} and g_n. Thus, this system has a tridiagonal character, just as did (6.26). However, this newer system has the *additional* property that its left-hand side is *homogeneous of degree three* as a function of the unknowns, meaning that if we write the system in the form,

$$F_k(g_1, \cdots, g_n) = 2\Delta_k, \quad k = 1, \cdots, n, \tag{6.39}$$

then for any value p, we have

$$F_k(pg_1, \cdots, pg_n) = p^3 F_k(g_1, \cdots, g_n). \tag{6.40}$$

This property may be used to make a somewhat more efficient version of Newton's method ([160]) to solve such a system numerically. We will return to this question in the next section.

When does the nonlinear system (6.39) even have a solution, when is it unique, and how may it be computed? According to [128], a rational interpolating spline s exists precisely when there is a fixed $v \in \{-1, 0, 1\}$ such that

$$sgn(\Delta_k) = v, \quad k = 1, \cdots, n, \tag{6.41}$$

where the right-hand side of (6.39), according to (6.36), (6.37), and (6.38), is defined by

$$\begin{array}{rcl} \Delta_1 & = & d_1 - y_1', \\ \Delta_k & = & d_k - d_{k-1}, \quad k = 2, \cdots, n - 1, \\ \Delta_n & = & y_n' - d_{n-1}. \end{array} \tag{6.42}$$

Moreover, it is then unique. If (6.39) has a solution $g_1, \cdots, g_n$ with $sgn(g_k) = v$, $k = 1, \cdots, n$, $v \in \{-1, 0, 1\}$, then s exists and $s''(x_k) = g_k^3$ ([128]). (Similar considerations can also be made for the system (6.26) ([47]).).

Condition (6.41) is not as completely restrictive as (6.7) or (6.8) but still sufficiently so that, for practical purposes, rational spline interpolants of the preceding form are rarely used, except when assumption (6.41) is satisfied and convex or concave data is to have its shape preserved by the interpolant. Generalizations of such rational spline interpolants are considered in [110, 129, 169, 170].

6.2. Adaptive Rational Spline Interpolants

In this section, we will attempt to overcome some of the restrictions of the previous section by replacing the rationals of the form (6.1) by cubic polynomials (4.2) on certain intervals or sequences of intervals. Such interruptions are necessary when the data requires the existence of inflection points, as these cannot be reproduced by the rational segments (6.1). We therefore ([130]) make use of cubic segments (4.2) on those sequences of intervals,

$$[x_i, x_{i+1}], [x_{i+1}, x_{i+2}], \cdots, [x_{j-1}, x_j], \qquad 1 \le i \le j - 1 \le n, \quad (6.43)$$

for which

$$\Delta_k \Delta_{k+1} \le 0, \qquad k = i, \cdots, j - 1, \tag{6.44}$$

where the quantities Δ_k, $k = 1, \cdots, n$, are defined by (6.42) and thus have the same signs as do the second divided difference approximations of the second derivative at x_k. For $j = i + 1$, (6.43) reduces to a single interval, but also for $j > i + 1$ we will refer to (6.43) with (6.44) as a *cubic segment*. *Adaptive rational spline interpolants* are thus composed of segments s_k of the form,

$$s_k(x) = \begin{cases} A_k + B_k(x - x_k) + \dfrac{C_k}{1 + D_k(x - x_k)} & \text{if } \Delta_k \Delta_{k+1} > 0 \\[2em] A_k + B_k(x - x_k) + C_k(x - x_k)^2 & \text{if } \Delta_k \Delta_{k+1} \le 0 \\[0.5em] \quad + D_k(x - x_k)^3 \end{cases}$$

$$\tag{6.45}$$

On non-cubic segments (6.43),

$$\Delta_k \Delta_{k+1} > 0, \qquad k = i, \cdots, j - 1, \tag{6.46}$$

and hence either $\Delta_k > 0$, $k = i, \cdots, j$ or $\Delta_k < 0$, $k = i, \cdots, j$. In this case, the earlier assertion, (6.41), guarantees the existence of rational spline interpolants on such segments.

On cubic segments (6.43), the values $\Delta_i, \cdots, \Delta_j$ change sign from one interval to the next, which means that if there are not two or more successive zeros in the sequence $\Delta_i, \cdots, \Delta_j$, then the *adaptive rational spline interpolant*, s, exists ([130]). As this can always be arranged by a slight modification of the data, there is no longer any theoretical restriction as regards applicability. Since by (6.15), inflection points can only occur on cubic segments, convexity or concavity of the data is reproduced in the interpolant (*shape preservation*) wherever this is possible.

We now turn to the problem of computing adaptive rational spline interpolants. The system of equations (6.37), (6.36), and (6.38) must be modified for knots lying inside or on the boundary of cubic segments. Now, for a cubic segment s_k,

$$s'_{k-1}(x_k) = B_{k-1} + 2h_{k-1}C_{k-1} + 3h^2_{k-1}D_{k-1}$$

and

$$s'_k(x_k) = B_k,$$

and so by (4.20),

$$s'_{k-1}(x_k) = d_{k-1} + \frac{1}{3}h_{k-1}g^3_{k-1} + \frac{1}{6}h_{k-1}g^3_k$$

and

$$s'_k(x_k) = d_k - \frac{1}{3}h_k g^3_k - \frac{1}{6}h_k g^3_{k+1},$$

where $2C_k = y''_k = g^3_k$. Then the C^1 continuity conditions (6.35) and end conditions, $s'_1(x_1) = y'_1$ and $s'_{n-1}(x_n) = y'_n$, result in the following system of equations:

$$\Delta_1 = \begin{cases} \frac{1}{2}h_1 g^2_1 g_2 & \text{if } \Delta_1 \Delta_2 > 0 \\[2mm] \frac{1}{3}h_1 g^3_1 + \frac{1}{6}h_1 g^3_2 & \text{if } \Delta_1 \Delta_2 \leq 0 \end{cases},$$

$$\Delta_k = \begin{cases} \frac{1}{2}h_{k-1}g_{k-1}g^2_k + \frac{1}{2}h_k g^2_k g_{k+1} & \text{if } \Delta_{k-1}\Delta_k > 0, \\ & \Delta_k \Delta_{k+1} > 0 \\[3mm] \frac{1}{6}h_{k-1}g^3_{k-1} + \frac{1}{3}h_{k-1}g^3_k + \frac{1}{2}h_k g^2_k g_{k+1} & \text{if } \Delta_{k-1}\Delta_k \leq 0, \\ & \Delta_k \Delta_{k+1} > 0 \\[3mm] \frac{1}{2}h_{k-1}g_{k-1}g^2_k + \frac{1}{3}h_k g^3_k + \frac{1}{6}h_k g^3_{k+1} & \text{if } \Delta_{k-1}\Delta_k > 0, \\ & \Delta_k \Delta_{k+1} \leq 0 \\[3mm] \frac{1}{6}h_{k-1}g^3_{k-1} + \frac{1}{3}(h_{k-1} + h_k)g^3_k + \frac{1}{6}h_k g^3_{k+1} & \text{if } \Delta_{k-1}\Delta_k \leq 0, \\ & \Delta_k \Delta_{k+1} \leq 0 \end{cases},$$

$$(6.47)$$

$$\Delta_n = \begin{cases} \frac{1}{2}h_{n-1}g_{n-1}g^2_n & \text{if } \Delta_{n-1}\Delta_n > 0 \\[2mm] \frac{1}{6}h_{n-1}g^3_{n-1} + \frac{1}{3}h_{n-1}g^3_n & \text{if } \Delta_{n-1}\Delta_n \leq 0 \end{cases}.$$

This nonlinear system (6.47) has the structure

$$\tilde{F}_k(g_1, \cdots, g_n) = \Delta_k, \qquad k = 1, \cdots, n, \tag{6.48}$$

and again the left-hand side is *homogeneous of degree three*, i. e.,

$$\tilde{F}_k(pg_1, \cdots, pg_n) = p^3 \tilde{F}_k(g_1, \cdots, g_n) \tag{6.49}$$

for any value p.

The existence theorem mentioned previously is proved by showing that a naturally associated *fixed point iteration* ([150]) has the contraction property on a certain domain ([130]). However, for practical calculations, it is suggested ([130]) that one use Newton's method, for which there is a little trick ([160]) based on the homogeneity property (6.49), which gives an implementation somewhat more efficient than the usual. Moreover, there are natural starting values that work quite well.

Let g denote the vector $(g_1, \cdots, g_n)^T$, $\tilde{F}$ and Δ the vectors of (6.48), and

$$J(g) = \left(\frac{\partial \tilde{F}_i}{\partial g_i} \right)_{i,j=1,\cdots,n} \tag{6.50}$$

the *Jacobian matrix* of (6.48). Then Newton's method is

$$g^{(t+1)} = g^{(t)} - \left[J(g^{(t)}) \right]^{-1} \left(\tilde{F}(g^{(t)}) - \Delta \right). \tag{6.51}$$

Since $g_k^3 = y_k''$, the natural choices for *starting values*, $g_k^{(0)}$, are cube roots of approximations to the second derivatives, i. e.,

$$g_1^{(0)} = \sqrt[3]{\frac{\Delta_1}{h_1}},$$

$$g_k^{(0)} = \sqrt[3]{\frac{2\Delta_k}{h_{k-1} + h_k}}, \quad k = 2, \cdots, n-1, \tag{6.52}$$

$$g_n^{(0)} = \sqrt[3]{\frac{\Delta_n}{h_{n-1}}}.$$

The homogeneity property (6.49) can be exploited in the following manner. By *Euler's identity* for *homogeneous functions* (here with *degree of homogeneity* $p = 3$), it follows that ([160])

$$J(g)g = 3\tilde{F}. \tag{6.53}$$

Hence, (6.51) reduces to

$$g^{(t+1)} = \frac{2}{3}g^{(t)} + \left[J(g^{(t)})\right]^{-1}\Delta. \qquad (6.54)$$

Notice that $\tilde{F}$ is no longer present. Now the Jacobian matrix $J(g)$ is almost never symmetric, although it is tridiagonal. Thus, the solution of the linear system,

$$J(g)x = \Delta,$$

required in (6.54), can be carried out by TRIDIU. Computing $J(g)$ from (6.47) is easy, but because of the numerous subcases we will not explicitly give the result here.

Test calculations show that the starting values (6.52) should be modified to

$$g_k^{(0)} = sgn(\Delta_k)$$

when $|\Delta_k| < \varepsilon_1$ (e.g., $\varepsilon_1 = .000001$). (In particular, for $\Delta_k = 0$, $g_k^{(0)} = 0$ is not allowed by (6.34); for reasons of rounding error, this modification should likewise be made for $\Delta_k \approx \varepsilon_1$, where ε_1 is a bit larger than the machine rounding unit.) Further, Newton's method, (6.51) or (6.54), did not always converge for the test examples we used, as the signs of the $g_k^{(0)}$ were not always maintained. We were, however, able to attain convergence emperically by introducing a damping factor $\lambda \leq 1$ in the correction term of (6.51). In the roughly 20 test examples we tried, $\lambda = \frac{1}{3}$ was always successful. For this particular choice of λ, (6.54) becomes

$$g^{(t+1)} = \frac{8}{9}g^{(t)} + \frac{1}{3}J(g^{(t)})^{-1}\Delta.$$

In the cases where we used $\lambda = 1$ or $\frac{1}{3} < \lambda < 1$, the number of iterations, and hence execution time, increased. (For other examples, $\lambda < \frac{1}{3}$ might be necessary.)

The subroutine RATSCH (Figs. 6.1 and 6.2) is based on such a modified Newton's method applied to (6.48). We did not encounter any convergence problems when computing the examples of this section. For $\varepsilon_2 = .0001$, the number of iterations, about 20, was higher than usual because of the damping; one case required up to 50 iterations. The first statement after DO 70 K=1,N would have to be modified if a $\lambda \neq 1/3$ were desired. Taking advantage of the homogeneity of degree three, in practice, saves up to 20% of the execution time. The spline formed from the segments (6.45) as computed by RATSCH may be evaluated using RATVAN (Fig. 6.3). The cases in (6.45) of whether the segment is rational (S(K)=.TRUE.) or cubic (S(K)=.FALSE.) are specified by the LOGICAL ARRAY S.

```
      SUBROUTINE RATSCH(N,X,Y,ITMAX,EPS1,EPS2,A,B,C,D,
     &                   E,G,H,S,ICASE,IFLAG)
      DIMENSION X(N),Y(N),A(N),B(N),C(N),D(N),E(N),G(N),H(N)
      INTERGER ICASE(N)
      LOGICAL S(N)
      IFLAG=0
      IF(N.LT.3) THEN
          IFLAG=1
          RETURN
      END IF
      N1=N-1
      EX=1./3.
      D1=B(1)
      DO 10 K=1,N1
          KP1=K+1
          H(K)=X(KP1)-X(K)
          D2=(Y(KP1)-Y(K))/H(K)
          D(K)=D2-D1
          E(K)=D(K)
          D1=D2
10    CONTINUE
      D(N)=B(N)-D2
      E(N)=D(N)
      DO 30 K=1,N1
          KP1=K+1
          P2=D(K)*D(KP1)
          IF(K.EQ.1) GOTO 20
          KM1=K-1
          IF(P1.GT.0.) THEN
              S(KM1)=.TRUE.
              IF(P2.GT.0.) THEN
                  ICASE(K)=1
              ELSE
                  ICASE(K)=3
              END IF
          ELSE
              S(KM1)=.FALSE.
              IF(P2.GT.0.) THEN
                  ICASE(K)=2
              ELSE
                  ICASE(K)=4
              END IF
          END IF
20        P1=P2
30    CONTINUE
      IF(S(1)) THEN
          ICASE(1)=1
      ELSE
          ICASE(1)=2
      END IF
      IF(P1.GT.0.) THEN
          S(N1)=.TRUE.
          ICASE(N)=1
      ELSE
          S(N1)=.FALSE.
          ICASE(N)=2
      END IF
```

(cont.)

```
      IF(ABS(D(1)).LE.EPS1) THEN
          G(1)=SIGN(1.,D(1))
      ELSE
          G(1)=((ABS(D(1)/H(1)))**EX)*SIGN(1.,D(1))
      END IF
      DO 40 K=2,N1
          IF(ABS(D(K)).LE.EPS1) THEN
              G(K)=SIGN(1.,D(K))
          ELSE
              G(K)=((ABS(2.*D(K)/(H(K-1)+H(K))))**EX)
     &                  *SIGN(1.,D(K))
          END IF
40    CONTINUE
      IF(ABS(D(N)).LE.EPS1) THEN
          G(N)=SIGN(1.,D(N))
      ELSE
          G(N)=((ABS(D(N)/H(N1)))**EX)*SIGN(1.,D(N))
      END IF
      IT=0
50    IT=IT+1
      IF(IT.GT.ITMAX) THEN
          IFLAG=13
          RETURN
      END IF
      H1=H(1)*G(1)
      H2=H(1)*G(2)
      IF(ICASE(1).EQ.1) THEN
          B(1)=H1*G(2)
          C(1)=H1*G(1)/2.
      ELSE
          B(1)=H1*G(1)
          C(1)=H2*G(2)/2.
      END IF
      DO 60 K=2,N1
          KM1=K-1
          KP1=K+1
          H1=H(KM1)*G(K)
          H2=H(K)*G(K)
          H3=H(KM1)*G(KM1)
          H4=H(K)*G(KP1)
          IF(ICASE(K).EQ.1) THEN
              A(KM1)=H1*G(K)/2.
              B(K)=H1*G(KM1)+H2*G(KP1)
              C(K)=H2*G(K)/2.
          END IF
          IF(ICASE(K).EQ.2) THEN
              A(KM1)=H3*G(KM1)/2.
              B(K)=H1*G(K)+H2*G(KP1)
              C(K)=H2*G(K)/2.
          END IF
          IF(ICASE(K).EQ.3) THEN
              A(KM1)=H1*G(K)/2.
              B(K)=H3*G(K)+H2*G(K)
              C(K)=H4*G(KP1)/2.
          END IF
          IF(ICASE(K).EQ.4) THEN
              A(KM1)=H3*G(KM1)/2.
```

(cont.)

```
                B(K)=(H1+H2)*G(K)
                C(K)=H4*G(KP1)/2.
          END IF
60    CONTINUE
      IF(ICASE(N).EQ.1) THEN
          A(N1)=H4*G(N)/2.
          B(N)=H4*G(N1)
      ELSE
          A(N1)=H2*G(N1)/2.
          B(N)=H4*G(N)
      END IF
      CALL TRIDIU(N,A,B,C,E,EPS1,IFLAG)
      IF(IFLAG.NE.0) THEN
          ITMAX=IT-1
          RETURN
      END IF
      RNORM1=0.
      RNORM2=0.
      DO 70 K=1,N
          GH=8.*G(K)/9.+E(K)/3.
          RNORM1=RNORM1+ABS(G(K))
          RNORM2=RNORM2+ABS(G(K)-GH)
          G(K)=GH
          E(K)=D(K)
70    CONTINUE
      IF(RNORM2.LT.EPS2*RNORM1) GOTO 80
      GOTO 50
80    E(1)=G(1)*G(1)*G(1)
      DO 90 K=1,N1
          K1=K+1
          E(K1)=G(K1)*G(K1)*G(K1)
          DXK=H(K)
          IF(S(K)) THEN
              H1=G(K)-G(K1)
              IF(ABS(H1).LT.EPS1) THEN
                  IFLAG=14
                  RETURN
              END IF
              D(K)=H1/(G(K1)*DXK)
              C(K)=E(K)/(2.*D(K)*D(K))
              A(K)=Y(K)-C(K)
              B(K)=(Y(K1)-A(K)-C(K)/(1.+D(K)*DXK))/DXK
          ELSE
              C(K)=E(K)/2.
              A(K)=Y(K)
              B(K)=(Y(K1)-Y(K))/DXK-DXK*(2.*E(K)+E(K1))/6.
              D(K)=(E(K1)-E(K))/(6.*DXK)
          END IF
90    CONTINUE
      ITMAX=IT
      RETURN
      END
```

Figure 6.1. Program listing of RATSCH.

Calling sequence:

CALL RATSCH(N,X,Y,ITMAX,EPS1,EPS2,A,B,C,D,E,G,
            H,S,ICASE,IFLAG)

Purpose:
RATSCH uses Newton's method to determine the coefficients A_k, B_k, C_k, and D_k, $k = 1, 2, \cdots, n-1$, of an adaptive rational spline interpolant. The spline interpolant consists of either a cubic on the interval $[x_k, x_{k+1}]$ if $\Delta_k \cdot \Delta_{k+1} \leq 0$, or a rational function of the form,

$$s_k(x) = A_k + B_k(x - x_k) + \frac{C_k}{1 + D_k(x - x_k)},$$

if $\Delta_k \cdot \Delta_{k+1} > 0$. Values for y_1' and y_n' must be given in B(1) and B(N), respectively.

Description of the parameters:

N	Number of given points. N$\geq$ 3 is required.
X	ARRAY(N): Upon calling must contain the abscissa values x_k, $k = 1, \cdots, n$, with $x_1 < x_2 < \cdots < x_n$.
Y	ARRAY(N): Upon calling must contain the ordinate values y_k, $k = 1, \cdots, n$.
ITMAX	Prescribed maximum number of iterations. Upon completion ITMAX holds the actual number of iterations used.
EPS1	Value used for accuracy test. Recommendation: EPS1=10^{-t} (t: number of available decimal digits). If the absolute value of a number is less than EPS1, it is interpreted as zero. EPS1 is used moreover as parameter in calling TRIDIU (see the description of TRIDIU).
EPS2	Value used in the stopping criterium for Newton's method. The iteration is stopped if for two consecutive approximate solutions, $g^{(t+1)}$ and $g^{(t)}$, $|g^{(t)} - g^{(t+1)}|_1 < $ EPS2 $\cdot \|g^{(t)}\|_1$, ($g_k := (y_k'')^{1/3}$). Recommendation: EPS2= 10^{4-t}.
A,B,C,D	ARRAY(N): Upon execution with IFLAG=0 contain the desired spline coefficients, K=1,2,$\cdots$,N-1.
E	ARRAY(N): Output: Upon execution contains the values y_k'', $k = 1, \cdots, n - 1$, of the second derivatives.
G,H	ARRAY(N): Work space.

S   LOGICAL ARRAY(N): Output: For $k = 1, 2, \cdots, n - 1$
            contains:
                    if $\Delta_k \cdot \Delta_{k+1} > 0$ then S(K)=.TRUE.
                    if $\Delta_k \cdot \Delta_{k+1} \leq 0$ then S(K)=.FALSE.
ICASE   INTEGER ARRAY(N): Output:
                    if $\Delta_1 \cdot \Delta_2 > 0$ then ICASE(1)=1
                    if $\Delta_1 \cdot \Delta_2 \leq 0$ then ICASE(1)=2
            Further, for $k = 2, 3, \cdots, n - 1$:
                    if $\Delta_{k-1} \cdot \Delta_k > 0$ and $\Delta_k \cdot \Delta_{k+1} > 0$
                    then ICASE(K)=1
                    if $\Delta_{k-1} \cdot \Delta_k \leq 0$ and $\Delta_k \cdot \Delta_{k+1} > 0$
                    then ICASE(K)=2
                    if $\Delta_{k-1} \cdot \Delta_k > 0$ and $\Delta_k \cdot \Delta_{k+1} \leq 0$
                    then ICASE(K)=3
                    if $\Delta_{k-1} \cdot \Delta_k \leq 0$ and $\Delta_k \cdot \Delta_{k+1} \leq 0$
                    then ICASE(K)=4
                    if $\Delta_{n-1} \cdot \Delta_n > 0$ then ICASE(N)=1
                    if $\Delta_{n-1} \cdot \Delta_n \leq 0$ then ICASE(N)=2
IFLAG   =0:   Normal execution.
        =1:   N$\geq$ 3 required.
        =2:   Error in solving the linear system (TRIDIU).
        =13:   Number of iterations exceeded ITMAX.
        =14:   If on a rational interval we have $y_k'' = y_{k+1}''$,
                then the spline coefficients cannot be computed
                (see the text).

Required subroutine: TRIDIU.

Figure 6.2. Description of RATSCH.

```
FUNCTION RATVAN(N,X,S,A,B,C,D,V,IFLAG)
DIMENSION X(N),A(N),B(N),C(N),D(N)
LOGICAL S(N)
DATA I/1/
IFLAG=0
IF(N.LT.2) THEN
    IFLAG=1
    RETURN
END IF
CALL INTONE(X,N,V,I,IFLAG)
IF(IFLAG.NE.0) RETURN
H=V-X(I)
IF(S(I)) THEN
    RATVAN=A(I)+B(I)*H+C(I)/(1.+D(I)*H)
ELSE
    RATVAN=A(I)+H*(B(I)+H*(C(I)+D(I)*H))
END IF
RETURN
END
```

FUNCTION RATVAN(N,X,S,A,B,C,D,V,IFLAG)

Purpose:
RATVAN is a FUNCTION subroutine for the calculation of a function value of an adaptive rational spline interpolant at a point $V \in [X(1), X(N)]$.

Description of the parameters:

N	Number of given points. $N \geq 2$ is required.
X	ARRAY(N): Abscissas.
S	LOGICAL ARRAY(N):
	S(K)=.FALSE.: The kth segment is cubic.
	S(K)=.TRUE.: The kth segment is rational.
A,B,C,D	ARRAY(N): Vectors of the spline coefficients.
V	Value of the point at which the spline function is to be evaluated.
IFLAG	=0: Normal execution.
	=1: $N \geq 2$ is required.
	=3: Error in interval determination (INTONE).

Required subroutine: INTONE.

Remark: The statement 'DATA I/1/' has the effect that I is set to 1 at the first call to RATVAN.

Figure 6.3. FUNCTION RATVAN and its description.

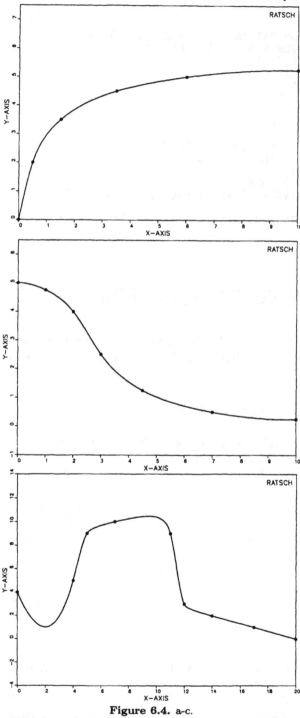

Figure 6.4. a-c.

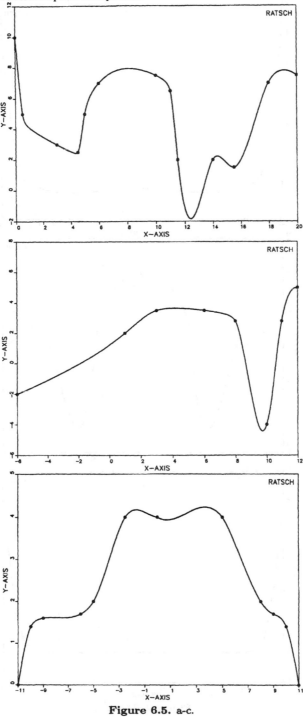

Figure 6.5. a-c.

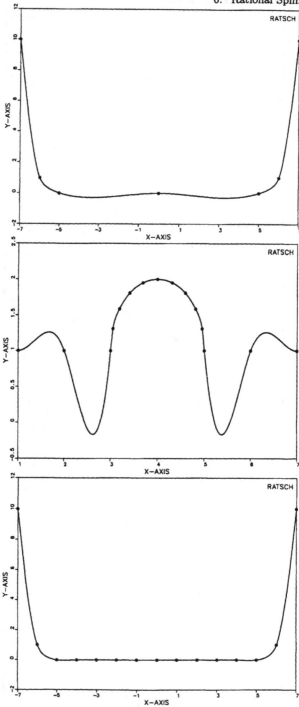

Figure 6.6. a-c.

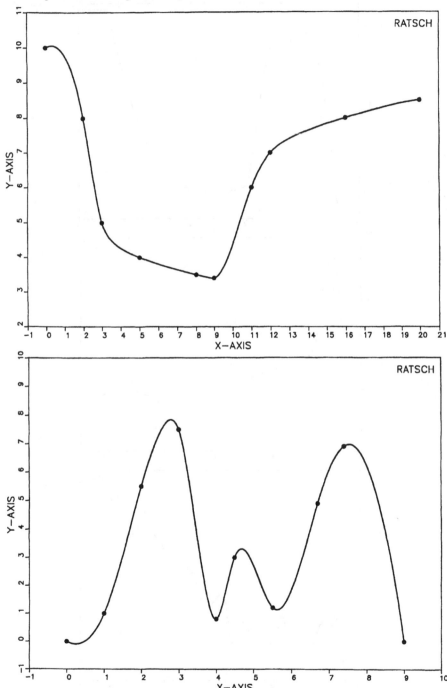

Figure 6.7. a, b.

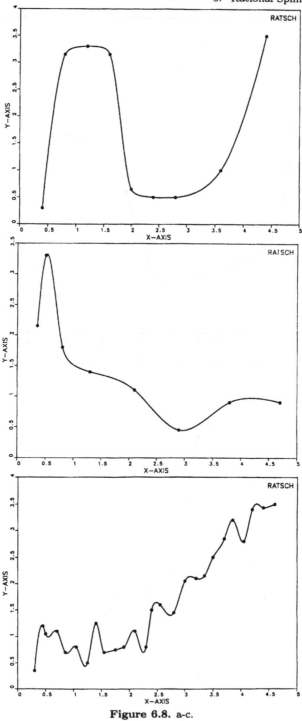

Figure 6.8. a-c.

In Figs. 6.4a–6.8c, we did not, as we had done previously, use end conditions $y_1' = d_1$, and $y_n' = d_n$, as then $\Delta_1 = \Delta_n = 0$, which would force the first and last segments to be cubic. Instead, we fit quadratics (using GRAD2) through the first three and last three points and used their derivatives at x_1 and x_n, respectively, for y_1' and y_n' (for comparison's sake, we use the same procedure in subsequent sections with the subroutines RATSAK, RATSP1, and RATGRE). The concave data of Fig. 6.4a was reproduced, of course, through only rational segments. In Fig. 6.4b, the third interval is cubic, as an inflection point is evidently needed. In 6.4c, the appearance of the curve in the first interval is somewhat annoying. This is caused by the fact that $y_1' = -2.75$, a value presumably too low. The curve resulting from the data of Fig. 6.5a is of an unacceptable appearance in intervals 9 and 10. This is caused by $d_9 = 0$. Figure 6.5b is very satisfactory. Among examples 6.5c–6.6c, only 6.6c is usable. In fact, we had substantially better results using C^1 shape- preserving polynomial methods. In 6.7a, the appearance of the curve in the first interval might be troublesome. This is caused by $y_1' = 1/3$. As to be expected, intervals 2 and 6 are cubic while the others are rational. Figures 6.8a–c show some very usable curves, although in 6.8c only intervals 1, 9, 10, 14, 18, and 25 are rational while all the others are cubic.

In summary, we may say that the strength of adaptive rational C^2 spline interpolation with free poles is most evident when the data requires only isolated cubic segments, i. e., when the data is piecewise concave or convex. In comparison to other methods, its computational expense is relatively high, although in absolute terms the execution time is quite bearable. For small values of the Δ_k, small perturbations of the data or rounding errors can result in large changes in the resulting curve.

6.3.   Rational Spline Interpolants with a Prescribable Pole

The work of [130] gives considerable advances in adaptive rational spline interpolants over that of [128]. The only drawback is that the computational expense is much higher, although not unbearably so, than that of the usual linear systems of equations of before. Quite early on, the idea of not letting the real poles vary freely was described in [154]. There, *two* of these are variably prescribed, which again leads to a linear system with tridiagonal coefficient matrix. Before we go into the details of this method in the next section, we want to study a newer (although already considered in [161]) variant ([122,123]). The idea is to fix a *single* real pole by means of a parameter, which, differing from [161], however, may possibly be different on each interval.

On $[x_k, x_{k+1}]$, we set

$$s_k(x) = A_k + B_k t + C_k t^2 + \frac{D_k t^3}{1 + p_k t}, \qquad t = \frac{x - x_k}{\Delta x_k}. \qquad (6.55)$$

(In [140], a similar "quadratic" form with three parameters is considered.)
Then

$$h_k s_k'(x) \quad = \quad B_k + 2C_k t + \frac{D_k \left(3t^2 + 2p_k t^3\right)}{(1 + p_k t)^2} \qquad (6.56)$$

and

$$h_k^2 s_k''(x) \quad = \quad 2C_k + 2D_k t \frac{p_k^2 t^2 + 3p_k t + 3}{(1 + p_k t)^3}. \qquad (6.57)$$

Now we will restrict ourselves to using the values y_k' of the first derivatives
as unknowns. We thus obtain from the conditions,

$$
\begin{aligned}
s_k(x_k) &= y_k &= A_k, \\
s_k(x_{k+1}) &= y_{k+1} &= y_k + B_k + C_k + \frac{D_k}{1 + p_k}, \\
h_k s_k'(x_k) &= h_k y_k' &= B_k,
\end{aligned}
$$

and

$$h_k s_k'(x_{k+1}) \quad = \quad h_k y_{k+1}' \quad = \quad B_k + 2C_k + \frac{D_k(3 + 2p_k)}{(1 + p_k)^2},$$

the desired coefficients,

$$
\begin{aligned}
A_k &= y_k, \\
B_k &= h_k y_k', \\
C_k &= (2p_k + 3)\Delta y_k - (p_k + 2)h_k y_k' - (p_k + 1)h_k y_{k+1}', \\
D_k &= (1 + p_k)^2 \left(-2\Delta y_k + h_k y_k' + h_k y_{k+1}'\right),
\end{aligned}
\qquad (6.58)
$$

as functions of the unknowns y_k', $k = 1, \cdots, n$. For this, it is assumed for
the time being that $p_k \neq -1$. However, if the assembled spline interpolant,
s, is to be twice continuously differentiable on $[x_1, x_n]$, then for each $k =
1, \cdots, n - 1$, s_k cannot have a pole anywhere in $[x_k, x_{k+1}]$. As the pole of
s_k lies at

$$\tilde{x} = x_k - \frac{h_k}{p_k},$$

this imposes the conditions

$$-1 < p_k < \infty, \qquad k = 1, \cdots, n - 1. \qquad (6.59)$$

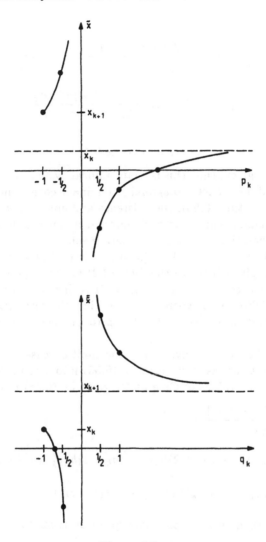

Figure 6.9.

(A sketch of this situation is given in the first half of Fig. 6.9.) The C^2 continuity conditions,

$$s''_{k-1}(x_k) = s''_k(x_k), \qquad k = 2, \cdots, n-1$$

yield

$$\frac{1}{1+p_{k-1}}\frac{1}{h_{k-1}}y'_{k-1} + \left(\frac{2+p_{k-1}}{1+p_{k-1}}\frac{1}{h_{k-1}} + (2+p_k)\frac{1}{h_k}\right)y'_k$$

$$+(1+p_k)\frac{1}{h_k}y'_{k+1} = (3+2p_k)\frac{d_k}{h_k} + \frac{3+p_{k-1}}{1+p_{k-1}}\frac{d_{k-1}}{h_{k-1}},$$

$$(6.60)$$

$$k = 2,\cdots,n-1.$$

If y'_1 and y'_n are given, then this is again a linear system with nonsymmetric but tridiagonal and strictly diagonally dominant coefficient matrix. Thus, under the assumption (6.59), the existence and uniqueness of such rational spline interpolants, where for each interval a single pole may be placed anywhere outside that interval, are guaranteed.

If $p_k = 0$, $k = 1,\cdots,n-1$ in (6.55), as is perfectly allowed by (6.59), then (6.55) simply reduces to the form (4.2) for cubic polynomials, except with $x - x_k$ replaced by $t = (x - x_k)/h_k$. It is therefore not surprising that for $p_k = 0$, (6.60) agrees with (4.7) and that the formulas (6.58) for C_k and D_k are identical to those of (4.6) up to the factors of $1/h_k$ and $1/h_k^2$, respectively.

In order to be able to investigate the limiting cases of $p_k \to -1$ and $p_k \to \infty$, we put the last two terms of (6.55) into a common denominator and substitute the values of C_k and D_k given by (6.58) to obtain

$$\frac{[C_k + (p_kC_k + D_k)\,t]\,t^2}{1+p_kt}$$

$$= \left[(2p_k + 3)\Delta y_k - (p_k+2)h_ky'_k - (p_k+1)h_ky'_{k+1} + (-(2+p_k)\Delta y_k \right.$$

$$\left. + h_ky'_k + (1+p_k)h_ky'_{k+1})\,t\right]t^2/(1+p_kt).$$

$$(6.61)$$

If $0 < t < 1$ then as $p_k \to \infty$, this expression tends to

$$\left[2\Delta y_k - h_ky'_k - h_ky'_{k+1} + \left(-\Delta y_k + h_ky'_{k+1}\right)t\right]t,$$

and as $p_k \to -1$ it tends to

$$(\Delta y_k - h_ky'_k)\,t^2.$$

In both cases, we obtain a quadratic in t, a fact that is not altered by the addition of $A_k + B_kt$.

For $t = 0$, we simply obtain $s_k(x_k) = A_k = y_k$, independently of p_k, while for $t = 1$ we get

$$A_k + B_k + C_k + \frac{D_k}{1+p_k} = y_{k+1}.$$

In both of the limiting cases under investigation (all the $p_k \to -1$ or all $p_k \to \infty$ simultaneously), our rational spline interpolant with poles chosen anywhere outside the interval tends to a continuous piecewise quadratic. This is consistent with the facts that by multiplying (6.60) by $1 + p_{k-1}$ and then letting $p_{k-1} \to -1$ and dividing it by $(2 + p_k)$ and then letting $p_k \to \infty$, the system reduces to

$$y'_{k-1} + y'_k = 2d_{k-1}, \qquad k = 2, \cdots, n-1,$$

and

$$y'_k + y'_{k+1} = 2d_k, \qquad k = 2, \cdots, n-1,$$

respectively, which we recognize as the equations of quadratic spline interpolation (3.7). (As the first system does not involve the prescribed values y'_n and the second does not involve y'_1, in certain cases the right- and leftmost segments, respectively, may not be quadratics.) For the effects of the signs of the p_k, we refer the reader to the examples that follow.

With regard to shape preservation, it is shown in [122], although explicitly only for equally spaced abscissas and $p_k = p$, $k = 1, \cdots, n-1$, that if $\Delta y_1 > \cdots > \Delta y_{n-1} > 0$, $y'_1 \geq 0$, and $0 \leq y'_n \leq 2\Delta y_{n-1}$, then s increases monotonically on $[x_1, x_n]$ for p sufficiently large. Similarly, if $0 < \Delta y_1 < \cdots < \Delta y_{n-1}$, $0 \leq y'_1 \leq 2\Delta y_1$ and $y'_n \geq 0$, then s also increases monotonically on $[x_1, x_n]$ for p sufficiently close to -1. Further, if $\Delta y_1 > \cdots > \Delta y_{n-1} > 0$, $\Delta y_{k+1} - 2\Delta y_k + \Delta y_{k-1} > 0$, $k = 2, \cdots, n-2$, and $s''(x_1) = s''(x_n) = 0$, then s is both monotonically increasing and concave for sufficiently large p. If $0 < \Delta y_1 < \cdots < \Delta y_{n-1}$, $\Delta y_{k+1} - 2\Delta y_k + \Delta y_{k-1} < 0$, $k = 2, \cdots, n-2$, and $s''(x_1) = s''(x_n) = 0$ then s is both monotonically increasing and convex for p sufficiently close to -1. These results for $p_k = p$ also hold if Δy_k is replaced by d_k in the preceding conditions ([122]). It is then intuitively clear that they will also hold as the p_k approach ∞ (respectively, -1) at the same rate. For *shape preservation*, we can therefore check, interval by interval, the local situation of the data whether to take $p_k \approx 0$ (when no monotonicity or convexity/concavity is present) or to take p_k large or near -1. In this way, we implicitly imitate the method of procedure used for adaptive rational spline interpolants.

The subroutine RATSAK (Figs. 6.10 and 6.11) forms the system (6.60) and solves it for given values of y'_1, y'_n, and p_k, $k = 1, \cdots, n-1$, satisfying condition (6.59). The coefficients of (6.55) are calculated from (6.58). RSKVAL (Fig. 6.12) can be used to evaluate s.

We first show by a simple example ($x_1 = 0, x_2 = 4, x_3 = 10, y_1 = 2, y_2 = 6, y_3 = 0, y'_1 = 1, y'_3 = -1$) the effect of the various sign patterns for neighboring values p_1 and p_2. In the four figures, 6.13a–6.14b, the values $p_1 = -.9, p_2 = 9$, then $p_1 = 9, p_2 = -.9$, then $p_1 = p_2 = -.9$ and finally

```
      SUBROUTINE RATSAK(N,X,Y,P,A,B,C,D,EPS,IFLAG)
      DIMENSION X(N),Y(N),P(N),A(N),B(N),C(N),D(N)
      IFLAG=0
      IF(N.LT.3) THEN
           IFLAG=1
           RETURN
      END IF
      N1=N-1
      N2=N-2
      B1=B(1)
      BN=B(N)
      DO 20 K=1,N1
           KP1=K+1
           KM1=K-1
           H2=1./(X(KP1)-X(K))
           D2=H2*H2*(Y(KP1)-Y(K))
           PK=P(K)
           IF(PR.LT.-1.+EPS) THEN
                IFLAG=12
                RETURN
           END IF
           P2=PK+1.
           IF(K.EQ.1) GOTO 10
           A(KM1)=H2/P2
           B(KM1)=(P1+1.)*H1/P1+(P2+1.)*H2
           C(KM1)=H2*P2
           D(KM1)=(2.*P2+1.)*D2+(P1+2.)*D1/P1
           IF(K.EQ.2) D(KM1)=D(KM1)-H1*B1/P1
           IF(K.EQ.N1) D(KM1)=D(KM1)-H2*BN*P2
10         H1=H2
           D1=D2
           P1=P2
20    CONTINUE
      CALL TRIDIU(N2,A,B,C,D,EPS,IFLAG)
      IF(IFLAG.NE.0) RETURN
      DO 30 K=1,N2
           A(K+1)=D(K)
30    CONTINUE
      A(1)=B1
      A(N)=BN
      DO 40 K=1,N1
           K1=K+1
           G=Y(K1)-Y(K)
           H=X(K1)-X(K)
           PK=P(K)+1.
           B(K)=H*A(K)
           C(K)=(2.*PK+1.)*G-(PK+1.)*H*A(K)-PK*H*A(K1)
           D(K)=PK*PK*(-2.*G+H*A(K)+H*A(K1))
40    CONTINUE
      RETURN
      END
```

Figure 6.10. Program listing of RATSAK.

Calling sequence:

CALL RATSAK(N,X,Y,P,A,B,C,D,EPS,IFLAG)

Purpose:
RATSAK determines coefficients Y_k, B_k, C_k, and D_k, $k = 1, 2, \cdots, n-1$, of a rational spline interpolant of the following form:

$$s_k(x) = y_k + B_k t + C_k t^2 + D_k \frac{t^3}{1 + p_k t},$$

where $t = (x - x_k)/(x_{k+1} - x_k)$, $k = 1, 2, \cdots, n - 1$. Values for y_1' and y_n' must be given in B(1) and B(N), respectively.

Description of the parameters:

N		Number of given points. $N \geq 3$ is required.
X		ARRAY(N): Upon calling must contain the abscissa values x_k, $k = 1, \cdots, n$, with $x_1 < x_2 < \cdots < x_n$.
Y		ARRAY(N): Upon calling must contain the ordinate values y_k, $k = 1, \cdots, n$.
P		ARRAY(N): Upon calling must contain the parameters p_k with $p_k > -1$, $k = 1, \cdots, n - 1$.
A		ARRAY(N): Output: Upon completion contains the values y_k', $k = 1, \cdots, n$, of the first derivatives.
EPS		see TRIDIU.
Y,B,C,D		ARRAY(N): Upon execution with IFLAG=0 contain the desired spline coefficients, K=1,2,$\cdots$,N−1.
IFLAG	=0:	Normal execution.
	=1:	$N \geq 3$ required.
	=2:	Error in solving the linear system (TRIDIU).
	=12:	$p_k > -1$ not satisfied for at least one k.

Required subroutine: TRIDIU.

Figure 6.11. Description of RATSAK.

```
FUNCTION RSKVAL(N,X,P,A,B,C,D,V,IFLAG)
DIMENSION X(N),P(N),A(N),B(N),C(N),D(N)
DATA I/1/
IFLAG=0
IF(N.LT.2) THEN
    IFLAG=1
    RETURN
END IF
CALL INTONE(X,N,V,I,IFLAG)
IF (IFLAG.NE.0) RETURN
T=(V-X(I))/(X(I+1)-X(I))
RSKVAL=A(I)+T*(B(I)+T*(C(I)+T*D(I)/(P(I)*T+1.)))
RETURN
END
```

FUNCTION RSKVAL(N,X,P,A,B,C,D,V,IFLAG)

Purpose:

RSKVAL is a FUNCTION subroutine for the calculation of a function value at a point $V \in [X(1),X(N)]$ of a rational spline interpolant of the following form:

$$s_k(x) = A_k + B_k t + C_k t^2 + D_k t^3/(1 + p_k t),$$

where $t = (x - x_k)/(x_{k+1} - x_k)$, $k = 1, 2, \cdots, n - 1$.

Description of the parameters:

N	Number of given points. $N \geq 2$ is required.
X	ARRAY(N): Abscissas.
P	ARRAY(N): Upon calling must contain the parameters p_k, with $p_k > -1$, $k = 1, \cdots, n - 1$.
A,B,C,D	ARRAY(N): Vectors of the spline coefficients.
V	Value of the point at which the spline interpolant is to be evaluated.
IFLAG	=0: Normal execution.
	=1: $N \geq 2$ required.
	=3: Error in interval determination (INTONE).

Required subroutine: INTONE.

Remark: The statement 'DATA I/1/' has the effect that I is set to 1 at the first call to RSKVAL.

Figure 6.12. RSKVAL and its description.

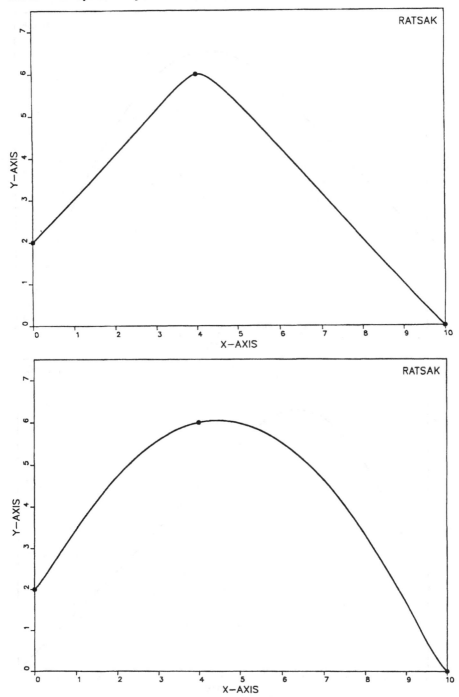

Figure 6.13. a, b.

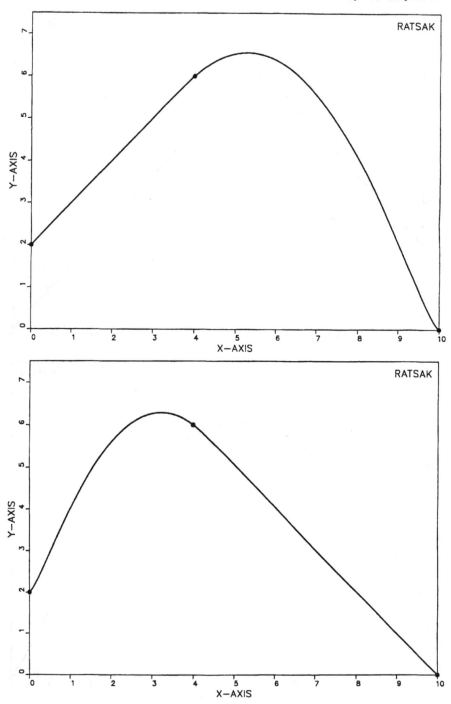

Figure 6.14. a, b.

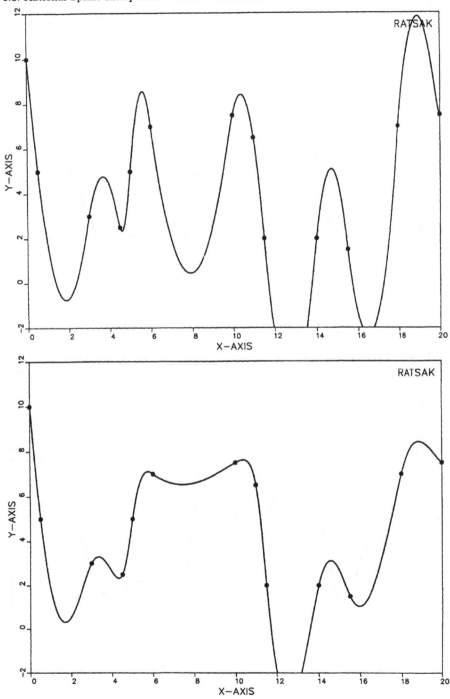

Figure 6.15. a, b.

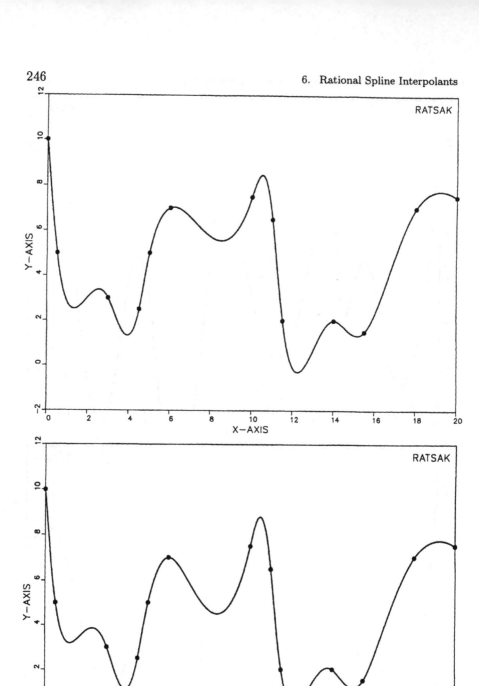

Figure 6.16. a, b.

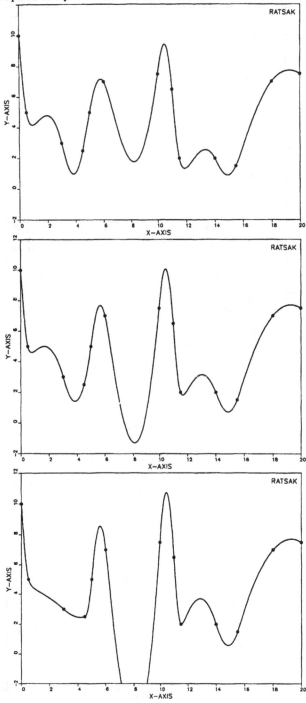

Figure 6.17. a-c.

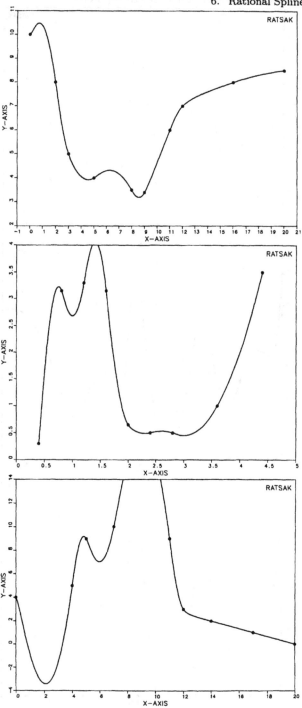

Figure 6.18. a–c.

$p_1 = p_2 = 9$ were chosen. From these, it is clear that we can control the shape of the curve by varying the location of the poles of s_1 and s_2.

A sequence of results with $p = -.9, -.5, .5, 1, 2.5, 5, 10$ and $p_k = p$, $k = 1, \cdots, n - 1$, applied to the same example are given in Figs. 6.15a–6.17c. Further standard examples with $p = p_k = 5$ are given in Figs. 6.18a–c. (For all of these, y_1' and y_n' were chosen, as for RATSCH.).

Obviously, it is very disadvantageous that with the form (6.55) we obtain parabolic segments as p_k tends both to -1 and to ∞. This could only be an advantage if the data is monotone and concave/convex. If this is not the case, then generally speaking, the results worsen as we move away from $p_k = 0$ (cubic spline interpolation).

6.4. Rational Spline Interpolants with Two Prescribable Real Poles

Instead of using the form (4.2) for cubic polynomials, we could just as well have taken

$$s_k(x) = A_k u + B_k t + C_k u^3 + D_k t^3, \qquad (6.62)$$

where

$$t = \frac{x - x_k}{\Delta x_k}, \qquad u = 1 - t.$$

We want to modify this form in a "rational" way so that in the limiting case of two extremely distant poles we get a straight line on $[x_k, x_{k+1}]$ instead of a quadratic as in the previous section. One way of accomplishing this ([154]) is to set

$$s_k(x) = A_k u + B_k t + \frac{C_k u^3}{1 + p_k t} + \frac{D_k t^3}{1 + q_k u}. \qquad (6.63)$$

It has two parameters, p_k and q_k, that determine two independent poles,

$$\tilde{x} = x_k - \frac{h_k}{p_k} \quad \text{and} \quad \overset{\approx}{x} = x_{k+1} + \frac{h_k}{q_k}. \qquad (6.64)$$

(A variant where either t or u only appear in the fractions of (6.33) is considered in [126]. It brings nothing new to the discussion.) If the assembled rational spline interpolant s is to be twice continuously differentiable, then the respective poles must lie outside $[x_k, x_{k+1}]$, i. e., we must have

$$p_k, q_k > -1, \qquad k = 1, \cdots, n - 1. \qquad (6.65)$$

A sketch of the location of the two poles is given in Fig. 6.9.

If $p_k = q_k$, then the distance between $\tilde{x}$ and x_k is the same as the distance between $\tilde{\tilde{x}}$ and x_{k+1}, and thus $\tilde{x}$ and $\tilde{\tilde{x}}$ lie symmetrically with respect to the midpoint of the interval $[x_k, x_{k+1}]$. We will be in the situation of a single pole when $\tilde{x} = \tilde{\tilde{x}}$, i.e., when $p_k q_k + p_k + q_k = 0$, or equivalently, when $p_k = -q_k/(q_k + 1)$ or $q_k = -p_k/(p_k + 1)$. In this case, we essentially reduce to the form (6.55), as then (6.63) becomes

$$s_k(x) = A_k u + B_k t + \frac{C_k u^3 + (1 + p_k)D_k t^3}{1 + p_k t}. \tag{6.66}$$

We may calculate the first two derivatives of (6.63) to obtain

$$h_k s_k'(x) = B_k - A_k + C_k \frac{2p_k u^3 - 3(1 + p_k)u^2}{(1 + p_k t)^2}$$

$$\tag{6.67}$$

$$- D_k \frac{2q_k t^3 - 3(1 + q_k)t^2}{(1 + q_k u)^2}$$

and

$$h_k^2 s_k''(x) = 2C_k \frac{p_k^2 u^3 - 3p_k(1 + p_k)u^2 + 3(1 + p_k)^2 u}{(1 + p_k t)^3}$$

$$\tag{6.68}$$

$$+ 2D_k \frac{q_k^2 t^3 - 3q_k(1 + q_k)t^2 + 3(1 + q_k)^2 t}{(1 + q_k u)^3}.$$

If we again take the y_k' as unknowns, we obtain the usual conditions,

$$y_k = A_k + C_k,$$
$$y_{k+1} = B_k + D_k,$$
$$h_k y_k' = B_k - A_k - (3 + p_k)C_k,$$
$$h_k y_{k+1}' = B_k - A_k + (3 + q_k)D_k.$$

From these, we may solve for

$$C_k = \frac{(3 + q_k)\Delta y_k - (2 + q_k)h_k y_k' - h_k y_{k+1}'}{(2 + p_k)(2 + q_k) - 1},$$

$$D_k = \frac{-(3 + p_k)\Delta y_k + h_k y_k' + (2 + p_k)h_k y_{k+1}'}{(2 + p_k)(2 + q_k) - 1},$$

$$A_k = y_k - C_k, \tag{6.69}$$

$$B_k = y_{k+1} - D_k, \qquad k = 1, \cdots, n - 1.$$

By (6.65), the denominators of C_k and D_k are nonzero.

Then, after some calculation, the C^2 continuity conditions become

$$\frac{q_{k-1}^2 + 3q_{k-1} + 3}{(2+p_{k-1})(2+q_{k-1}) - 1}\frac{1}{h_{k-1}}y_{k-1}'$$

$$+ \left[\frac{(q_{k-1}^2 + 3q_{k-1} + 3)(2+p_{k-1})}{(2+p_{k-1})(2+q_{k-1}) - 1}\frac{1}{h_{k-1}} + \frac{(p_k^2 + 3p_k + 3)(2+q_k)}{(2+p_k)(2+q_k) - 1}\frac{1}{h_k}\right]y_k'$$

$$+ \frac{p_k^2 + 3p_k + 3}{(2+p_k)(2+q_k) - 1}\frac{1}{h_k}y_{k+1}'$$

$$= \frac{(q_{k-1}^2 + 3q_{k-1} + 3)(3+p_{k-1})}{(2+p_{k-1})(2+q_{k-1}) - 1}\frac{d_{k-1}}{h_{k-1}} + \frac{(p_k^2 + 3p_k + 3)(3+q_k)}{(2+p_k)(2+q_k) - 1}\frac{d_k}{h_k},$$

$$\text{(6.70)}$$

$$k = 2, \cdots, n - 1.$$

If y_1' and y_n' are given, then we are again dealing with a linear system with a nonsymmetric but tridiagonal and strictly diagonally dominant coefficient matrix. Existence and uniqueness of the corresponding spline interpolants are then guaranteed whenever (6.65) holds.

If all the poles are symmetric, i.e., $p_k = q_k$, $k = 1, \cdots, n - 1$, then the coefficient matrix of (6.70) is symmetric. If $p_k = q_k = p$, $k = 1, \cdots, n - 1$, then it simplifies even more drastically to

$$\frac{1}{h_{k-1}}y_{k-1}' + (2+p)\left(\frac{1}{h_{k-1}} + \frac{1}{h_k}\right)y_k' + \frac{1}{h_k}y_{k+1}'$$

$$\text{(6.71)}$$

$$= (3+p)\left(\frac{d_{k-1}}{h_{k-1}} + \frac{d_k}{h_k}\right), \qquad k = 2, \cdots, n - 1.$$

As expected, for $p = 0$, this reduces to (4.7).

The limiting case of $p \to -1$ is no longer tractable, as an elimination such as was done for (6.61) is no longer possible. However, as $p \to \infty$, C_k and D_k do tend towards zero and hence, as desired, on (x_k, x_{k+1}) we get the linear segment $s_k(x) = y_k u + y_{k+1}t$. Further, (6.71) then reduces to

$$y_k' = \frac{h_{k-1}d_k + h_k d_{k-1}}{h_{k-1} + h_k}, \qquad \text{(6.72)}$$

which, by (3.57), is the slope at $x = x_k$ of the quadratic through points $k - 1, k$, and $k + 1$. Hence, for large values of p_j and q_j, $j = k - 1, k$, s_k approaches, in the interior of $[x_k, x_{k+1}]$, the polygonal path with the same ordinates, while at x_k and x_{k+1} s itself is still a twice continously differentiable curve with slope approximately (6.72).

For the sake of completeness and also because it may at times be desirable to be able to specify values for y_1'' and y_n'' (e. g., zero), we will now as an alternative use the values y_k'' of the second derivatives as unknowns. From (6.68), it follows that

$$h_k^2 y_k'' \;=\; 2(p_k^2 + 3p_k + 3)C_k$$

and

$$h_k^2 y_{k+1}'' \;=\; 2(q_k^2 + 3q_k + 3)D_k.$$

These, together with the interpolation conditions, imply that

$$C_k \;=\; \frac{1}{2(p_k^2 + 3p_k + 3)} h_k^2 y_k'',$$

$$D_k \;=\; \frac{1}{2(q_k^2 + 3q_k + 3)} h_k^2 y_{k+1}'',$$ \hfill (6.73)

$$A_k \;=\; y_k - C_k,$$

$$B_k \;=\; y_{k+1} - D_k.$$

The condition $s_k'(x_{k-1}) = s_k'(x_k)$ then results, as usual, in a system,

$$\frac{3h_{k-1}}{p_{k-1}^2 + 3p_{k-1} + 3} y_{k-1}''$$

$$+ \left[\frac{3(2 + q_{k-1})h_{k-1}}{q_{k-1}^2 + 3q_{k-1} + 3} + \frac{3(2 + p_k)h_k}{p_k^2 + 3p_k + 3} \right] y_k'' \qquad (6.74)$$

$$+ \frac{3h_k}{q_k^2 + 3q_k + 3} y_{k+1}'' \;=\; 6(d_k - d_{k-1}),$$

$$k = 2, \cdots, n - 1.$$

This linear system of equations has a nonsymmetric, tridiagonal coefficient matrix that is *column-wise* strictly diagonally dominant and hence is also nonsingular (see the appendix).

The special case of $p_k = q_k = p$, $k = 1, \cdots, n - 1$, simplifies to

$$h_{k-1}y_{k-1}'' + (2 + p)(h_{k-1} + h_k)y_k'' + h_k y_{k+1}''$$

$$\;=\; 2(p^2 + 3p + 3)(d_k - d_{k-1}), \qquad k = 2, \cdots, n - 1.$$ \hfill (6.75)

When $p = 0$, this corresponds to the system (4.21) with $y_k'' = 2C_k$.
The limiting case of $p \to \infty$ has $y_k'' \to \infty$, although still $C_k, D_k \to 0$. For $p \to -1$, from (6.73), it follows that

$$C_k = \frac{1}{2}h_k^2 y_k'' \quad \text{and} \quad D_k = \frac{1}{2}h_k^2 y_{k+1}'', \qquad (6.76)$$

and (6.75) becomes

$$h_{k-1}y_{k-1}'' + (h_{k-1} + h_k)y_k'' + h_k y_{k+1}'' = 2(d_k - d_{k-1}),$$

$$k = 2, \cdots, n-1, \qquad (6.77)$$

a linear system that is still uniquely solvable. By (6.76) and (6.73), the form (6.63) reduces to

$$s_k(x) = y_k u + y_{k+1} t + \frac{1}{2}h_k^2 y_k''(u^2 - u) + \frac{1}{2}h_k^2 y_{k+1}''(t^2 - t), \qquad (6.78)$$

i. e., a quadratic as $p_k, q_k \to -1$.

The case of differing values $p_k \neq q_k$, $p_k \neq p$, $q_k \neq p$, $k = 1, \cdots, n-1$, really requires further investigation, since the values y_k' in (6.70) and y_k'' in (6.75) depend on *all* of the p_k and q_k, $k = 1, \cdots, n-1$. This statement also applies to the last and next to last sections. When discussing exponential spline interpolants, we will give references to where such investigations are explicitly carried out. For practical purposes, one will try out various parameters anyway, adjusting them until the curve has an acceptable shape, preferably interactively on a computer screen.

The subroutine RATSP1 (Figs. 6.19 and 6.20) solves the system (6.70) for given values of p_k and q_k, $k = 1, \cdots, n-1$, as well as given boundary values y_1' and y_n'. The special version RATPS1 (Figs. 6.21 and 6.22) handles the case of $p_k = q_k = p$, i. e., the system (6.71). Similarly, the subroutines RATSP2 (Figs. 6.23 and 6.24) and RATPS2 (Figs. 6.25 and 6.26) treat the case when y_1'' and y_n'' are given. In the first case, the coefficients of (6.63) are calculated by (6.69) and in the second by (6.73). RATVAL (Fig. 6.27) may be used to evaluate the assembled spline.

The results of RATSP2 for the same little example as before (Figs. 6.13a–6.14b for RATSAK), with the combinations $(p_1 = 9, q_1 = -.9, p_2 = 9, q_2 = -.9)$, $(p_1 = 9, q_1 = -.9, p_2 = -.9, q_2 = 9)$, $(p_1 = q_1 = p_2 = q_2 = 9)$, and $(p_1 = q_1 = p_2 = q_2 = -.9)$, are given in the four figures, 6.28a–6.29b. Figures 6.30a–6.34 were calculated with RATSP1, with y_1' and y_n' given as previously. Figures 6.30a and 6.30b contrast the results for $p_k = q_k = p = -.5$, and $p_k = q_k = 5$. Evidently, just as for RATSAK, $p \to -1$ is almost always uninteresting unless the data is monotone and convex/concave. Figure 6.30c was calculated with $p = 5$; for intervals 2 and 4, $p_2 = q_2$ and $p_4 = q_4$

```
      SUBROUTINE RATSP1(N,X,Y,P,Q,A,B,C,D,Y1,EPS,IFLAG)
      DIMENSION X(N),Y(N),P(N),Q(N),A(N),B(N),C(N),D(N),Y1(N)
      IFLAG=0
      IF(N.LT.3) THEN
          IFLAG=1
          RETURN
      END IF
      E1=EPS-1.
      N1=N-1
      N2=N-2
      Y11=Y1(1)
      Y1N=Y1(N)
      DO 20 K=1,N1
          KP1=K+1
          KM1=K-1
          PK=P(K)
          QK=Q(K)
          IF(PK.LT.E1.OR.EQ.LT.E1) THEN
              IFLAG=12
              RETURN
          END IF
          PK2=PK*(PK+3.)+3.
          QK2=QK*(QK+3.)+3.
          P22=2.+PK
          Q22=2.+QK
          H=1./(X(KP1)-X(K))
          A(K)=1./(P22*Q22-1.)
          G2=H*A(K)
          R2=H*G2*(Y(KP1)-Y(K))
          IF(K.EQ.1) GOTO 10
          B(KM1)=QK2*G2
          C(KM1)=QK1*P21*G1+PK2*Q22*G2
          D(KM1)=PK2*G2
          Y1(KM1)=R1*QK1*(1.+P21)+R2*PK2*(1.+Q22)
          IF(K.EQ.2) Y1(KM1)=Y1(KM1)-QK1*G1*Y11
          IF(K.EQ.N1) Y1(KM1)=Y1(KM1)-PK2*G2*Y1N
10        P21=P22
          QK1=QK2
          G1=G2
          R1=R2
20    CONTINUE
      CALL TRIDIU(N2,B,C,D,Y1,EPS,IFLAG)
      IF(IFLAG.NE.0) RETURN
      DO 30 K=N2,1,-1
          Y1(K+1)=Y1(K)
30    CONTINUE
      Y1(1)=Y11
      Y1(N)=Y1N
      DO 40 K=1,N1
          K1=K+1
          H=A(K)*(Y(K1)-Y(K))
          Z=A(K)*(X(K1)-X(K))
          P2=2.+P(K)
          Q2=2.+Q(K)
          D(K)=-(1.+P2)*H+Z*(P2*Y1(K1)+Y1(K))
          C(K)=(1.+Q2)*H-Z*(Y1(K1)+Q2*Y1(K))
```

(*cont.*)
```
         B(K)=Y(K1)-D(K)
         A(K)=Y(K)-C(K)
40    CONTINUE
      RETURN
      END
```

Figure 6.19. Program listing of RATSP1.

Calling sequence:

CALL RATSP1(N,X,Y,P,Q,A,B,C,D,Y1,EPS,IFLAG)

Purpose:
RATSP1 determines coefficients A_k, B_k, C_k, and D_k, $k = 1, 2, \cdots, n-1$, of a rational spline interpolant of the following form:

$$s_k(x) = A_k u + B_k t + C_k u^3/(p_k t + 1) + D_k t^3/(q_k u + 1),$$

where $t = (x - x_k)/(x_{k+1} - x_k)$ and $u = 1 - t$ for $k = 1, \cdots, n - 1$. Values for y_1' and y_n' must be given in Y1(1) and Y1(N), respectively.

Description of the parameters:

N	Number of given points. N$\geq$ 3 is required.
X	ARRAY(N): Upon calling must contain the abscissa values x_k, $k = 1, \cdots, n$, with $x_1 < x_2 < \cdots < x_n$.
Y	ARRAY(N): Upon calling must contain the ordinate values y_k, $k = 1, \cdots, n$.
P,Q	ARRAY(N): Upon calling must contain the parameters $p_k > -1$ and $q_k > -1$ for $k = 1, \cdots, n - 1$.
A,B,C,D	ARRAY(N): Upon execution with IFLAG=0 contain the desired spline coefficients, K = 1, 2, $\cdots$, N − 1.
Y1	ARRAY(N): Output: Upon completion contains the values y_k', $k = 1, \cdots, n$, of the first derivatives.
EPS	see TRIDIU.
IFLAG	=0: Normal execution.
	=1: N$\geq$ 3 required.
	=2: Error in solving the linear system (TRIDIU).
	=12: $p_k > -1$ or $q_k > -1$ violated for some k.

Required subroutine: TRIDIU.

Figure 6.20. Description of RATSP1.

```
      SUBROUTINE RATPS1(N,X,Y,P,A,B,C,D,Y1,EPS,IFLAG)
      DIMENSION X(N),Y(N),A(N),B(N),C(N),D(N),Y1(N)
      IFLAG=0
      IF(N.LT.3) THEN
           IFLAG=1
           RETURN
      END IF
      E1=EPS-1.
      IF(P.LT.E1) THEN
           IFLAG=12
           RETURN
      END IF
      N1=N-1
      N2=N-2
      Y11=Y1(1)
      Y1N=Y1(N)
      P2=P+2
      P3=P+3
      DO 20 K=1,N1
           KP1=K+1
           KM1=K-1
           H2=1./(X(KP1)-X(K))
           R2=H2*H2*(Y(KP1)-Y(K))
           IF(K.EQ.1) GOTO 10
           B(KM1)=P2*(H1+H2)
           C(KM1)=H2
           Y1(KM1)=P3*(R1+R2)
           IF(K.EQ.2) Y1(KM1)=Y1(KM1)-H1*Y11
           IF(K.EQ.N1) Y1(KM1)=Y1(KM1)-H2*Y1N
10         H1=H2
           R1=R2
20    CONTINUE
      CALL TRIDIS(N2,B,C,Y1,EPS,IFLAG)
      IF(IFLAG.NE.0) RETURN
      DO 30 K=N2,1,-1
           Y1(K+1)=Y1(K)
30    CONTINUE
      Y1(1)=Y11
      Y1(N)=Y1N
      PP=1./(P2*P2-1.)
      DO 40 K=1,N1
           K1=K+1
           DX=X(K1)-X(K)
           H=P3*(Y(K1)-Y(K))
           D(K)=(-H+DX*(Y1(K)+P2*Y1(K1)))*PP
           C(K)=(H-DX*(P2*Y1(K)+Y1(K1)))*PP
           B(K)=Y(K1)-D(K)
           A(K)=Y(K)-C(K)
40    CONTINUE
      RETURN
      END
```

Figure 6.21. Program listing of RATPS1.

Calling sequence:

CALL RATPS1(N,X,Y,P,A,B,C,D,Y1,EPS,IFLAG)

Purpose:
RATPS1 determines coefficients A_k, B_k, C_k, and D_k, $k = 1, 2, \cdots, n-1$, of a rational spline interpolant of the following form:

$$s_k(x) = A_k u + B_k t + C_k \frac{u^3}{pt+1} + D_k \frac{t^3}{pu+1},$$

where $t = (x - x_k)/(x_{k+1} - x_k)$ and $u = 1 - t$ for $k = 1, \cdots, n-1$. Values for y_1' and y_n' must be given in Y1(1) and Y1(N), respectively.

Description of the parameters:

N,X,Y,A,B,C,D,Y1 as in RATSP1.
P   Upon calling must contain the parameter $p > -1$.
EPS   see TRIDIS.
IFLAG   =0:   Normal execution.
               =1:   N$\geq$ 3 required.
               =2:   Error in solving the linear system (TRIDIS).
               =12: $p > -1$ not satisfied.

Required subroutine: TRIDIS.

Figure 6.22. Description of RATPS1.

```
      SUBROUTINE RATSP2(N,X,Y,P,Q,A,B,C,D,Y2,EPS,IFLAG)
      DIMENSION X(N),Y(N),P(N),Q(N),A(N),B(N),C(N),D(N),Y2(N)
      IFLAG=0
      IF(N.LT.3) THEN
           IFLAG=1
           RETURN
      END IF
      N1=N-1
      N2=N-2
      Y21=Y2(1)
      Y2N=Y2(N)
      DO 20 K=1,N1
           KP1=K+1
           KM1=K-1
           PK=P(K)
           QK=Q(K)
           H=X(KP1)-X(K)
           P2=PK+2.
           Q2=QK+2.
           PK2=3./(PK*(PK+3.)+3.)*H
           QK2=3./(QK*(QK+3.)+3.)*H
           A(K)=H*H/2.
           R2=(Y(KP1)-Y(K))/H
           IF(K.EQ.1) THEN
                   PK1=PK2
                   GOTO 10
           END IF
           B(KM1)=PK2
           C(KM1)=QK1*Q1+PK2*P2
           D(KM1)=QK2
           Y2(KM1)=6.*(R2-R1)
           IF(K.EQ.2) Y2(KM1)=Y2(KM1)-PK1*Y21
           IF(K.EQ.N1) Y2(KM1)=Y2(KM1)-QK2*Y2N
  10       Q1=Q2
           QK1=QK2
           R1=R2
  20  CONTINUE
      CALL TRIDIU(N2,B,C,D,Y2,EPS,IFLAG)
      IF(IFLAG.NE.0) RETURN
      DO 30 K=N2,1,-1
           Y2(K+1)=Y(K)
  30  CONTINUE
      Y2(1)=Y21
      Y2(N)=Y2N
      DO 40 K=1,N1
           K1=K+1
           AK=A(K)
           PK=P(K)
           QK=Q(K)
           D(K)=AK*Y2(K1)/(QK*(QK+3.)+3.)
           C(K)=AK*Y2(K)/(PK*(PK+3.)+3.)
           B(K)=Y(K1)-D(K)
           A(K)=Y(K)-C(K)
  40  CONTINUE
      RETURN
      END
```

Figure 6.23. Program listing of RATSP2.

Calling sequence:

CALL RATSP2(N,X,Y,P,Q,A,B,C,D,Y2,EPS,IFLAG)

Purpose:

RATSP2 determines coefficients A_k, B_k, C_k, and D_k, $k = 1, 2, \cdots, n-1$, of a rational spline interpolant of the following form:

$$s_k(x) = A_k u + B_k t + C_k \frac{u^3}{p_k t + 1} + D_k \frac{t^3}{q_k u + 1},$$

where $t = (x - x_k)/(x_{k+1} - x_k)$ and $u = 1 - t$ for $k = 1, \cdots, n-1$. Values for y_1'' and y_n'' must be given in Y2(1) and Y2(N), respectively.

Description of the parameters:

N,X,Y,P,Q,A,B,C,D,EPS,IFLAG as in RATSP1.

Y2   ARRAY(N): Output: Contains upon execution with
                   IFLAG=0 the values y_k'', $k = 1, \cdots, n$, of the
                   second derivatives.

Required subroutine: TRIDIU.

Figure 6.24. Description of RATSP2.

```
      SUBROUTINE RATPS2(N,X,Y,P,A,B,C,D,Y2,EPS,IFLAG)
      DIMENSION X(N),Y(N),A(N),B(N),C(N),D(N),Y2(N)
      IFLAG=0
      IF(N.LT.3) THEN
          IFLAG=1
          RETURN
      END IF
      P2=P+2
      PP=2.*P*(P+3.)+6.
      N1=N-1
      N2=N-2
      Y21=Y2(1)
      Y2N=Y2(N)
      DO 20 K=1,N1
          KP1=K+1
          KM1=K-1
          H2=X(KP1)-X(K)
          R2=(Y(KP1)-Y(K))/H2
          IF(K.EQ.1) GOTO 10
          B(KM1)=P2*(H1+H2)
          C(KM1)=H2
          Y2(KM1)=PP*(R2-R1)
          IF(K.EQ.2) Y2(KM1)=Y2(KM1)-H1*Y21
          IF(K.EQ.N1) Y2(KM1)=Y2(KM1)-H2*Y2N
10        H1=H2
          R1=R2
20    CONTINUE
      CALL TRIDIS(N2,B,C,Y2,EPS,IFLAG)
      IF(IFLAG.NE.0) RETURN
      DO 30 K=N2,1,-1
          Y2(K+1)=Y2(K)
30    CONTINUE
      Y2(1)=Y21
      Y2(N)=Y2N
      DO 40 K=1,N1
          K1=K+1
          DX=X(K1)-X(K)
          H=DX*DX/PP
          D(K)=H*Y2(K1)
          C(K)=H*Y2(K)
          B(K)=Y(K1)-D(K)
          A(K)=Y(K)-C(K)
40    CONTINUE
      RETURN
      END
```

Figure 6.25. Program listing of RATPS2.

Calling sequence:

CALL RATPS2(N,X,Y,P,A,B,C,D,Y2,EPS,IFLAG)

Purpose:
RATPS2 determines coefficients A_k, B_k, C_k, and D_k, $k = 1, 2, \cdots, n-1$, of a rational spline interpolant of the following form:

$$s_k(x) = A_k u + B_k t + C_k \frac{u^3}{pt+1} + D_k \frac{t^3}{pu+1},$$

where $t = (x - x_k)/(x_{k+1} - x_k)$ and $u = 1 - t$ for $k = 1, \cdots, n-1$. Values for y_1'' and y_n'' must be given in Y2(1) and Y2(N), respectively.

Description of the parameters:

N,X,Y,P,A,B,C,D,EPS,IFLAG as in RATPS1.
Y2   ARRAY(N): Output: Upon execution contains the values
            y_k'', $k = 1, \cdots, n - 1$, of the second derivatives.

Required subroutine: TRIDIS.

Figure 6.26. Description of RATPS2.

```
FUNCTION RATVAL(N,X,P,Q,A,B,C,D,V,IFLAG)
DIMENSION X(N),P(N),Q(N),A(N),B(N),C(N),D(N)
DATA I/1/
IFLAG=0
IF(N.LT.2) THEN
    IFLAG=1
    RETURN
END IF
CALL INTONE(X,N,V,I,IFLAG)
IF(IFLAG.NE.0) RETURN
T=(V-X(I))/(X(I+1)-X(I))
U=1.-T
RATVAL=A(I)*U+B(I)*T+C(I)*U*U*U/(P(I)*T+1.)+
&       D(I)*T*T*T/(Q(I)*U+1.)
RETURN
END
```

FUNCTION RATVAL(N,X,P,Q,A,B,C,D,V,IFLAG)

Purpose:

RATVAL is a FUNTION subroutine for the calculation of a function value at a point V$\in$ [X(1),X(N)] of a rational spline interpolant of the following form:

$$s_k(x) = A_k u + B_k t + C_k u^3/(p_k t + 1) + D_k t^3/(q_k u + 1),$$

where $t = (x - x_k)/(x_{k+1} - x_k)$ and $u = 1 - t$ for $k = 1, \cdots, n - 1$.

Description of the parameters:

N	Number of given points. N$\geq$ 2 is required.
X	ARRAY(N): Abscissas.
P,Q	ARRAY(N): Upon calling must contain the parameters $p_k > -1$ and $q_k > -1$ for $k = 1, \cdots, n - 1$.
A,B,C,D	ARRAY(N): Vectors of the spline coefficients.
V	Value of the point at which the spline interpolant is to be evaluated.
IFLAG	=0: Normal execution.
	=1: N$\geq$ 2 required.
	=3: Error in interval determination (INTONE).

Required subroutine: INTONE.

Remark: The statement 'DATA I/1/' has the effect that I is set to 1 at the first call to RATVAL.

Figure 6.27. FUNCTION RATVAL and its description.

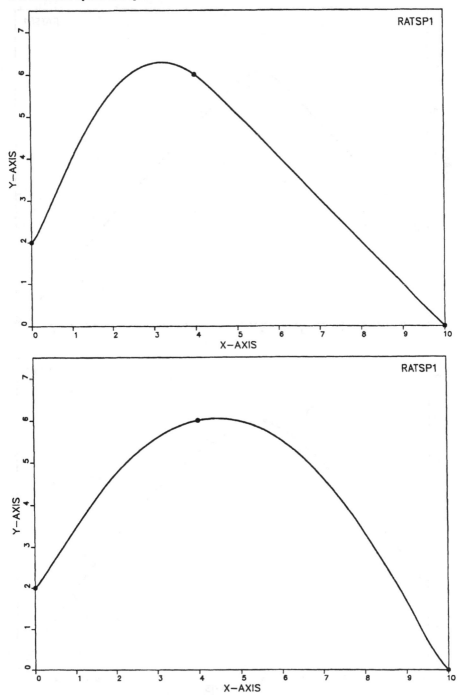

Figure 6.28. a, b.

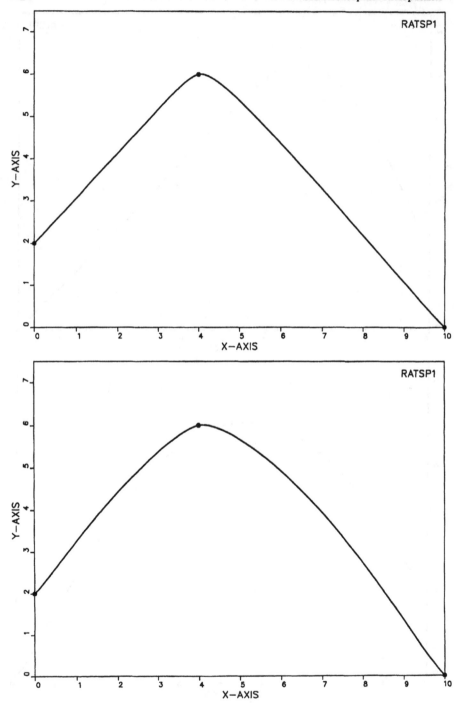

Figure 6.29. a, b.

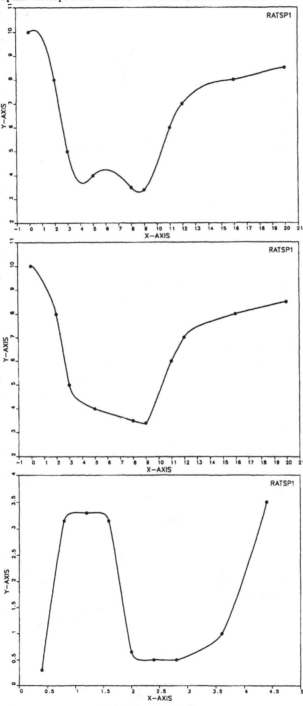

Figure 6.30. a-c.

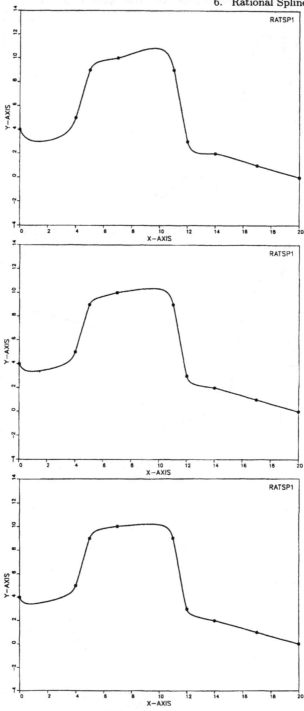

Figure 6.31. a-c.

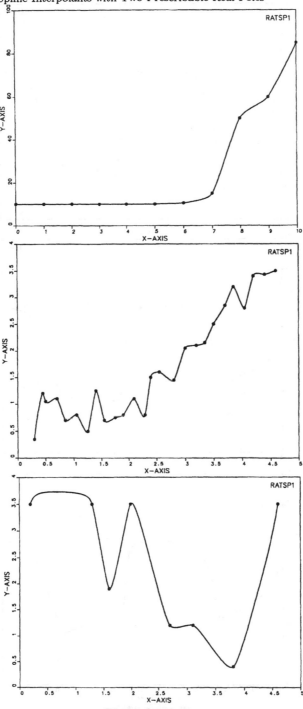

Figure 6.32. a-c.

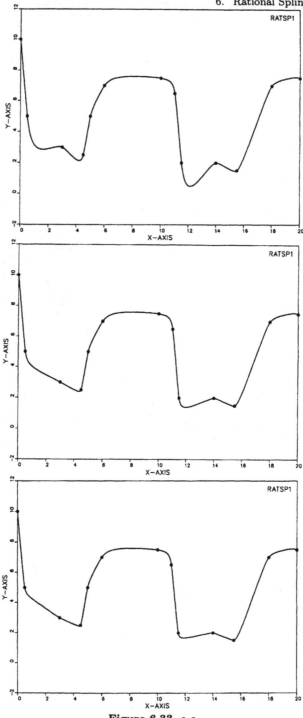

Figure 6.33. a-c.

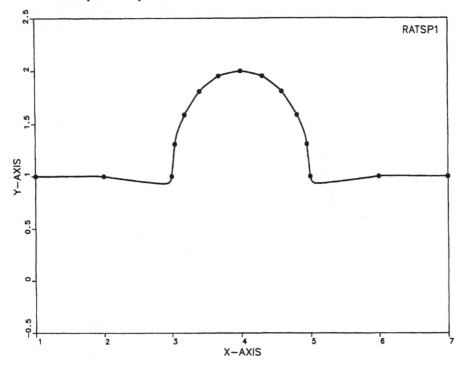

Figure 6.34.

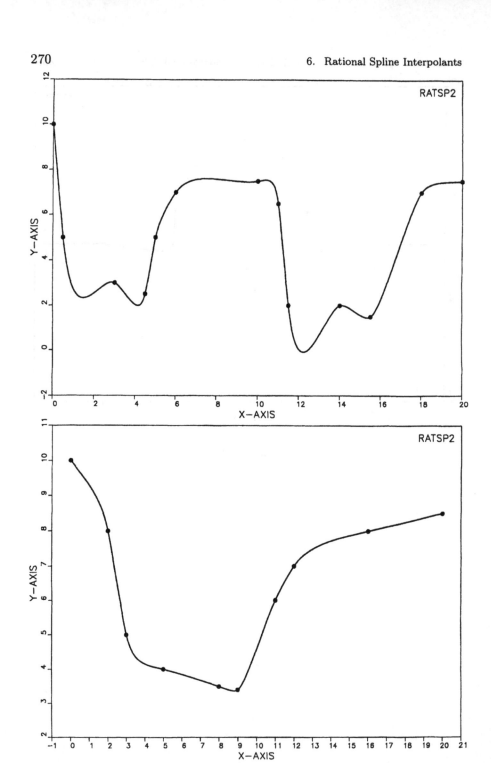

Figure 6.35. a, b.

should have been chosen to be somewhat larger. The results for $p = 5$, $p = 9$ and $p_1 = p_4 = p_6 = 10$, $p_2 = p_3 = p_5 = p_7 = p_8 = p_9 = 5$ $(q_k = p_k)$ are contrasted in Figs. 6.31a–c. In particular, this illustrates, on intervals 1,4, and 6, how increasing the values "straightens" the curve on the corresponding intervals while keeping it C^2. For Figs. 6.32a and b, p was again $p = 5$. In Fig. 6.32c, $p_1 = q_1 = 15$ and $p_k = q_k = 5$ for $k \geq 2$. Figures 6.33a–c show the results for $p = 5$, $p_k = 5$ except $p_2 = p_3 = p_9 = 15$ and $p_{10} = 10$, and $p_k = 5$ except $p_2 = p_3 = p_9 = 25$ and $p_{10} = 20$ $(q_k = p_k)$. From these, it is very clear how useful varying the parameters by interval can be. Finally, Fig. 6.34 was calculated with $p_k = 5$ except $p_1 = p_{14} = 10$ and $p_2 = p_3 = 50$ $(p_k = q_k)$. Given that the curve is to be C^2, it is difficult to improve upon this result. Figures 6.35a and b show the curves resulting from RATSP2, with $y_1'' = y_n'' = 0$ and $p_k = q_k = 2.5$ for two of our standard examples. These rational spline interpolants have met with better success than others; also, for a practical application see [152]. Different values for p_k and q_k have not yet been taken advantage of at all.

As the examples show, already often with $p_k = q_k = p$ and decidedly with $p_k = q_k$ varying by interval, data that is positive, monotone, or convex/concave on a sequence of intervals has these properties reproduced by the interpolant. Nevertheless, there is still the question whether the parameters p_k and q_k can be chosen *automatically* instead of just by trial and error.

One possibility arises from considering the second derivative (6.68) in the form $(p_k \neq 0)$,

$$h_k^2 s_k''(x) = 2C_k \frac{p_k^2 u \left[\left(u - \frac{3(1+p_k)}{2p_k} \right)^2 + \frac{3(1+p_k)^2}{4p_k^2} \right]}{(1+p_k t)^3}$$

$$+2D_k \frac{q_k^2 t \left[\left(t - \frac{3(1+q_k)}{2q_k} \right)^2 + \frac{3(1+q_k)^2}{4q_k^2} \right]}{(1+q_k u)^3}.$$

As the coefficients of C_k and D_k are nonnegative functions on $[0, 1]$, the sign of s_k'' will be constant on $[x_k, x_{k+1}]$ if C_k and D_k have the same sign or if one of them vanishes. By (6.73), this is the case if $sgn(y_k'') = sgn(y_{k+1}'')$, and hence if the solutions of (6.74) have this property, s_k will be convex/concave on $[x_k, x_{k+1}]$. In the expressions (6.69) for C_k and D_k, we can consider the y_k' as being given or having been calculated in some manner, i.e., consider a *Hermite rational spline interpolant*. Then, for example, in the special case of $p_k = q_k$, we can calculate a value p' so that $C_k = 0$ and a value p'' so that $D_k = 0$. If $\max(p', p'') > -1$ (as required by (6.65)), then we set

$p_k = \max(p', p'')$ and otherwise we "adaptively" set $p_k = 0$. In this manner, one of the coefficients C_k and D_k becomes zero wherever this is possible, and consequently convexity or concavity of the interpolant is attained when possible. Moreover, evaluation of the spline is also somewhat simplified. We will not pursue this problem any further here but refer the reader to the section after the next one, where it is investigated for a different type of interpolant, although for all of the data simultaneously.

6.5. Periodic Rational Spline Interpolants with Two Prescribable Real Poles

To obtain a *periodic* rational spline interpolant (rational as in the sense of the previous section), we proceed in exactly the same manner as we did for cubic splines. If besides $y_n = y_1$, $y_n' = y_1'$ is also to be satisfied, then these conditions should be substituted into both sides of the $(n-1)$st equation of (6.70). The other equations, $k = 2, \cdots, n-2$, are unaffected. In addition, the condition $s_1''(x_1) = s_{n-1}''(x_n)$ results in

$$\frac{p_{n-1}^2 + 3p_{n-1} + 3}{(2 + p_{n-1})(2 + q_{n-1}) - 1} \frac{1}{h_{n-1}} y_1'$$

$$+ \frac{q_{n-2}^2 + 3q_{n-2} + 3}{(2 + p_{n-2})(2 + q_{n-2}) - 1} \frac{1}{h_{n-2}} y_{n-2}'$$

$$+ \left[\frac{(q_{n-2}^2 + 3q_{n-2} + 3)(2 + p_{n-2})}{(2 + p_{n-2})(2 + q_{n-2}) - 1} \frac{1}{h_{n-2}} \right.$$

$$\left. + \frac{(p_{n-1}^2 + 3p_{n-1} + 3)(2 + q_{n-1})}{(2 + p_{n-1})(2 + q_{n-1}) - 1} \frac{1}{h_n} \right] y_{n-1}' \qquad (6.79)$$

$$= \frac{(q_{n-2}^2 + 3q_{n-2} + 3)(3 + p_{n-2})}{(2 + p_{n-2})(2 + q_{n-2}) - 1} \frac{d_{n-2}}{h_{n-2}}$$

$$+ \frac{(p_{n-1}^2 + 3p_{n-1} + 3)(3 + q_{n-1})}{(2 + p_{n-1})(2 + q_{n-1}) - 1} \frac{d_{n-1}}{h_{n-1}}.$$

Here, d_{n-1} on the right should be taken as $d_{n-1} = (y_1 - y_{n-1})/(x_1 - x_{n-1})$. The coefficient matrix of the resulting linear system of $(n-1)$ equations in the $(n-1)$ unknowns, $y_1', y_2', \cdots, y_{n-1}'$, is nonsymmetric, cyclically tridiagonal, and strictly diagonally dominant. Hence, the existence and uniqueness of the desired spline interpolant are guaranteed. The subroutine RATPER

```
      SUBROUTINE RATPER(N,X,Y,P,Q,A,B,C,D,Y1,EPS,IFLAG)
      DIMENSION X(N),Y(N),P(N),Q(N),A(N),B(N),C(N),D(N),Y1(N)
      IFLAG=0
      IF(N.LT.4) THEN
          IFLAG=1
          RETURN
      END IF
      E1=EPS-1.
      N1=N-1
      Y(N)=Y(1)
      QN1=Q(N1)
      P21=P(N1)+2.
      QK1=QN1*(QN1+3.)+3.
      H=1./(X(N)-X(N1))
      G1=H/(P21*(Q(N1)+2.)-1.)
      R1=H*G1*(Y(N)-Y(N1))
      DO 10 K=1,N1
          KP1=K+1
          PK=P(K)
          QK=Q(K)
          IF(PK.LT.E1.OR.QK.LT.E1) THEN
              IFLAG=12
              RETURN
          END IF
          PK2=PK*(PK+3.)+3.
          QK2=QK*(QK+3.)+3.
          P22=2.+PK
          Q22=2.+QK
          H=1./(X(KP1)-X(K))
          A(K)=1./(P22*Q22-1.)
          G2=H*A(K)
          R2=H*G2*(Y(KP1)-Y(K))
          B(K)=QK1*G1
          C(K)=QK1*P21*G1+PK2*Q22*G2
          D(K)=PK2*G2
          Y1(K)=R1*QK1*(1.+P21)+R2*PK2*(1.+Q22)
          P21=P22
          QK1=QK2
          G1=G2
          R1=R2
10    CONTINUE
      CALL TRIPEU(N1,B,C,D,Y1,EPS,IFLAG)
      IF(IFLAG.NE.0) RETURN
      Y1(N)=Y1(1)
      DO 20 K=1,N1
          K1=K+1
          H=A(K)*(Y(K1)-Y(K))
          Z=A(K)*(X(K1)-X(K))
          P2=2.+P(K)
          Q2=2.+Q(K)
          D(K)=-(1.+P2)*H+Z*(P2*Y1(K1)+Y1(K))
          C(K)=(1.+Q2)*H-Z*(Y(K1)+Q2*Y1(K))
          B(K)=Y(K1)-D(K)
          A(K)=Y(K)-C(K)
20    CONTINUE
      RETURN
      END
```

Figure 6.36. Program listing of RATPER.

Calling sequence:

CALL RATPER(N,X,Y,P,Q,A,B,C,D,Y1,EPS,IFLAG)

Purpose:
RATPER determines coefficients A_k, B_k, C_k, and D_k, $k = 1, 2, \cdots, n-1$,
of a periodic rational spline interpolant of the following form:

$$s_k(x) = A_k u + B_k t + C_k \frac{u^3}{p_k t + 1} + D_k \frac{t^3}{q_k u + 1},$$

where $t = (x - x_k)/(x_{k+1} - x_k)$ and $u = 1 - t$ for $k = 1, \cdots, n - 1$.

Description of the parameters:

N,X,P,Q,A,B,C,D,Y1 as in RATSP1 (N$\geq$ 4 required).
Y   ARRAY(N): Upon calling must contain the ordinate values
                 y_k, $k = 1, \cdots, n$. $y_n = y_1$ is required and is
                 enforced by the program.
EPS   see TRIPEU.
IFLAG =0:   Normal execution.
        =1:   N$<$ 4 not allowed.
        =2:   Error in solving the linear system (TRIPEU).
        =12: $p_k > -1$ or $q_k > -1$ violated for some k.
Required subroutine: TRIPEU.

Figure 6.37. Description of RATPER.

```
SUBROUTINE RATPPR(N,X,Y,P,A,B,C,D,Y1,EPS,IFLAG)
DIMENSION X(N),Y(N),A(N),B(N),C(N),D(N),Y1(N)
IFLAG=0
IF(N.LT.4) THEN
     IFLAG=1
     RETURN
END IF
E1=EPS-1.
IF(P.LT.E1) THEN
     IFLAG=12
     RETURN
END IF
Y(N)=Y(1)
N1=N-1
P2=P+2.
P3=P+3.
H1=1./(X(N)-X(N1))
R1=H1*H1*(Y(N)-Y(N1))
DO 10 K=1,N1
     KP1=K+1
     H2=1./(X(KP1)-X(K))
```

(cont.)

```
        R2=H2*H2*(Y(KP1)-Y(K))
        B(K)=P2*(H1+H2)
        C(K)=H2
        Y1(K)=P3*(R1+R2)
        H1=H2
        R1=R2
10  CONTINUE
        CALL TRIPES(N1,B,C,Y1,EPS,IFLAG)
        IF(IFLAG.NE.0) RETURN
        Y1(N)=Y1(1)
        PP=1./(P2*P2-1.)
        DO 20 K=1,N1
        K1=K+1
        DX=X(K1)-X(K)
        H=P3*(Y(K1)-Y(K))
        D(K)=(-H+DX*(Y1(K)+P2*Y1(K1)))*PP
        C(K)=(H-DX*(P2*Y1(K)+Y1(K1)))*PP
        B(K)=Y(K1)-D(K)
        A(K)=Y(K)-C(K)
20  CONTINUE
        RETURN
        END
```

Figure 6.38. Program listing of RATPPR.

Calling sequence:
CALL RATPPR(N,X,Y,P,A,B,C,D,Y1,EPS,IFLAG)

Purpose:
RATPPR determines coefficients A_k, B_k, C_k, and D_k, $k = 1, 2, \cdots, n-1$, of a periodic rational spline interpolant of the following form:

$$s_k(x) = A_k u + B_k t + C_k u^3/(pt + 1) + D_k t^3/(pu + 1),$$

where $t = (x - x_k)/(x_{k+1} - x_k)$ and $u = 1 - t$ for $k = 1, \cdots, n - 1$.

Description of the parameters:

N,X,P,A,B,C,D,Y1 as in RATPS1 (N$\geq$ 4 required).
Y   ARRAY(N): Upon calling must contain the ordinate values
                     y_k, $k = 1, \cdots, n$. $y_n = y_1$ is required and is
                     enforced by the program.
EPS   see TRIPES.
IFLAG   =0:   Normal execution.
           =1:   N< 4 not allowed.
           =2:   Error in solving the linear system (TRIPES).
           =12: $p > -1$ violated.
Required subroutine: TRIPES.

Figure 6.39. Description of RATPPR.

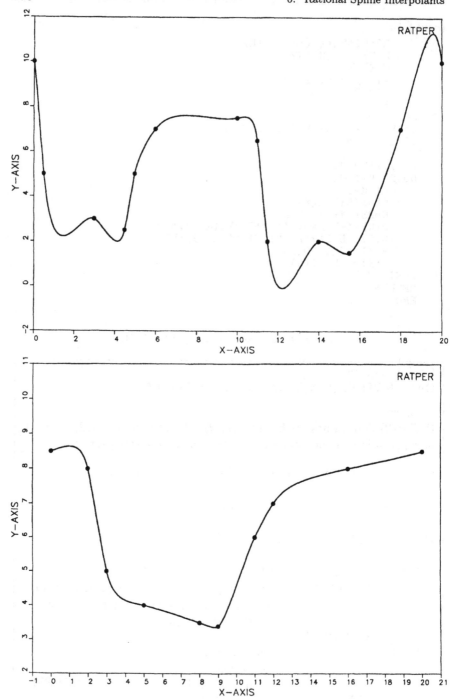

Figure 6.40. a, b.

(Figs. 6.36 and 6.37) solves this system using TRIPEU and calculates the coefficients by (6.69). The special case of $p_k = q_k = p$, $k = 1, \cdots, n - 1$, results in a drastic simplification that is exploited in RATPPR (Figs. 6.38 and 6.39). Again, RATVAL may be used for evaluation.

Figures 6.40a and b show the results of RATPER for $p_k = q_k = 2.5$ on two of our standard examples.

6.6.   Rational Spline Interpolants with Two Prescribable Real or Complex Poles

In [47,48], the form or result (4.95), used in cubic Hermite splines, is modified by dividing the cubic part by $1 + (r_k - 3)tu$. This results in

$$s_k(x) = y_k u + y_{k+1} t + \frac{tu\left[(t - u)\Delta y_k - h_k(ty'_{k+1} - uy'_k)\right]}{1 + (r_k - 3)tu}. \qquad (6.80)$$

For the time being, r_k is an arbitrary real parameter.

When $r_k = 3$, we recover cubic Hermite splines. When $r_k \neq 3$, the zeros of the denominators are

$$t = \frac{1}{2}\left(1 \pm \sqrt{\frac{r_k + 1}{r_k - 3}}\right). \qquad (6.81)$$

Since these are in the interval (0,1) for $-\infty < r_k \leq -1$, we must now restrict ourselves to

$$r_k > -1, \qquad k = 1, \cdots, n - 1, \qquad (6.82)$$

in order for s_k to be continuous on $[x_k, x_{k+1}]$. For $-1 < r_k < 3$, the zeros are complex. Nevertheless, we will not be able to make use of this form as $r_k \to -1$. The reason is that by (6.81), when $r_k = -1$, (6.80) has a double pole at $t = 1/2$. This causes large spikes in the graph of the interpolant near the interval midpoint as $r_k \to -1$. As $r_k \to \infty$, the poles tend to $t = 0$ and $t = 1$, i. e., the interpolation points x_k and x_{k+1}, and also the form (6.80) tends to a linear segment on the interior of the interval $[x_k, x_{k+1}]$.

We must first verify that the rationally modified form (6.80) still solves the Hermite interpolation problem. To this end, we temporarily make the following abbreviations:

$$\alpha = \Delta y_k - h_k y'_k, \quad \beta = h_k y'_{k+1} - \Delta y_k, \quad \text{and } \gamma = r_k - 3, \qquad (6.83)$$

where, as already has been done for t, we suppress the index k on the left-hand sides. Further, we let

$$Q(t) = \frac{\left[(\alpha - \beta)t - \alpha\right](t - t^2)}{1 + \gamma(t - t^2)} \qquad (6.84)$$

denote the last term of (6.80). Since $Q(0) = Q(1) = 0$, the interpolation conditions, $s_k(x_k) = y_k$ and $s_k(x_{k+1}) = y_{k+1}$, are satisfied. We may calculate

$$h_k Q'(t) = \frac{\left[1 + \gamma(t - t^2)\right](\alpha - \beta)(t - t^2) + \left[(\alpha - \beta)t - \alpha\right](1 - 2t)}{\left[1 + \gamma(t - t^2)\right]^2}, \quad (6.85)$$

from which it follows that $Q'(0) = -\alpha/h_k$ and $Q'(1) = \beta/h_k$. Hence, since $s_k'(x) = d_k + Q'(t)$, we see that $s_k'(x_k) = y_k'$ and $s_k'(x_{k+1}) = y_{k+1}'$. If the y_k' are given, then we have a C^1 rational Hermite spline interpolant s with segments s_k given by (6.80).

In the next section, we will choose the r_k as some appropriate functions of the y_k' so that the resulting spline has certain shape-preserving properties. Here, just as in the last section, we first want to consider the parameters r_k as given and then determine the y_k' so that the resulting spline is twice continuously differentiable. To this end, we calculate

$$h_k^2 Q''(t) = \frac{\left\{ \begin{array}{l} -2[(\alpha - 2\beta)t - (2\alpha - \beta)u](1 + \gamma(t - t^2)) \\ +2\gamma(\beta t + \alpha u)(1 - 2t)^2 \end{array} \right\}}{\left[1 + \gamma(t - t^2)\right]^3}. \quad (6.86)$$

From this, we may easily deduce that the C^2 continuity conditions, $s_{k-1}''(x_k) = s_k''(x_k)$, $k = 2, \cdots, n - 1$, become

$$\frac{1}{h_{k-1}} y_{k-1}' \left[(r_{k-1}) \frac{1}{h_{k-1}} + (r_k - 1) \frac{1}{h_k} \right] y_k' + \frac{1}{h_k} y_{k+1}'$$

$$= \frac{r_{k-1}}{h_{k-1}} d_{k-1} + \frac{r_k}{h_k} d_k, \qquad k = 2, \cdots, n - 1.$$

$$(6.87)$$

When $r_k = 3$, we recover (4.7) with $B_k = y_k'$. For given values of y_1' and y_n', the system (6.87) is strictly diagonally dominant only for $-1 < r_k < 0$ and $r_k > 2$. For $r_k = 0$ and $r_k = 2$, the coefficient matrix is only weakly diagonally dominant, but it is, however, irreducible. Hence, existence and uniqueness of such rational spline interpolants for $0 < r_k < 2$ cannot be shown by this technique. As $r_k \to \infty$, (6.72) again holds.

The subroutine RATGRE (Figs. 6.41 and 6.42) allows, nevertheless, any $r_k > -1$, $k = 1, \cdots, n - 1$. For $0 < r_k < 2$, the solvability of the system (6.87) is determined by TRIDIS. The reader is reminded, however, that TRIDIS does no pivoting. With the resulting values of the y_k', the s_k of (6.80) can be evaluated using GREVAL (Figs. 6.43 and 6.44).

Again, we first give the results for our little example. The effects of the choices $r_1 = -.9$, $r_2 = 9$, $r_1 = 9$, $r_2 = -.9$, $r_1 = r_2 = -.9$, and $r_1 = r_2 = 9$ are shown in Figs. 6.45a–6.46b.

```
      SUBROUTINE RATGRE(N,X,Y,R,Y1,A,B,EPS,IFLAG)
      DIMENSION X(N),Y(N),R(N),Y1(N),A(N),B(N)
      IFLAG=0
      IF(N.LT.3) THEN
          IFLAG=1
          RETURN
      END IF
      N1=N-1
      N2=N-2
      Y11=Y1(1)
      Y1N=Y1(N)
      DO 20 K=1,N1
          KP1=K+1
          KM1=K-1
          H2=1./(X(KP1)-X(K))
          R2=R(K)
          IF(R2.LE.-1.) THEN
              IFLAG=12
              RETURN
          END IF
          S2=R2*H2*H2*(Y(KP1)-Y(K))
          IF(K.EQ.1) GOTO 10
          A(KM1)=H1*(R1-1.)+H2*(R2-1.)
          B(KM1)=H2
          Y1(KM1)=S1+S2
          IF(K.EQ.2) Y1(KM1)=Y1(KM1)-H1*Y11
          IF(K.EQ.N1) Y1(KM1)=Y1(KM1)-H2*Y1N
10        H1=H2
          R1=R2
          S1=S2
20    CONTINUE
      CALL TRIDIS(N2,A,B,Y1,EPS,IFLAG)
      IF(IFLAG.NE.0) RETURN
      DO 30 K=N2,1,-1
          Y1(K+1)=Y1(K)
30    CONTINUE
      Y1(1)=Y11
      Y1(N)=Y1N
      RETURN
      END
```

Figure 6.41. Program listing of RATGRE.

Calling sequence:

CALL RATGRE(N,X,Y,R,Y1,A,B,EPS,IFLAG)

Purpose:

RATGRE determines values y'_k, $k = 2, \cdots, n - 1$, for a rational spline interpolant of the following form (Gregory, J.A.):

$$s_k(x) = y_k u + y_{k+1} t + t \frac{(t - u)\Delta y_k - \Delta x_k(t y'_{k+1} - u y'_k)}{1 + (r_k - 3)tu},$$

where $t = (x - x_k)/\Delta x_k$ and $u = 1 - t$, $k = 1, 2, \cdots, n - 1$.

Description of the parameters:

N,X,Y as in RATSP1. $N \geq 3$ is required.

R	ARRAY(N):	Upon calling must contain the parameters $r_k > -1$, $k = 1, \cdots, n - 1$.
Y1	ARRAY(N):	Upon calling Y1(1) and Y1(N) must contain values for y'_1 and y'_n, respectively. Output: Upon execution with IFLAG=0 contains the values y'_k, $k = 1, \cdots, n$, for the first derivatives.
A,B	ARRAY(N):	Work space.
EPS	see TRIDIS.	
IFLAG	=0:	Normal execution.
	=1:	$N \geq 3$ required.
	=2:	Error in solving the linear system (TRIDIS).
	=12:	$r_k > -1$ violated for some k.

Required subroutine: TRIDIS.

Remark: The linear system to be solved is not strictly diagonally dominant for $0 \leq r_k \leq 2$.

Figure 6.42. Description of RATGRE.

```
FUNCTION GREVAL(N,X,Y,Y1,R,V,IFLAG)
DIMENSION X(N),Y(N),R(N),Y1(N)
DATA I/1/
IFLAG=0
IF(N.LT.2) THEN
    IFLAG=1
    RETURN
END IF
CALL INTONE(X,N,V,I,IFLAG)
IF(IFLAG.NE.0) RETURN
T=(V-X(I))/(X(I+1)-X(I))
U=1.-T
I1=I+1
GREVAL=Y(I)*U+Y(I1)*T+T*U*((T-U)*(Y(I1)-Y(I))-
&        (X(I1)-X(I))*(T*Y1(I1)-U*Y1(I)))/
&        (1.+(R(I)-3.)*T*U)
RETURN
END
```

Figure 6.43. Program listing of GREVAL.

FUNCTION GREVAL(N,X,Y,Y1,R,V,IFLAG)

Purpose:
GREVAL is a FUNCTION subroutine for the calculation of a function value at a point V∈ [X(1),X(N)] of a rational spline interpolant of the form:

$$s_k(x) = y_k u + y_{k+1} t + tu[(t-u)\Delta y_k - \Delta x_k(ty'_{k+1} - uy'_k)]/[1+(r_k-3)tu],$$

where $t = (x - x_k)/\Delta x_k$ and $u = 1 - t$, $k = 1, 2, \cdots, n - 1$.

Description of the parameters:

N,X,Y,Y1,R as in RATGRE.
V   Value of the point at which the spline interpolant is
         to be evaluated.
IFLAG   =0:   Normal execution.
         =1:   N≥ 2 required.
         =3:   Error in interval determination (INTONE).

Required subroutine: INTONE.

Remark: The statement 'DATA I/1/' has the effect that I is set to 1 at the first call to GREVAL.

Figure 6.44. Description of GREVAL.

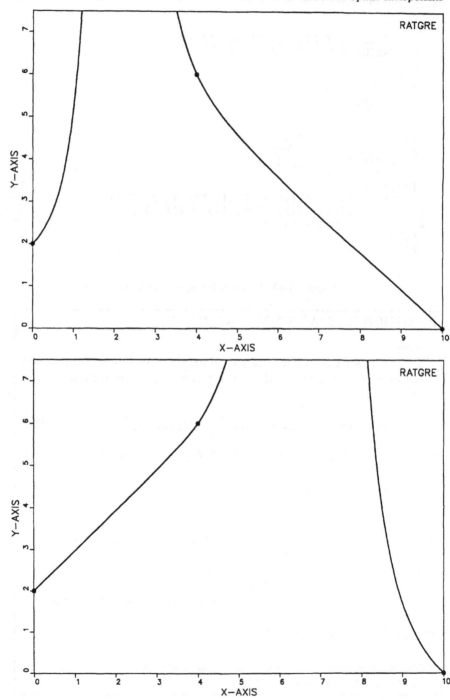

Figure 6.45. a, b.

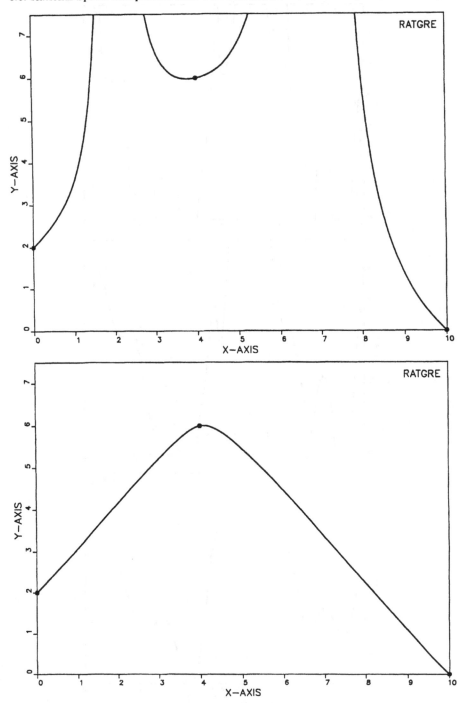

Figure 6.46. a, b.

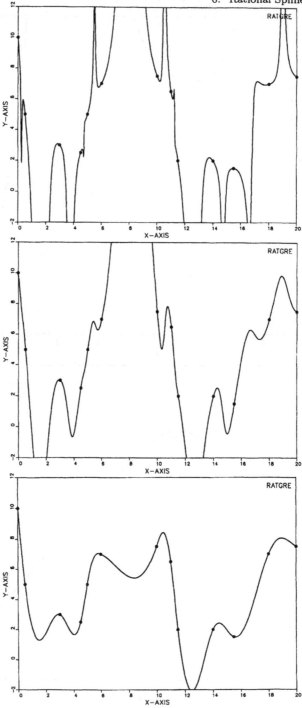

Figure 6.47. a-c.

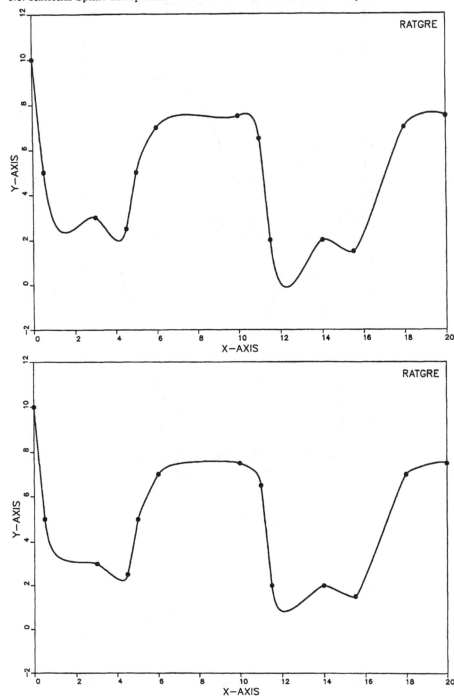

Figure 6.48. a, b.

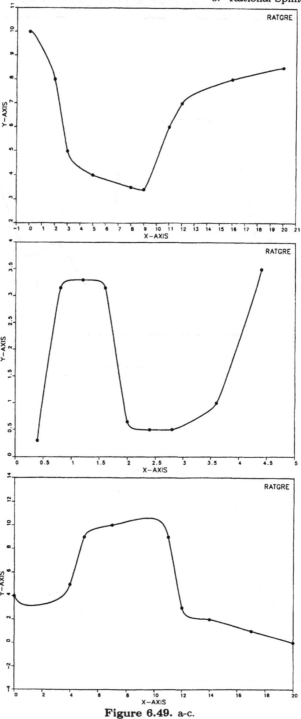

Figure 6.49. a-c.

Figures 6.47a–6.48b show a sequence of results for a single data set with $r_k = r$, $k = 1, \cdots, n - 1$ and $r = -.9, .5, 2.5, 5, 10$ (cf. 6.15a–6.17b). The boundary values y'_1 and y'_n were chosen as for RATSCH. Further, Figs. 6.49a–c show the curves resulting from $r_k = r = 9$ applied to some other standard examples.

Taking these figures into consideration and understanding the effect of the location of the poles, one would normally only use RATGRE with $r_k \geq 3$. This restriction is similar to that of $p_k, q_k \geq 0$ for the form of (6.63). For the purposes of shape preservation, it is also possible to consider values $r_k \approx 2$, since then the resulting curves are similar to quadratics. As we have already indicated, in the next section we will choose the r_k as certain functions of the y'_k in order to guarantee shape preservation. Finally, we remark that in the denominator of $Q(t)$, we could just as well have chosen other nonzero quadratics, e. g., $r_k^2 + ut$.

6.7. Shape Preservation of Monotone Data

We now consider (6.80) of the previous section but in the equivalent form,

$$s_k(x) = \frac{y_{k+1}t^3 + \left(r_k y_{k+1} - h_k y'_{k+1}\right) t^2 u + \left(r_k y_k + h_k y'_k\right) tu^2 + y_k u^3}{t^3 + r_k(t^2 u + ut^2) + u^3}.$$

(6.88)

Some lengthy calculations ([23]) show that

$$s'_k(x) = \frac{y'_{k+1}t^4 + \alpha t^3 u + \beta t^2 u^2 + \gamma tu^3 + y'_k u^4}{[1 + (r_k - 3)tu]^2},$$

(6.89)

where we temporarily set

$$
\begin{aligned}
\alpha &= 2(r_k d_k - y'_k), \\
\beta &= (r_k^2 + 3)d_k - r_k(y'_k + y'_{k+1}), \\
\gamma &= 2(r_k d_k - y'_{k+1}).
\end{aligned}
$$

(6.90)

We assume now *monotonically increasing ordinates*, i. e.,

$$y_1 \leq y_2 \leq \cdots \leq y_n \quad \Leftrightarrow \quad d_k \geq 0, \quad k = 1, \cdots, n - 1.$$

(6.91)

(The case of *monotonically decreasing ordinates* can be dealt with in a similiar manner.) Then s preserves the monotonicity if $s'_k(x) \geq 0$ on $[x_k, x_{k+1}]$. For this, a necessary condition is that $y'_k, y'_{k+1} \geq 0$, while a sufficient condition is, by (6.89), that the values (6.90) are nonnegative. (Further conditions are discussed in [113].) This, in turn, in the case that $d_k > 0$, is guaranteed by

$$r_k \geq \frac{y'_k + y'_{k+1}}{d_k}. \tag{6.92}$$

In particular, the choice of ([23])

$$r_k = 1 + \frac{y'_k + y'_{k+1}}{d_k} \tag{6.93}$$

considerably simplifies (6.80) or (6.88) to

$$s_k(x) = y_k + \frac{\Delta y_k(d_k t^2 + y'_k tu)}{d_k + (y'_k + y'_{k+1} - 2d_k)tu}, \quad (d_k > 0), \tag{6.94}$$

i. e., to a quotient of two quadratics ([49]). The nonnegativity of the derivative,

$$s'_k(x) = \frac{d_k^2 \left[y'_k u^2 + 2d_k tu + y'_{k+1} t^2 \right]}{\left[d_k + (y'_k + y'_{k+1} - 2d_k)tu \right]^2}, \tag{6.95}$$

then follows immediately from (6.91). If $d_k = 0$, then we must have $y'_k = y'_{k+1} = 0$, and then from (6.80) it follows that, in this case, $s_k(x) \equiv y_k$.

Thus, we obtain a monotone C^1 spline interpolant of this type when, for example, the y'_k are computed according to (4.68) or (4.74), i. e., by the subroutine GRAD2B. However, for reasons previously discussed, it is usually better ([49]) to use GRAD2R. The subroutine RMONC1 (Figs. 6.50 and 6.51) gives the user the choice of which to use. It also handles monotonically decreasing data. The resulting spline may be evaluated by using GREVAM (Fig. 6.52).

Figures 6.53a–6.57b contrast the results of RMONC1 using GRAD2B and GRAD2R, applied to a collection of five examples (taken in part from the relevant literature). In these, it is clear that GRAD2R is to be preferred.

If the y'_k are not, as just before, given, but are to be determined so that the resulting spline is C^2, then we can substitute the r_k of (6.93) into the C^2 conditions (6.87). Assuming in addition that the ordinates are strictly monotonically increasing, i. e.,

$$y_1 < y_2 < \cdots < y_n \quad \Leftrightarrow \quad d_k > 0, \quad k = 1, \cdots, n - 1, \tag{6.96}$$

this results in the nonlinear system of equations ([22]),

$$y'_k \left[-\left(\frac{1}{h_{k-1}} + \frac{1}{h_k} \right) + \frac{1}{\Delta y_{k-1}} y'_{k-1} + \left(\frac{1}{\Delta y_{k-1}} + \frac{1}{\Delta y_k} \right) y'_k + \frac{1}{\Delta y_k} y'_{k+1} \right]$$

$$= \frac{d_{k-1}}{h_{k-1}} + \frac{d_k}{h_k}, \quad k = 2, \cdots, n - 1,$$

$$\tag{6.97}$$

```
      SUBROUTINE RMONC1(N,X,Y,METHOD,EPS,Y1,ICASE,IFLAG)
      DIMENSION X(N),Y(N),Y1(N)
      IFLAG=0
      ICASE=0
      IF(N.LT.3) THEN
          IFLAG=1
          RETURN
      END IF
      D1=Y(2)-Y(1)
      DO 10 K=2,N-1
          D2=Y(K+1)-Y(K)
          IF(D1*D2.LT.0.) THEN
              IFLAG=17
              RETURN
          END IF
          IF(ABS(D1).LE.EPS) D1=D2
10    CONTINUE
      IF(D1.GT.0.) THEN
          ICASE=1
      ELSE
          ICASE=2
      END IF
      IF(METHOD.EQ.1) THEN
          CALL GRAD2B(N,X,Y,Y1,EPS,IFLAG)
      ELSE
          CALL GRAD2R(N,X,Y,Y1,EPS,IFLAG)
      END IF
      RETURN
      END
```

Figure 6.50. Program listing of RMONC1.

Calling sequence:

CALL RMONC1(N,X,Y,METHOD,EPS,Y1,ICASE,IFLAG)

Purpose:

For the form:

$$
s_k(x) = \begin{cases} y_k + \Delta y_k \dfrac{d_k t^2 + y_k' tu}{d_k + (y_k' + y_{k+1}' - 2d_k)tu} & \text{if } y_k \neq y_{k+1} \\ y_k & \text{if } y_k = y_{k+1} \end{cases} ,
$$

where $t = (x - x_k)/\Delta x_k$, $u = 1 - t$, $k = 1, 2, \cdots, n - 1$, and $d_k = \Delta y_k/\Delta x_k$, the y_k' are calculated so as to produce a monotonically increasing (respectively, decreasing) once continuously differentiable spline interpolant. The user has the option of two different methods (see GRAD2B and GRAD2R).

Description of the parameters:

N,X as in RATSP1 ($N \geq 3$ required).

Y	ARRAY(N):	Vector of the monotonically increasing (or monotonically decreasing) ordinate values.
METHOD=1:		The y_k' are calculated using GRAD2B.
=2:		The y_k' are calculated using GRAD2R (cf. the program descriptions of GRAD2B and GRAD2R).
EPS		see GRAD2B or GRAD2R.
Y1	ARRAY(N):	Output: Upon execution with IFLAG=0 contains the values y_k', $k = 1, \cdots, n$, of the first derivatives.
ICASE	=1:	The ordinates are monotonically increasing.
	=2:	The ordinates are monotonically decreasing.
IFLAG	=0:	Normal execution.
	=1:	$N \geq 3$ required.
	=17:	The ordinates are not monotone.

Required subroutines: GRAD2B and GRAD2R.

Figure 6.51. Description of RMONC1.

```
FUNCTION GREVAM(N,X,Y,Y1,V,EPS,IFLAG)
DIMENSION X(N),Y(N),Y1(N)
DATA I/1/
IFLAG=0
IF(N.LT.2) THEN
```

(cont.)

```
      IFLAG=1
      RETURN
   END IF
   CALL INTONE(X,N,V,I,IFLAG)
   IF(IFLAG.NE.0) RETURN
   I1=I+1
   DY=Y(I1)-Y(I)
   IF(ABS(DY).LT.EPS) THEN
      GREVAM=Y(I)
      RETURN
   END IF
   DX=X(I1)-X(I)
   H=DY/DX
   T=(V-X(I))/DX
   U=1.-T
   GREVAM=Y(I)+DY*((H*T+Y1(I)*U)*T)/(H+(Y1(I)
 &       +Y1(I1)-2.*H)*T*U)
   RETURN
   END
```

FUNCTION GREVAM(N,X,Y,Y1,V,EPS,IFLAG)

Purpose:
GREVAM is a FUNCTION subroutine for the calculation of a function value at a point V∈ [X(1),X(N)] of a rational spline interpolant of the form:

$$s_k(x) = \begin{cases} y_k + \Delta y_k \dfrac{d_k t^2 + y'_k tu}{d_k + (y'_k + y'_{k+1} - 2d_k)tu} & \text{if } y_k \neq y_{k+1} \\ y_y & \text{if } y_k = y_{k+1} \end{cases},$$

where $t = (x - x_k)/\Delta x_k$, $u = 1 - t$, $k = 1, 2, \cdots, n - 1$, and $d_k = \Delta y_k/\Delta x_k$.

Description of the parameters:

N,X,Y,Y1,V,IFLAG as in GREVAL (N≥ 3 required).
EPS   Value for the accuracy test. Recommendation: EPS= 10^{-t} (t : number of available decimal digits). If the absolute value of the difference Δy_k is smaller than EPS, then it is interpreted as zero.

Required subroutine: INTONE.

Remark: The statement 'DATA I/1/' has the effect that I is set to 1 at the first call to GREVAM.

Figure 6.52. FUNCTION GREVAM and its description.

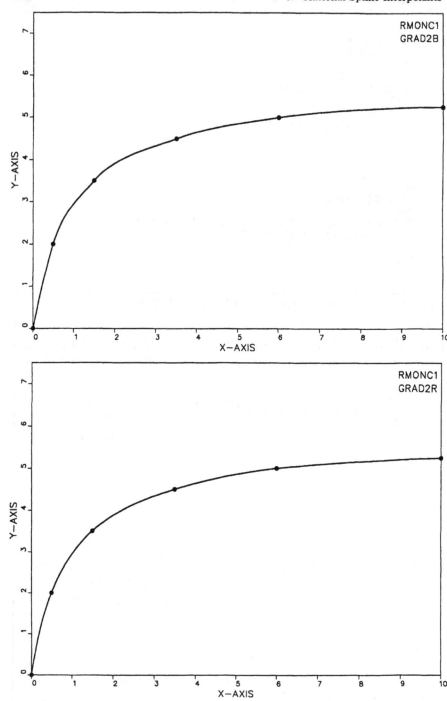

Figure 6.53. a, b.

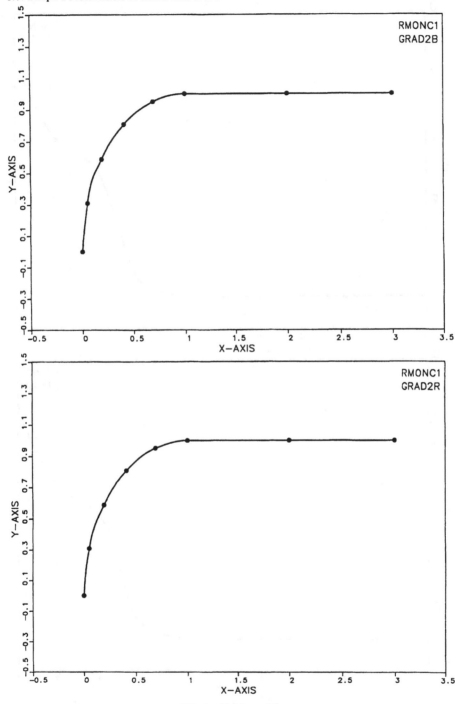

Figure 6.54. a, b.

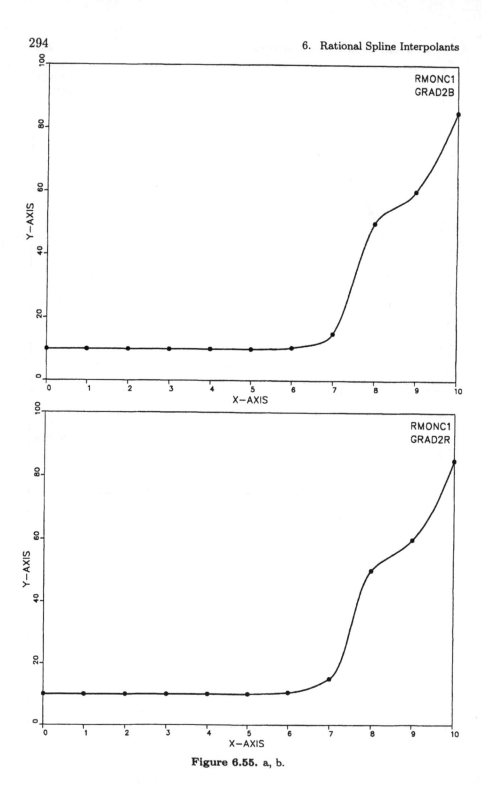

Figure 6.55. a, b.

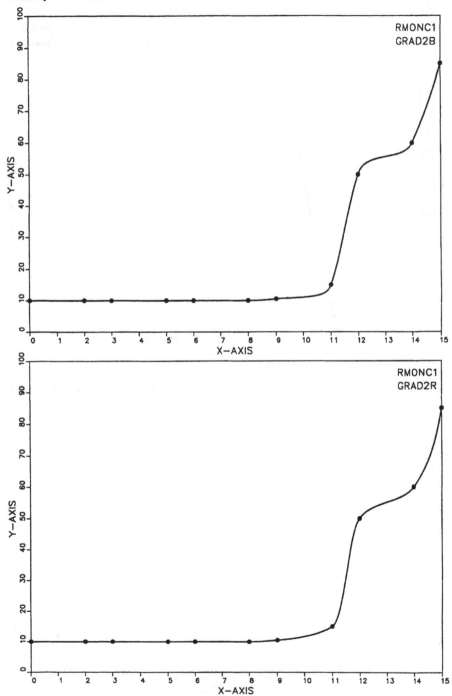

Figure 6.56. a, b.

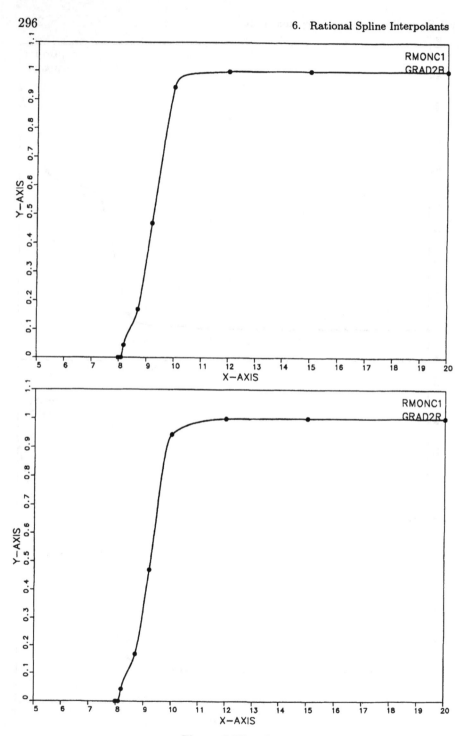

Figure 6.57. a, b.

which for given $y_1', y_n' \geq 0$ must be solved for $y_k' > 0$, $k = 2, \cdots, n-1$. (For strictly decreasing ordinates, we would have $y_1', y_n' \leq 0$ and require $y_k' < 0$, $k = 2, \cdots, n-1$.)

Setting

$$a_k = \frac{1}{\Delta y_k}, \quad c_k = \frac{1}{h_{k-1}} + \frac{1}{h_k}, \quad b_k = \frac{d_{k-1}}{h_{k-1}} + \frac{d_k}{h_k}, \tag{6.98}$$

(6.97) simplifies to

$$y_k' \left[-c_k + a_{k-1}y_{k-1}' + (a_{k-1} + a_k)y_k' + a_k y_{k+1}' \right] = b_k.$$

This quadratic equation in y_k' may be solved for

$$
\begin{aligned}
y_k' \quad = \quad & \frac{1}{2(a_{k-1} + a_k)} [(c_k - a_{k-1}y_{k-1}' - a_k y_{k+1}') \\
& + \sqrt{\left(c_k - a_{k-1}y_{k-1}' - a_k y_{k+1}'\right)^2 + 4(a_{k-1} + a_k)b_k}],
\end{aligned}
\tag{6.99}
$$

$$k = 2, \cdots, n-1.$$

The denominator that so arises is nonzero by assumption (6.96). Moreover, the square root is also real. Because of the requirement that $y_k' > 0$, the sign of the square root must be chosen to be positive (negative for decreasing ordinates). One obvious way to solve this system is to use a Gauss-Seidel iteration,

$$
\begin{aligned}
(y_k')^{(i+1)} \quad = \quad & \frac{1}{2(a_{k-1} + a_k)} \left[\left(c_k - a_{k-1}(y_{k-1}')^{(i+1)} - a_k(y_{k+1}')^{(i)} \right) \right. \\
& \left. + \sqrt{(c_k - a_{k-1}(y_{k-1}')^{(i+1)} - a_k(y_{k+1}')^{(i)})^2 + 4(a_{k-1} + a_k)b_k} \right],
\end{aligned}
$$

$$k = 2, \cdots, n-1, \quad i = 0, 1, 2, \cdots. \tag{6.100}$$

We use

$$(y_k')^{(0)} = \sqrt{\frac{b_k}{a_{k-1} + a_k}}, \quad k = 2, \cdots, n-1, \tag{6.101}$$

as starting values. They come from setting $c_k - a_{k-1}y_{k-1}' - a_k y_{k+1}' = 0$ in (6.99). For the end slopes, we use

$$(y_1')^{(i+1)} = (y_1')^{(i)} = y_1' \quad \text{and} \quad (y_n')^{(i+1)} = (y_n')^{(i)} = y_n'.$$

Constructive convergence proofs of the Gauss-Seidel method as well as the corresponding Jacobi method ([150]) (this is actually done first) are given in [22]. This then also shows the existence and uniqueness of such spline interpolants. The Gauss-Seidel iteration converges asymptotically faster than does the Jacobi iteration ([22]).

The Gauss-Seidel method is implemented in the subroutine RMONC2 (Figs. 6.58 and 6.59). The convergence test used is described in the program description. GREVAM can again be used for spline evaluation.

Because of the assumption (6.96) on the strict montonicity of the data, we are unable to present all of the previous examples of monotone data for purposes of comparison. In Figs. 6.60a–c we have set $y_1' = d_1$ and $y_{n-1}' = d_{n-1}$. Figure 6.60a corresponds to 6.53a and 6.53b, while 6.60b corresponds to 6.57a and 6.57b. Evidently, the C^2 requirement results in no visible improvement over the C^1 results. However, this could be caused by the inexactness of the graphs. Figure 6.60c is a new example.

When instead of assumption (6.84), we assume, for example,

$$y_1 \leq \cdots \leq y_j, \quad y_j \geq \cdots \geq y_n,$$

then we may use the C^1 interpolation method described previously on each of the two interval segments. We would then also have to set $y_j' = 0$. Of course, we could then similarly handle cases of more than two interval segments of changing monotonicity.

6.8.   Shape Preservation of Convex or Concave Data

Instead of monotone data (6.91), we now assume strictly convex data ([21, 23, 47, 48, 113]),

$$d_1 < d_2 < \cdots < d_{n-1}. \tag{6.102}$$

(The case of concave data, where the inequality signs are reversed, can be handled in a similar manner and actually most easily by multiplying the y_k by -1.) In order to obtain a strictly convex interpolant, it is necessary that the values y_k' of the first derivative at x_k satisfy

$$y_1' < d_1 < y_2' < d_2 < \cdots < d_{k-1} < y_k' < d_k < \cdots < y_n'. \tag{6.103}$$

Now the segments s_k of (6.80), or equivalently (6.88), are convex precisely when $s_k''(x) \geq 0$ on $[x_k, x_{k+1}]$. If this is true for $k = 1, \cdots, n-1$, then s is also convex. A lengthy calculation shows ([23]) that

$$s_k''(x) = \frac{2}{h_k} \frac{\alpha t^3 + \beta t^2 u + \gamma t u^2 + \delta u^3}{[1 + (r_k - 3)tu]^3}, \tag{6.104}$$

```
      SUBROUTINE RMONC2(N,X,Y,EPS1,EPS2,ITMAX,Y1,A,B,C,
     &                  ICASE,IT,IFLAG)
      DIMENSION X(N),Y(N),Y1(N),A(N),B(N),C(N)
      IFLAG=0
      IF(N.LT.3) THEN
          IFLAG=1
          RETURN
      END IF
      ICASE=0
      IT=0
      N1=N-1
      D1=Y(2)-Y(1)
      IF(ABS(D1).LT.EPS1) THEN
          IFLAG=18
          RETURN
      END IF
      DO 10 K=2,N1
          D2=Y(K+1)-Y(K)
          IF(ABS(D2).LT.EPS1) THEN
              IFLAG=18
              RETURN
          END IF
          IF(D1*D2.LT.0.) THEN
              IFLAG=17
              RETURN
          END IF
 10   CONTINUE
      IF(D1.GE.0.) THEN
          ICASE=1
          CC=1.
      ELSE
          ICASE=2
          Y1(1)=-Y1(1)
          Y1(N)=-Y1(N)
          CC=-1.
      END IF
      IF(Y1(1).LT.0..OR.Y1(N).LT.0.) THEN
          IFLAG=20
          RETURN
      END IF
      A(1)=CC/D1
      DO 30 K=1,N1
          K1=K+1
          H2=1./(X(K1)-X(K))
          G2=CC*(Y(K1)-Y(K))
          IF(K.EQ.1) GOTO 20
          A(K)=1./G2
          C(K)=H1+H2
          B(K)=H1*H1*G1+H2*H2*G2
 20       H1=H2
          G1=G2
 30   CONTINUE
 40   IT=IT+1
      IF(IT.GT.ITMAX) THEN
          IFLAG=13
          RETURN
      END IF
```

(cont.)

```
      H=Y1(1)
      RNORM1=0.
      RNORM2=0.
      DO 50 K=2,N1
          KM1=K-1
          KP1=K+1
          H1=2.*(A(KM1)+A(K))
          H2=C(K)-A(KM1)*H-A(K)*Y1(KP1)
          IF(IT.EQ.1) H2=0.
          H=(H2+SQRT(H2*H2+2.*H1*B(K)))/H1
          IF(IT.GT.1) RNORM1=RNORM1+ABS(H-Y1(K))
          RNORM2=RNORM2+ABS(H)
          Y1(K)=H
50    CONTINUE
      IF(IT.EQ.1) GOTO 40
      IF(RNORM1.LE.EPS2*RNORM2) GOTO 60
      GOTO 40
60    IF(ICASE.EQ.2) THEN
          DO 70 K=1,N
              Y1(K)=-Y1(K)
70        CONTINUE
      END IF
      RETURN
      END
```

Figure 6.58. Program listing of RMONC2.

Calling sequence:

CALL RMONC2(N,X,Y,EPS1,EPS2,ITMAX,Y1,A,B,C,
                      ICASE,IT,IFLAG)

Purpose:
For the form:

$$s_k(x) = y_k + \Delta y_k \frac{d_k t^2 + y'_k t u}{d_k + (y'_k + y'_{k+1} - 2d_k)tu},$$

where $t = (x - x_k)/\Delta x_k$, $u = 1 - t$, $k = 1, 2, \cdots, n - 1$, and $d_k = \Delta y_k/\Delta x_k$, the $y'_k > 0$ (respectively, $y'_k < 0$) are calculated so as to produce a strictly monotonically increasing (resp., decreasing) twice continuously differentiable spline interpolant for given $y_1 < \cdots < y_n$ (resp., $y_1 > \cdots > y_n$), and given $y'_1, y'_n \geq 0$ (resp., $y'_1, y'_n \leq 0$).

Description of the parameters:

N,X as in RATSP1 (N$\geq$ 3 required).
Y   ARRAY(N): Vector of the monotonically increasing (or
                      monotonically decreasing) ordinate values.

EPS1	Value for the accuracy test. Recommendation: EPS1$= 10^{-t}$ (t : number of available decimal digits). If the absolute value of a number is smaller than EPS1, then it is interpreted as zero.
EPS2	Value used for testing convergence of the Gauss-Seidel iteration. If two consecutive iterates, $(y')^{(i)}$ and $(y')^{(i+1)}$, satisfy $\|(y')^{(i+1)} - (y')^{(i)}\|_1 \leq$ EPS2 $\cdot \|(y')^{(i+1)}\|_1$ then the iteration is halted. Recommendation: EPS2$= 10^{-t/2+1}$.
ITMAX	Prescribed value for the maximum number of iterations.
Y1	ARRAY(N): Input: Values for y_1' and y_n' must be given in Y1(1) and Y1(N), respectively. Output: Upon execution with IFLAG=0 contains the values y_k', $k = 1, \cdots, n$, of the first derivatives.
A,B,C	ARRAY(N): Work space.
ICASE	=1: The ordinates are monotonically increasing.
	=2: The ordinates are monotonically decreasing.
IT	Number of iterations actually used.
IFLAG	=0: Normal execution.
	=1: N$\geq$ 3 required.
	=13: ITMAX number of iterations exceeded.
	=17: The ordinates are not strictly monotone.
	=18: For at least one k, $y_k = y_{k+1}$ in the sense that $|\Delta y_k| <$EPS1.
	=20: $y_1', y_n' \geq 0$ not satisfied for $y_1 < \cdots < y_n$ or $y_1', y_n' \leq 0$ not satisfied for $y_1 > \cdots > y_n$.

Figure 6.59. Description of RMONC2.

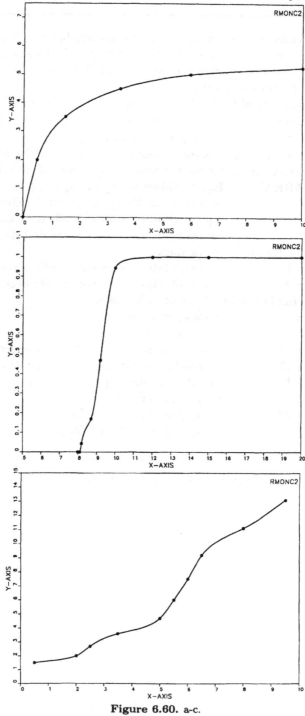

Figure 6.60. a-c.

where we have set

$$
\begin{aligned}
\alpha &= r_k(y'_{k+1} - d_k) - y'_{k+1} + y'_k, \\
\beta &= 3(y'_{k+1} - d_k), \\
\gamma &= 3(d_k - y'_k), \\
\delta &= r_k(d_k - y'_k) - y'_{k+1} + y'_k,
\end{aligned}
\tag{6.105}
$$

and have again suppressed the index k on the left. By (6.104), convexity requires that

$$
\alpha \geq 0 \quad \text{and} \quad \delta \geq 0.
\tag{6.106}
$$

This is, however, also sufficient, since given y'_k satisfying (6.103), $\beta > 0$ and $\gamma > 0$ are automatically true. From (6.106) with (6.105) and (6.103), it follows that s_k is convex precisely when

$$
r_k \geq \max \left(\frac{y'_{k+1} - y'_k}{y'_{k+1} - d_k}, \frac{y'_{k+1} - y'_k}{d_k - y'_k} \right).
\tag{6.107}
$$

Setting

$$
\begin{aligned}
M_k &= \max \left(y'_{k+1} - d_k, d_k - y'_k \right), \\[4pt]
m_k &= \min \left(y'_{k+1} - d_k, d_k - y'_k \right),
\end{aligned}
\tag{6.108}
$$

(6.107) reduces to

$$
r_k \geq 1 + \frac{M_k}{m_k}, \qquad k = 1, \cdots, n - 1,
\tag{6.109}
$$

seeing that $M_k + m_k = y'_{k+1} - y'_k$.

On the basis of some further theoretical considerations, [23] suggests the use of

$$
r_k = 2 + \frac{M_k}{m_k}
\tag{6.110}
$$

or

$$
r_k = 1 + \frac{M_k}{m_k} + \frac{m_k}{M_k} = 1 + \frac{y'_{k+1} - d_k}{d_k - y'_k} + \frac{d_k - y'_k}{y'_{k+1} - d_k}.
\tag{6.111}
$$

This is consistent with our earlier remark that the r_k should usually be chosen so that $r_k \geq 3$. The choice of r_k according to (6.111) eliminates the technically unpleasant use of max and min.

We leave it to the reader to write his own FUNCTION that will evaluate a convex (concave) C^1 interpolant of the form (6.80) for given values y_k and y'_k satisfying (6.103) and r_k satisfying one of (6.110) or (6.111) (or even chosen by some other means consistent with (6.109)). (The y'_k can again be computed by GRAD2B, or even better, by GRAD2R. Alternatively, one might simply set $y'_k = \frac{1}{2}(d_{k-1} + d_k)$, $k = 2, \cdots, n - 1$.)

We now want to attain twice continuous differentiability. Using the y_k' as unknowns, restricted, however, to satisfy (6.103), choosing the r_k according to (6.111) and substituting these into (6.87) yields ([47, 48]) the nonlinear system,

$$\frac{(d_k - y_k')^2}{(y_k' - d_{k-1})^2} = \frac{h_k}{h_{k-1}} \frac{y_{k+1}' - d_k}{d_{k-1} - y_{k-1}'}, \qquad k = 2, \cdots, n-1. \qquad (6.112)$$

Astonishingly, this is the same as (6.26), which was obtained from the completely different form (6.1). Hence, the resulting values of the y_k' are the same even though the forms (6.1) and (6.88) with r_k according to (6.111) are obviously rather different.

If we again suppose that (6.103) holds, then $d_k - y_k' > 0$ and $y_k' - d_{k-1} > 0$, and hence we may take the square root of (6.112) to obtain

$$\frac{d_k - y_k'}{y_k' - d_{k-1}} = +\sqrt{\frac{h_k}{h_{k-1}} \frac{y_{k+1}' - d_k}{d_{k-1} - y_{k-1}'}}. \qquad (6.113)$$

Finally, from this we obtain the fixed-point expression,

$$y_k' = \frac{\sqrt{h_{k-1}(d_{k-1} - y_{k-1}')}\, d_k + \sqrt{h_k(y_{k+1}' - d_k)}\, d_{k-1}}{\sqrt{h_{k-1}(d_{k-1} - y_{k-1}')} + \sqrt{h_k(y_{k+1}' - d_k)}}, \qquad (6.114)$$

$$k = 2, \cdots, n-1.$$

If values for the first derivative at x_1 and x_n satisfying

$$y_1' < d_1, \qquad y_n' > d_{n-1}, \qquad (6.115)$$

are given, then this results in a system of $n-2$ nonlinear equations in $n-2$ unknowns, $y_2', \cdots, y_{n-1}'$. It can be shown ([47]) that there exists exactly one solution of (6.114) satisfying (6.103), and hence the desired rational spline interpolant is uniquely determined.

The system (6.114) can again be solved by a Jacobi or a Gauss-Seidel iterative method, although the convergence of these methods has not yet been explicitly proven. We have implemented Gauss-Seidel, i. e.,

$$(y_k')^{(i+1)} = \frac{\sqrt{h_{k-1}(d_{k-1} - (y_{k-1}')^{(i+1)})}\, d_k + \sqrt{h_k((y_{k+1}')^{(i)} - d_k)}\, d_{k-1}}{\sqrt{h_{k-1}(d_{k-1} - (y_{k-1}')^{(i+1)})} + \sqrt{h_k((y_{k+1}')^{(i)} - d_k)}},$$

$$(6.116)$$

$k = 2, \cdots, n - 1; \quad i = 0, 1, 2, \cdots,$ in the subroutine RCONC2 (Figs. 6.61 and 6.62). As starting values, we use

$$(y_k')^{(0)} = \frac{1}{2}(d_{k-1} + d_k), \qquad k = 2, \cdots, n - 1 \qquad (6.117)$$

and for the end slopes, we set

$$(y_1')^{(i+1)} = (y_1')^{(i)} = y_1' \quad \text{and} \quad (y_n')^{(i+1)} = (y_n')^{(i)} = y_n', \qquad (6.118)$$

assuming of course that (6.115) holds. The convergence test is described in the program description. We did not encounter any convergence problems in the examples we computed. With the stopping criterium as recommended in RCONC2, the number of iterations needed in our examples was between 5 and 15.

Because of (6.103), attention must be paid, when calculating examples, to the conditions, $y_1' < d_1$ and $y_n' > d_{n-1}$, in the case of convex data. In Fig. 6.63a, $y_1' = -y_n' = 10$ were chosen. The end conditions seem to strongly influence the curve near the boundary, and hence they must be chosen with some care. For nonsymmetric data, as in Fig. 6.63b, this is even more important.

If instead of assumption (6.102) we have for example,

$$d_1 < \cdots < d_j, \qquad d_j > d_{j+1} > \cdots > d_{n-1},$$

then the corresponding C^2 segments may be joined to be once continuously differentiable by, e.g., setting $y_{j+1}' = \frac{1}{2}(d_j + d_{j+1})$. It is also possible to proceed in an adaptive manner, as was done in the second section of this chapter, by using cubic segments on intervals requiring an inflection point.

6.9.   Rational Histosplines

Suppose that we are given abscissas $x_1 < \cdots < x_n$ and rectangles with bases x_k to x_{k+1} and heights $\tilde{y}_k$, $k = 1, \cdots, n - 1$. Further, suppose we are given function values $\tilde{y}_0$ and $\tilde{y}_n$ at x_1 and x_n. As before, a histospline requires a certain degree of differentiability, the area conditions (5.23), and the boundary conditions (5.24).

Now in order to construct a *once continuously differentiable rational histospline*, we could proceed in analogy with the quadratic form of Chapter 3. For example, if we would like to have a single control parameter, then we might think of starting with

$$\tilde{s}_k = \tilde{A}_k + \tilde{B}_k t + \frac{\tilde{C}_k t^2}{1 + p_k t}, \qquad t = \frac{x - x_k}{h_k}. \qquad (6.119)$$

```
      SUBROUTINE RCONC2(N,X,Y,EPS,ITMAX,Y1,R,ICASE,IT,IFLAG)
      DIMENSION X(N),Y(N),Y1(N),R(N)
      IFLAG=0
      IF(N.LT.3) THEN
          IFLAG=1
          RETURN
      END IF
      IT=0
      ICASE=1
      CC=1.
      N1=N-1
      DO 20 K=1,N1
          K1=K+1
          H2=CC*(Y(K1)-Y(K))/(X(K1)-X(K))
          IF(K.EQ.1) GOTO 10
          IF(K.EQ.2.AND.H1.GT.H2) THEN
              CC=-1.
              ICASE=2
              H1=CC*H1
              H2=CC*H2
          END IF
          IF(H1.GE.H2) THEN
              IFLAG=19
              RETURN
          END IF
          Y1(K)=(H1+H2)/2.
10        R(K)=H2
          H1=H2
20    CONTINUE
      R(1)=CC*R(1)
      Y1(1)=CC*Y1(1)
      Y1(N)=CC*Y1(N)
      IF(Y1(1).GE.R(1).OR.Y1(N).LE.R(N1)) THEN
          IFLAG=21
          RETURN
      END IF
30    IT=IT+1
      IF(IT.GT.ITMAX) THEN
          IFLAG=13
          RETURN
      END IF
      H=Y1(1)
      RNORM1=0.
      RNORM2=0.
      DO 50 K=1,N1
          KP1=K+1
          H2=X(KP1)-X(K)
          IF(K.EQ.1) GOTO 40
          KM1=K-1
          H3=SQRT(H1*(R(KM1)-H))
          H4=SQRT(H2*(Y1(KP1)-R(K)))
          H=(H3*R(K)+H4*R(KM1))/(H3+H4)
          RNORM1=RNORM1+ABS(Y1(K))
          RNORM2=RNORM2+ABS(Y1(K)-H)
          Y1(K)=H
40        H1=H2
```

(*cont.*)

```
50   CONTINUE
     IF(RNORM2.LT.EPS*RNORM1) GOTO 60
     GOTO 30
60   Y1(1)=CC*Y1(1)
     DO 70 K=1,N1
        K1=K+1
        Y1(K1)=CC*Y1(K1)
        H1=CC*R(K)
        H=(Y1(K1)-H1)/(H1-Y1(K))
        R(K)=1.+H+1./H
70   CONTINUE
     RETURN
     END
```

Figure 6.61. Program listing of RCONC2.

Calling sequence:

CALL RCONC2(N,X,Y,EPS,ITMAX,Y1,R,ICASE,IT,IFLAG)

Purpose:

For the form:

$$s_k(x) = y_k u + y_{k+1} t + tu \frac{(t-u)\Delta y_k - \Delta x_k (ty'_{k+1} - uy'_k)}{1 + (r_k - 3)tu},$$

where $t = (x - x_k)/\Delta x_k$, $u = 1 - t$, $k = 1, 2, \cdots, n - 1$, and $r_k = 1 + z_k + 1/z_k$ with $z_k = (y'_{k+1} - d_k)/(d_k - y'_k)$, the y'_k with $d_{k-1} < y'_k < d_k$, $k = 2, \cdots, n - 1$, are calculated so as to produce a convex, twice continuously differentiable spline interpolant for given convex data satisfying $d_{k+1} > d_k$, $k = 1, \cdots, n - 2$ and given values $y'_1 < d_1$ and $y'_n > d_{n-1}$. Alternatively, a concave function can also be calculated for $d_{k+1} < d_k$, $k = 1, \cdots, n - 2$, and given $y'_1 > d_1$ and $y'_n < d_{n-1}$.

Description of the parameters:

N,X as in RATSP1 (N$\geq$ 3 required).

Y   ARRAY(N): Vector containing the y-values. The y_k must satisfy either the convexity or the concavity condition.

EPS   Value used for testing convergence of the Gauss-Seidel iteration. If two consecutive iterates, $(y')^{(i)}$ and $(y')^{(i+1)}$, satisfy $\|(y')^{(i+1)} - (y')^{(i)}\|_1 \leq \text{EPS2} \cdot \|(y')^{(i+1)}\|_1$, then the iteration is halted. Recommendation: EPS2$= 10^{-t/2+1}$ (t : number of available decimal digits).

ITMAX   Prescribed value for the maximum number of iterations.

Y1   ARRAY(N): Input: Values for y_1' and y_n' must be given
          in Y1(1) and Y1(N), respectively. Output:
          Upon execution with IFLAG=0 contains the
          values y_k', $k = 1, \cdots, n$, of the first derivatives.

R   ARRAY(N): Output: Upon execution with IFLAG=0
          contains the computed values for the
          parameters r_k, $k = 1, \cdots, n - 1$.

ICASE   =1:   Convex initial data.

          =2:   Concave initial data.

IT   Number of iterations actually used.

IFLAG   =0:   Normal execution.

          =1:   N$\geq$ 3 required.

          =13:   ITMAX number of iterations exceeded.

          =19:   Convexity or concavity condition violated.

          =21: $y_1' < d_1$ and $y_n' > d_{n-1}$ violated in the case
                    of convexity, or $y_1' > d_1$ and $y_n' < d_{n-1}$ in the
                    case of concavity.

Figure 6.62. Description of RCONC2.

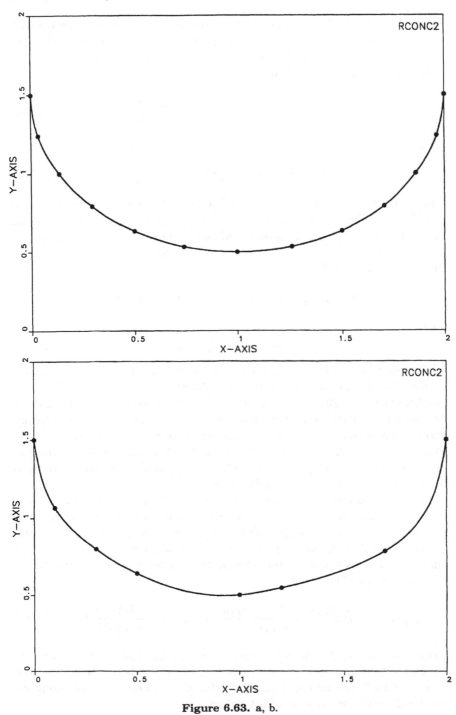

Figure 6.63. a, b.

However, the form (6.119) does not, unfortunately, come from a form such
as (6.55) (like quadratics are the derivatives of cubics), and hence the considerations of the next to last section of Chapter 5 do not apply.

Hence, because of (6.56), we set ([123])

$$\tilde{s}_k = \tilde{A}_k + \tilde{B}_k t + \tilde{C}_k \frac{3t^2 + 2p_k t^3}{(1 + p_k t)^2}. \qquad (6.120)$$

Here,

$$\tilde{A}_k = \frac{1}{h_k} B_k, \quad \tilde{B}_k = \frac{2}{h_k} C_k, \quad \tilde{C}_k = \frac{1}{h_k} D_k, \qquad (6.121)$$

where B_k, C_k, and D_k are those of (6.58). Notice that now s_k has a
double pole. The coefficients B_k, C_k, and D_k can easily be obtained from
RATSAK by computing a rational spline interpolant s, of the form (6.55),
for (x_k, y_k), $k = 1, \cdots, n$, and with end conditions $s'(x_1) = y_1'$, $s'(x_n) = y_n'$,
and where, as in (5.25),

$$y_1 = 0, \quad y_{k+1} = y_k + h_k \tilde{y}_k, \quad k = 1, \cdots, n-1, \qquad (6.122)$$

and

$$y_1' = \tilde{y}_0, \quad y_n' = \tilde{y}_n. \qquad (6.123)$$

The subroutine RHIST1 (Figs. 6.64 and 6.65) first calculates the y_k according to (6.122), sets (6.123), calls RATSAK, and then calculates the
coefficients of (6.120) according to (6.121). Evaluation of (6.120) can be
carried out by RH1VAL (Figs. 6.66 and 6.67). Since, as we have already
shown, the forms (6.55) tend to quadratics as $p_k \to -1$ or $p_k \to \infty$, it
follows that (6.120) tends to a linear for these limiting cases. From the
examples, Figs. 6.68a ($p_k = -.9$) and 6.68b ($p_k = 9$), it is clear that these
limiting cases are not particularly useful.

Figures 6.69a and b show the results for $p_k = 0$ (quadratics) and $p_k = 1$.
It was not possible to find values for the p_k that resulted in a curve that
reproduced the monotonicity, unimodality, and positivity of the histogram.

Now recall that the form (6.63) also tends to a quadratic as $p_k \to -1$ but
that it tends to a linear on the interior of the interval as $p_k \to \infty$. Hence,
if because of (6.67), we set

$$\tilde{s}_k(x) = \tilde{A}_k + \tilde{B}_k \frac{2p_k u^3 - 3(1 + p_k)u^2}{(1 + p_k t)^2} + \tilde{C}_k \frac{2q_k t^3 - 3(1 + q_k)t^2}{(1 + q_k u)^2}, \qquad (6.124)$$

(which has two double poles), then we expect that these tend to constants
(linears with slope zero) on the interiors of the intervals $[x_k, x_{k+1}]$, i.e., the
given rectangle can be approximated arbitrarily closely by a once continuously differentiable area-matching spline.

The coefficients of (6.124) are obtained from

$$\tilde{A}_k = \frac{1}{h_k}(B_k - A_k), \quad \tilde{B}_k = \frac{1}{h_k}C_k, \quad \tilde{C}_k = -\frac{1}{h_k}D_k, \qquad (6.125)$$

where the A_k, B_k, C_k, and D_k, $k = 1, \cdots, n - 1$, are given by (6.69), and the s_k are computed from (6.63) with data (6.122) and (6.123). The control parameters $p_k, q_k > -1$ determine, as before, two real poles outside $[x_k, x_{k+1}]$. In this case, however, they are both double poles.

The corresponding subroutine RHIST2 (Figs. 6.70 and 6.71) is exactly analogous to RHIST1. It makes use of RATSP1 instead of RATSAK. RH2VAL (Figs. 6.72 and 6.73) can be used for evaluation.

Figures 6.74a and b show the results for $p_k = q_k = -.9$ and $p_k = q_k = 9$ for the same examples as before. They illustrate how large values of the parameters may be used to good effect. Figure 6.75a has $p_k = q_k = 1$. The resulting curve does not yet have the desired shape-preserving properties. However, in Fig. 6.75b, also with $p_k = q_k = 1$ except for $p_4 = p_5 = p_7 = p_{13} = p_{14} = 10$ $(p_k = q_k)$, we have an example of a parameter set that does guarantee the monotonicity, unimodality, and positivity. This is essentially due to the fact that for large values of the parameters, the curve is approximately constant on the interior of the corresponding intervals. For this reason RHIST2, is almost always to be preferred to RHIST1.

We remark that it would also have been possible to proceed analogously with the derivative of the form (6.80) and even with other nonlinear forms ([110]), e. g., with the exponential spline of the next chapter.

```
      SUBROUTINE RHIST1(N,X,Y,P,A,B,C,D,E,EPS,IFLAG)
      DIMENSION X(N),Y(N),P(N),A(N),B(N),C(N),D(N),E(N)
      IFLAG=0
      IF(N.LT.3) THEN
          IFLAG=1
          RETURN
      END IF
      N1=N-1
      E(1)=0
      DO 10 K=1,N1
          K1=K+1
          E(K1)=E(K)+(X(K1)-X(K))*Y(K)
10    CONTINUE
      CALL RATSAK(N,X,E,P,A,B,C,D,EPS,IFLAG)
      IF(IFLAG.NE.0) RETURN
      DO 20 K=1,N1
          H=1./(X(K+1)-X(K))
          A(K)=H*B(K)
          B(K)=2.*H*C(K)
          C(K)=H*D(K)
20    CONTINUE
      RETURN
      END
```

Figure 6.64. Program listing of RHIST1.

Calling sequence:

CALL RHIST1(N,X,Y,P,A,B,C,D,E,EPS,IFLAG)

Purpose:

Given abscissas $x_1 < x_2 < \cdots < x_n$, an initial value y_0, an end value y_n, as well as rectangle heights y_k and parameters $p_k > -1$, $k = 1, \cdots, n-1$, RHIST1 determines coefficients A_k, B_k and C_k, $k = 1, \cdots, n-1$, of a rational histospline of the following form (a single pole):

$$s_k(x) = A_k + B_k t + C_k \frac{3t^2 + 2p_k t^3}{(1 + p_k t)^2},$$

where $t = (x - x_k)/h_k$, $k = 1, 2, \cdots, n-1$. Values for y_0 and y_n must be given in B(1) and B(N), respectively. The spline function goes through both endpoints, (x_1, y_0) and (x_n, y_n), and has the property that its area above the interval $[x_k, x_{k+1}]$, $k = 1, 2, \cdots, n-1$, equals that of the corresponding rectangle, i.e., equals $h_k y_k$.

Description of the parameters:

N	Number of given points. $N \geq 3$ is required.
X	ARRAY(N): Upon calling must contain the abscissa values x_k, $k = 1, \cdots, n$, with $x_1 < x_2 < \cdots < x_n$.
Y	ARRAY(N): Y(1),Y(2),$\cdots$,Y(N$-$1) must contain the rectangle heights y_k, $k = 1, \cdots, n-1$.
P	ARRAY(N): Upon calling must contain the parameters p_k, with $p_k > -1$, $k = 1, \cdots, n-1$.
A,B,C	ARRAY(N): Upon execution with IFLAG=0 contain the desired spline coefficients, K= $1, 2, \cdots$,N$-$1.
D,E	ARRAY(N): Work space.
EPS	Value for the accuracy test (see TRIDIU).
IFLAG	=0: Normal execution.
	=1: $N \geq 3$ required.
	=2: Error in solving the linear system (TRIDIU).
	=12: $p_k > -1$ violated for some k (RATSAK).

Required subroutines: RATSAK, TRIDIU.

Figure 6.65. Description of RHIST1.

```
FUNCTION RH1VAL(N,X,P,A,B,C,V,IFLAG)
DIMENSION X(N),P(N),A(N),B(N),C(N)
DATA I/1/
IFLAG=0
IF(N.LT.2) THEN
    IFLAG=1
    RETURN
END IF
CALL INTONE(X,N,V,I,IFLAG)
IF(IFLAG.NE.0) RETURN
T=(V-X(I))/(X(I+1)-X(I))
H=P(I)*T+1.
RH1VAL=A(I)+T*(B(I)+C(I)*T*(3.+2.*P(I)*T)/(H*H))
RETURN
END
```

Figure 6.66. Program listing of RH1VAL.

FUNCTION RH1VAL(N,X,P,A,B,C,V,IFLAG)

Purpose:
RH1VAL is a FUNCTION subroutine for the calculation of a function value at a point V∈ [X(1),X(N)] of a rational spline interpolant of the following form:

$$s_k(x) = A_k + B_k t + C_k \frac{3t^2 + 2p_k t^3}{(1 + p_k t)^2},$$

where $t = (x - x_k)/h_k$, $k = 1, 2, \cdots, n - 1$.

Description of the parameters:

N,X,P,A,B,C as in RHIST1.
V   Value of the point at which the spline interpolant is to
           be evaluated.
IFLAG   =0:   Normal execution.
        =1: $N \geq 2$ required.
        =3:   Error in interval determination (INTONE).

Required subroutine: INTONE.

Remark: The statement 'DATA I/1/' has the effect that I is set to 1 at the first call to RH1VAL.

Figure 6.67. Description of RH1VAL.

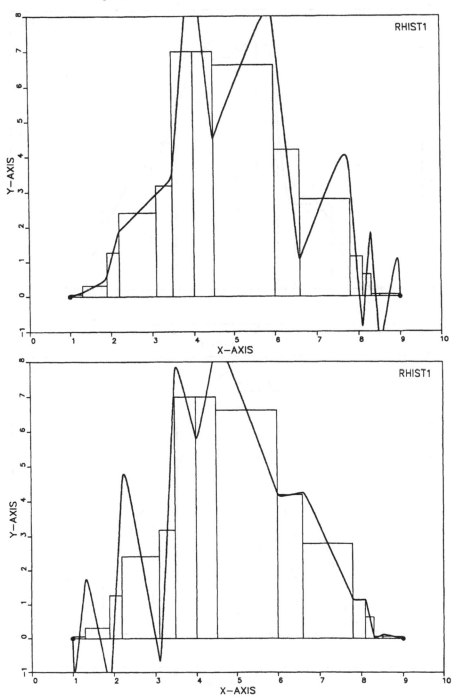

Figure 6.68. a, b.

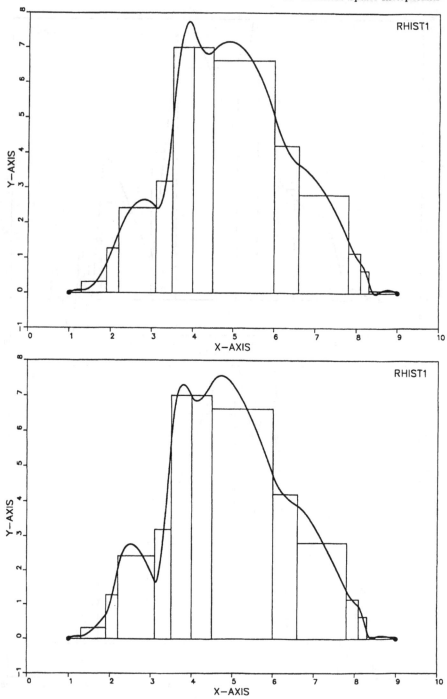

Figure 6.69. a, b.

```
      SUBROUTINE RHIST2(N,X,Y,P,Q,A,B,C,D,Y1,E,EPS,IFLAG)
      DIMENSION X(N),Y(N),P(N),Q(N),A(N),B(N),C(N),D(N),
     &          Y1(N),E(N)
      IFLAG=0
      IF(N.LT.3) THEN
          IFLAG=1
          RETURN
      END IF
      N1=N-1
      E(1)=0
      DO 10 K=1,N1
          K1=K+1
          E(K1)=E(K)+(X(K1)-X(K))*Y(K)
10    CONTINUE
      CALL RATSP1(N,X,E,P,Q,A,B,C,D,Y1,EPS,IFLAG)
      IF(IFLAG.NE.0) RETURN
      DO 20 K=1,N1
          H=1./(X(K+1)-X(K))
          A(K)=H*(B(K)-A(K))
          B(K)=H*C(K)
          C(K)=-H*D(K)
20    CONTINUE
      RETURN
      END
```

Figure 6.70. Program listing of RHIST2.

Calling sequence:

CALL RHIST2(N,X,Y,P,Q,A,B,C,D,Y1,E,EPS,IFLAG)

Purpose:

Given abscissas $x_1 < x_2 < \cdots < x_n$, an initial value y_0, an end value y_n, as well as rectangle heights y_k and parameters $p_k, q_k > -1$, $k = 1, \cdots, n - 1$, RHIST2 determines coefficients A_k, B_k, and C_k, $k = 1, \cdots, n - 1$, of a rational histospline of the following form (two poles):

$$s_k(x) = A_k + B_k \frac{2p_k u^3 - 3u^2(1 + p_k)}{(p_k t + 1)^2} + C_k \frac{2q_k t^3 - 3t^2(1 + q_k)}{(q_k u + 1)^2},$$

where $t = (x - x_k)/h_k$ and $u = 1 - t$, $k = 1, 2, \cdots, n - 1$. The spline function goes through both endpoints, (x_1, y_0) and (x_n, y_n), and has the property that its area above the interval $[x_k, x_{k+1}]$, $k = 1, 2, \cdots, n - 1$, equals that of the corresponding rectangle, i.e., equals $h_k y_k$. Values for y_0 and y_n must be given in Y1(1) and Y1(N), respectively. $N \geq 3$ is required.

Description of the parameters:

N,X,Y,A,B,C,D,E,EPS as in RHIST1.

P,Q   ARRAY(N): Upon calling must contain the parameters
          $p_k > -1$ and $q_k > -1$ for $k = 1, \cdots, n - 1$.
Y1   ARRAY(N): Work space.
IFLAG =0:   Normal execution.
     =1: $N \geq 3$ required.
     =2:   Error in solving the linear system (TRIDIU).
     =12: $p_k > -1$ or $q_k > -1$ violated for some k
          (RATSP1).

Required subroutines: RATSP1, TRIDIU.

Figure 6.71. Description of RHIST2.

```
FUNCTION RH2VAL(N,X,P,Q,A,B,C,V,IFLAG)
DIMENSION X(N),P(N),Q(N),A(N),B(N),C(N)
DATA I/1/
IFLAG=0
IF(N.LT.2) THEN
     IFLAG=1
     RETURN
END IF
CALL INTONE(X,N,V,I,IFLAG)
IF(IFLAG.NE.0) RETURN
T=(V-X(I))/(X(I+1)-X(I))
U=1.-T
H1=P(I)*T+1.
H2=Q(I)*U+1.
RH2VAL=A(I)+B(I)*U*U*(2.*P(I)*U-3.*(1.+P(I)))/(H1*H1)+
&        C(I)*T*T*(2.*Q(I)*T-3.*(1.+Q(I)))/(H2*H2)
RETURN
END
```

Figure 6.72. Program listing of RH2VAL.

FUNCTION RH2VAL(N,X,P,Q,A,B,C,V,IFLAG)

Purpose:
RH2VAL is a FUNCTION subroutine for the calculation of a function value at a point $V \in [X(1),X(N)]$ of a rational spline interpolant of the following form:

$$s_k(x) = A_k + B_k \frac{2p_k u^3 - 3u^2(1 + p_k)}{(p_k t + 1)^2} + C_k \frac{2q_k t^3 - 3t^2(1 + q_k)}{(q_k u + 1)^2},$$

where $t = (x - x_k)/h_k$ and $u = 1 - t$, $k = 1, 2, \cdots, n - 1$.

Description of the parameters:

N,X,P,Q,A,B,C as in RHIST2. V,IFLAG as in RH1VAL.

Required subroutine: INTONE.

Remark: The statement 'DATA I/1/' has the effect that I is set to 1 at the first call to RH2VAL.

Figure 6.73. Description of RH2VAL.

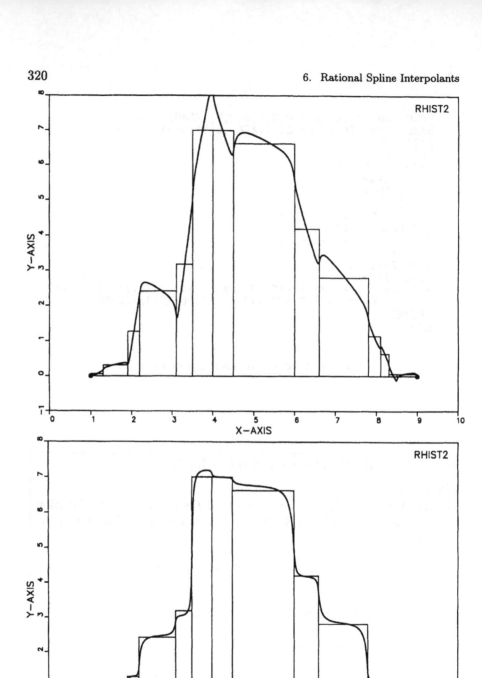

Figure 6.74. a, b.

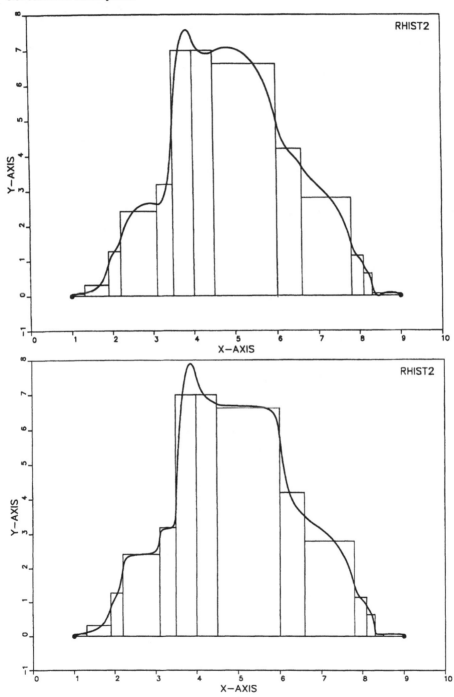

Figure 6.75. a, b.

7

Exponential Spline Interpolants

7.1. First Derivatives as Unknowns

Historically, exponential spline interpolants ([148, 149, 162]) were considered before rational splines with parameter-controlled poles were. Their purpose is to be able to achieve a visually pleasing interpolating curve through the appropriate choice of certain *tension parameters*, $p_k \geq 0$, which have the effect of "tightening" the curve on the corresponding intervals. The special case of all the $p_k = 0$ results in ordinary cubic spline interpolation. Since they make use of exponential functions, their computational expense is considerably greater than that of rational splines whose complexity, as we saw in the last chapter, is already quite high. For this reason, exponential splines are not usually considered to be a competitive alternative.

Various forms have been considered in the literature. For example, the authors of [10, 11, 111, 118, 119, 120, 172] set

$$s_k(x) = A_k t + B_k u + \frac{C_k}{p_k^2}\left[\frac{\sinh(p_k t)}{\sinh(p_k)} - t\right] + \frac{D_k}{p_k^2}\left[\frac{\sinh(p_k u)}{\sinh(p_k)} - u\right], \quad (7.1)$$

which has the advantage of simplifying the calculations involved in using the second derivatives as unknowns. In [164], the form,

$$s_k(x) \;=\; A_k + B_k(x - x_k) + C_k \frac{2}{p_k^2}\left[\cosh(p_k(x - x_k)) - 1\right]$$

$$\tag{7.2}$$

$$+ D_k \frac{6}{p_k^3}\left[\sinh(p_k(x - x_k)) - p_k(x - x_k)\right],$$

is suggested, but so far this has not gained further acceptance.
We will stick to the original form ([148, 149, 162]) of

$$s_k(x) = A_k u + B_k t + C_k \phi_k(u) + D_k \phi_k(t), \tag{7.3}$$

where

$$\phi_k(t) = \frac{\sinh(p_k t) - t \sinh(p_k)}{\sinh(p_k) - p_k}, \tag{7.4}$$

but also leave open the possibility of other choices for the ϕ_k.
    Since by definition,

$$\sinh(x) = \frac{e^x - e^{-x}}{2} = x + \frac{x^3}{3!} + \frac{x^5}{5!} + \frac{x^7}{7!} + \frac{x^9}{9!} + \cdots \tag{7.5}$$

and

$$\cosh(x) = \frac{e^x + e^{-x}}{2} = 1 + \frac{x^2}{2!} + \frac{x^4}{4!} + \frac{x^6}{6!} + \frac{x^8}{8!} + \cdots, \tag{7.6}$$

all of the preceding forms are equivalent in the sense that they give cubic
polynomials as $p_k \to 0$ and linears as $p_k \to \infty$. We verify this fact for (7.3).
Substituting the power series expansion (7.5) into (7.4), we see that

$$\phi_k(t) = \frac{\left[p_k t + \dfrac{(p_k t)^3}{3!} + \dfrac{(p_k t)^5}{5!} + \cdots\right] - t\left[p_k + \dfrac{p_k^3}{3!} + \dfrac{p_k^5}{5!} + \cdots\right]}{\left[p_k + \dfrac{p_k^3}{3!} + \dfrac{p_k^5}{5!} + \cdots\right] - p_k}.$$

This expression may be rewritten in either of the forms,

$$\phi_k(t) = \frac{\dfrac{1}{3!}(t^3 - t) + \dfrac{p_k^2}{5!}(t^5 - t) + \dfrac{p_k^4}{7!}(t^7 - t) + \cdots}{\dfrac{1}{3!} + \dfrac{p_k^2}{5!} + \dfrac{p_k^4}{7!} + \cdots} \tag{7.7}$$

or

$$\phi_k(t) = -t + \frac{t^3\left[\dfrac{1}{3!} + \dfrac{(p_k t)^2}{5!} + \dfrac{(p_k t)^4}{7!} + \cdots\right]}{\dfrac{1}{3!} + \dfrac{p_k^2}{5!} + \dfrac{p_k^4}{7!} + \cdots}. \tag{7.8}$$

From (7.7), we see that

$$\lim_{p_k \to 0} \phi_k(t) = t^3 - t \qquad \text{for } 0 \le t \le 1, \tag{7.9}$$

and from (7.8) that

$$\lim_{p_k \to \infty} \phi_k(t) = -t \qquad \text{for } 0 \le t < 1, \tag{7.10}$$

and

$$\lim_{p_k \to \infty} \phi_k(t) = 0 \qquad \text{for } t = 1. \tag{7.11}$$

Hence, as $p_k \to 0$, we get

$$s_k(x) = A_k u + B_k t + C_k(u^3 - u) + D_k(t^3 - t), \tag{7.12}$$

a variant of (4.2). As $p_k \to \infty$, we obtain

$$\lim_{p_k \to \infty} s_k(x) = \begin{cases} (A_k - C_k)u + (B_k - D_k)t & \text{if } 0 \le t < 1 \\ A_k u + B_k t & \text{if } t = 1 \end{cases}. \tag{7.13}$$

In order to guarantee continuity, we must have

$$\lim_{p_k \to \infty} C_k = \lim_{p_k \to \infty} D_k = 0. \tag{7.14}$$

We will later show that this is true in the case that the y_k' (and, respectively, the y_k'') remain bounded.

Since the hyperbolic sine, (7.5), is an odd function, the form (7.3) remains the same when p_k is replaced by $-p_k$. Hence, we may restrict ourselves to

$$p_k \ge 0, \qquad k = 1, \cdots, n - 1. \tag{7.15}$$

The properties of the form (7.3) with (7.4), described previously, correspond to those of (6.80) with the r_k restricted to be $r_k \ge 3$.

In what follows, instead of just (7.4), we will allow the ϕ_k to be essentially arbitrary functions. Since (7.3) must interpolate, we will require that

$$\phi_k(0) = \phi_k(1) = 0, \qquad k = 1, \cdots, n - 1. \tag{7.16}$$

For the sake of brevity, we set

$$\phi_0' = \left. \frac{d\phi_k}{dt} \right|_{t=0}, \qquad \phi_1' = \left. \frac{d\phi_k}{dt} \right|_{t=1}. \tag{7.17}$$

We may calculate

$$h_k s_k'(x) = B_k - A_k - C_k \frac{d\phi_k(u)}{dt} + D_k \frac{d\phi_k(t)}{dt}. \tag{7.18}$$

Further, the interpolation conditions imply that

$$A_k = y_k, \qquad B_k = y_{k+1}. \tag{7.19}$$

Thus, we have

$$
\begin{aligned}
h_k y_k' &= \Delta y_k - C_k \phi_1' + D_k \phi_0', \\[1em]
h_k y_{k+1}' &= \Delta y_k - C_k \phi_0' + D_k \phi_1'.
\end{aligned}
\tag{7.20}
$$

If

$$\phi_0' \neq \pm \phi_1', \tag{7.21}$$

then these may be solved for the remaining coefficients,

$$C_k = \frac{1}{(\phi_1')^2 - (\phi_0')^2} \left[(\phi_1' - \phi_0') \, \Delta y_k + h_k \left(\phi_0' y_{k+1}' - \phi_1' y_k' \right) \right], \tag{7.22}$$

$$D_k = \frac{1}{(\phi_1')^2 - (\phi_0')^2} \left[-(\phi_1' - \phi_0') \Delta y_k - h_k (\phi_0' y_k' - \phi_1' y_{k+1}') \right],$$

in terms of the y_k'.

Since

$$h_k^2 s_k''(x) = C_k \frac{d^2 \phi_k(u)}{dt^2} + D_k \frac{d^2 \phi_k(t)}{dt^2}, \tag{7.23}$$

the C^2 continuity conditions $s_{k-1}''(x_k) = s_k''(x_k)$ yield

$$
\frac{\alpha_{k-1}}{h_{k-1}} y_{k-1}' + \left(\frac{\beta_{k-1}}{h_{k-1}} + \frac{\beta_k}{h_k} \right) y_k' + \frac{\alpha_k}{h_k} y_{k+1}'
$$

$$
\tag{7.24}
$$

$$
= \gamma_{k-1} \frac{d_{k-1}}{h_{k-1}} + \gamma_k \frac{d_k}{h_k}, \qquad k = 2, \cdots, n-1,
$$

where we have set

$$\phi_0'' = \frac{d^2 \phi_k}{dt^2} \bigg|_{t=0}, \qquad \phi_1'' = \frac{d^2 \phi_k}{dt^2} \bigg|_{t=1}, \tag{7.25}$$

$$\alpha_k = -\frac{\phi_0' \phi_1'' + \phi_1' \phi_0''}{(\phi_1')^2 - (\phi_0')^2}, \tag{7.26}$$

$$\beta_k = \frac{\phi_0' \phi_0'' + \phi_1' \phi_1''}{(\phi_1')^2 - (\phi_0')^2}, \tag{7.27}$$

and

$$\gamma_k = \frac{\phi_1'' - \phi_0''}{\phi_1' + \phi_0'}. \tag{7.28}$$

For the special case of (7.4), we have

$$\frac{d\phi_k}{dt} = \frac{p_k \cosh(p_k t) - \sinh(p_k)}{\sinh(p_k) - p_k} \tag{7.29}$$

and

$$\frac{d^2\phi_k}{dt^2} = \frac{p_k^2 \sinh(p_k t)}{\sinh(p_k) - p_k}, \tag{7.30}$$

from which it follows that

$$\phi_0' = -1, \quad \phi_1' = b_k := \frac{p_k \cosh(p_k) - \sinh(p_k)}{\sinh(p_k) - p_k},$$

$$\tag{7.31}$$

$$\phi_0'' = 0, \quad \phi_1'' = a_k := \frac{p_k^2 \sinh(p_k)}{\sinh(p_k) - p_k}.$$

The coefficients (7.26) through (7.28) in the linear system (7.24) then simplify to

$$\alpha_k = \frac{a_k}{b_k^2 - 1}, \quad \beta_k = b_k \alpha_k, \quad \gamma_k = (b_k + 1)\alpha_k = \frac{a_k}{b_k - 1}. \tag{7.32}$$

We now show that for given values of y_1' and y_n', this system has a symmetric strictly diagonally dominant coefficient matrix. For this purpose, we consider power series expansions of the a_k and the b_k of (7.31). These are

$$a_k = \frac{p_k^2 \left(p_k + \dfrac{p_k^3}{3!} + \dfrac{p_k^5}{5!} + \dfrac{p_k^7}{7!} + \dfrac{p_k^9}{9!} + \cdots \right)}{\dfrac{p_k^3}{3!} + \dfrac{p_k^5}{5!} + \dfrac{p_k^7}{7!} + \dfrac{p_k^9}{9!} + \cdots}$$

$$\tag{7.33}$$

$$= \frac{1 + \dfrac{p_k^2}{6} + \dfrac{p_k^4}{120} + \dfrac{p_k^6}{5040} + \cdots}{\dfrac{1}{6} + \dfrac{p_k^2}{120} + \dfrac{p_k^4}{5040} + \dfrac{p_k^6}{362880} + \cdots}$$

and

$$b_k = \frac{p_k \left(1 + \dfrac{p_k^2}{2!} + \dfrac{p_k^4}{4!} + \dfrac{p_k^6}{6!} + \cdots \right) - \left(p_k + \dfrac{p_k^3}{3!} + \dfrac{p_k^5}{5!} + \dfrac{p_k^7}{7!} + \cdots \right)}{\dfrac{p_k^3}{3!} + \dfrac{p_k^5}{5!} + \dfrac{p_k^7}{7!} + \dfrac{p_k^9}{9!} + \cdots}$$

$$\text{(7.34)}$$

$$= \frac{\frac{1}{3} + \frac{1}{30}p_k^2 + \frac{1}{840}p_k^4 + \frac{1}{45360}p_k^6 + \cdots}{\frac{1}{6} + \frac{1}{120}p_k^2 + \frac{1}{5040}p_k^4 + \frac{1}{362880}p_k^6 + \cdots}.$$

Clearly,

$$a_k \geq 6 \quad \text{and} \quad b_k \geq 2, \tag{7.35}$$

and hence, by (7.32),

$$\alpha_k > 0 \quad \text{and} \quad \beta_k \geq 2\alpha_k, \tag{7.36}$$

from which the strict diagonal dominance follows. Further, we see that

$$\lim_{p_k \to 0} \alpha_k = 2, \quad \lim_{p_k \to 0} \beta_k = 4, \quad \lim_{p_k \to 0} \gamma_k = 6, \tag{7.37}$$

which implies that (7.24) becomes the linear system (4.7) of cubic spline interpolants as $p_k \to 0$, $k = 1, \cdots, n-1$.

The coefficients C_k and D_k, given by (7.22), may in the special case of (7.4) be expressed as

$$C_k = \frac{\Delta y_k}{b_k - 1} - \frac{h_k}{b_k^2 - 1}(y'_{k+1} + b_k y'_k),$$

$$\tag{7.38}$$

$$D_k = -\frac{\Delta y_k}{b_k - 1} + \frac{h_k}{b_k^2 - 1}(y'_k + b_k y'_{k+1}).$$

Since

$$\lim_{p_k \to \infty} a_k = \lim_{p_k \to \infty} b_k = \infty, \tag{7.39}$$

(7.14) follows in the case that the solutions y'_k, $k = 2, \cdots, n-1$, of (7.24) remain bounded. If all the p_k, $k = 1, \cdots, n-1$, go uniformly to infinity, then (7.24) reduces to (6.72). If the p_k go to infinity at differing rates or if some of them remain fixed, then further investigation is required. Such investigations are carried out in [111] using the second derivatives as unknowns.

The formation and solution of the linear system (7.24) as well as the subsequent evaluation of the coefficients of the exponential spline interpolant according to (7.19) and (7.38) require the calculation of the a_k and b_k of (7.31) to be as numerically accurate as possible, since as $p_k \to 0$, these expressions become numerically unstable. For $p_k \leq$ BIG, the subroutine ABKVAL (Figs. 7.1 and 7.2) breaks this computation down into two cases:

```
SUBROUTINE ABKVAL(P,AK,BK,BIG,IFLAG)
IFLAG=0
IF(P.LT.0..OR.P.GT.BIG) THEN
    IFLAG=11
    RETURN
END IF
IF (P.GT..1) THEN
    SINHP=SINH(P)
    H=1./(SINHP-P)
    AK=H*P*P*SINHP
    BK=H*(P*COSH(P)-SINHP)
ELSE
    P2=P*P
    ABN=1./(((P2/362880.+1./5040.)*P2+1./120.)*P2+1./6.)
    AZ=((P2/5040.+1./120.)*P2+1./6.)*P2+1.
    BZ=((P2/45360.+1./840.)*P2+1./30.)*P2+1./3.
    AK=ABN*AZ
    BK=ABN*BZ
END IF
RETURN
END
```

Figure 7.1. Program listing of ABKVAL.

Calling sequence:

CALL ABKVAL(P,AK,BK,BIG,IFLAG)

Purpose:

For a given value P with $0 \leq P \leq BIG$, ABKVAL computes the values,

$$AK= \frac{P^2 \sinh(P)}{\sinh(P)-P} \text{ and } BK= \frac{P\cosh(P)-\sinh(P)}{\sinh(P)-P}.$$

Description of the parameters:

P	Input parameter for the calculation of AK and BK.
AK,BK	Calculated values.
BIG	The program tests whether P≤BIG. We recommend that BIG be set to M/2, where M is the largest decimal exponent available in the computer's floating point representation.
IFLAG	=0: Normal execution.
	=11: 0≤P≤BIG not satisfied.

Figure 7.2. Description of ABKVAL.

```
      SUBROUTINE EXPSP1(N,X,Y,P,A,B,C,D,EPS,BIG,IFLAG)
      DIMENSION X(N),Y(N),P(N),A(N),B(N),C(N),D(N)
      IFLAG=0
      IF(N.LT.3) THEN
         IFLAG=1
         RETURN
      END IF
      N1=N-1
      N2=N-2
      B1=B(1)
      BN=B(N)
      DO 20 K=1,N1
         KP1=K+1
         KM1=K-1
         H=1./(X(KP1)-X(K))
         CALL ABKVAL (P(K),A(K),B(K),BIG,IFLAG)
         IF (IFLAG.NE.0) RETURN
         D2=A(K)*(Y(KP1)-Y(K))*H*H/(B(K)-1.)
         A(K)=A(K)*H/(B(K)*B(K)-1.)
         C2=B(K)*A(K)
         IF(K.EQ.1) GOTO 10
         C(KM1)=C1+C2
         D(KM1)=D1+D2
         IF(K.EQ.2) D(KM1)=D(KM1)-A(KM1)*B1
         IF(K.EQ.N1) D(KM1)=D(KM1)-A(K)*BN
         A(KM1)=A(K)
10       C1=C2
         D1=D2
20    CONTINUE
      CALL TRIDIS(N2,C,A,D,EPS,IFLAG)
      IF(IFLAG.NE.0) RETURN
      DO 30 K=1,N2
         A(K+1)=D(K)
30    CONTINUE
      A(1)=B1
      A(N)=BN
      DO 40 K=1,N1
         K1=K+1
         H=(X(K1)-X(K))/(B(K)*B(K)-1.)
         H1=(Y(K1)-Y(K))/(B(K)-1.)
         D(K)=-H1+H*(A(K)+B(K)*A(K1))
         C(K)=H1-H*(A(K1)+B(K)*A(K))
         B(K)=Y(K1)
40    CONTINUE
      RETURN
      END
```

Figure 7.3. Program listing of EXPSP1.

Calling sequence:

CALL EXPSP1(N,X,Y,P,A,B,C,D,EPS,BIG,IFLAG)

Purpose:

EXPSP1 determines coefficients Y_k, B_k, C_k, and D_k, $k = 1, 2, \cdots, n-1$, of an exponential spline interpolant,

$$s_k(x) = y_k u + B_k t + C_k \phi_k(u) + D_k \phi_k(t),$$

where $t = (x - x_k)/(x_{k+1} - x_k)$, $u = 1 - t$ and $\phi_k(u) = (\sinh(p_k u) - u \sinh(p_k))/(\sinh(p_k) - p_k)$. Values for y_1' and y_n' must be given in B(1) and B(N), respectively.

Description of the parameters:

N	Number of given points. N$\geq$ 3 required.
X	ARRAY(N): Upon calling must contain the abscissa values x_k, $k = 1, \cdots, n$, with $x_1 < x_2 < \cdots < x_n$.
Y	ARRAY(N): Upon calling must contain the ordinate values y_k, $k = 1, \cdots, n$.
P	ARRAY(N): Upon calling must contain the factors p_k with $0 \leq p_k \leq$ BIG, $k = 1, \cdots, n-1$.
A	ARRAY(N): Output: Upon execution contains the values y_k', $k = 1, \cdots, n-1$, of the first derivatives.
EPS	see TRIDIS.
BIG	see ABKVAL.
Y,B,C,D	ARRAY(N): Upon execution with IFLAG=0, contain the desired spline coefficients, K $= 1, 2, \cdots,$ N $- 1$.
IFLAG	=0: Normal execution.
	=1: N$\geq$ 3 is required.
	=2: Error in solving the linear system (TRIDIS).
	=11: $0 \leq p_k \leq$ BIG violated for some k (ABKVAL).

Required subroutines: ABKVAL, TRIDIS.

Figure 7.4. Description of EXPSP1.

```
FUNCTION EXPVAL(N,X,P,A,B,C,D,V,IFLAG)
DIMENSION X(N),P(N),A(N),B(N),C(N),D(N)
DATA I/1/
IFLAG=0
IF(N.LT.2) THEN
    IFLAG=1
    RETURN
END IF
CALL INTONE(X,N,V,I,IFLAG)
IF(IFLAG.NE.0) RETURN
DX=V-X(I)
H=X(I+1)-X(I)
T=DX/H
U=1.-T
PP=P(I)
PT=PP*T
PU=PP*U
P2=PP*PP
IF(PP.LE..1.OR.PU.LE..1.OR.PT.LE..1) THEN
    H1=1./6.
    H2=1./120.
    H3=1./5040.
    H4=1./362880.
    HN=H1+P2*(H2+P2*(H3+P2*H4))
END IF
IF(PP.GT..1) THEN
    SINHP=SINH(PP)
ELSE
    SINHP=PP*(1.+P2*HN)
END IF
IF(PU.GT..1) THEN
    SHPU=(SINH(PU)-U*SINHP)/(SINHP-PP)
ELSE
    PU2=PU*PU
    HZ=H1+PU2*(H2+PU2*(H3+PU2*H4))
    SHPU=U*(-1.+U*U*(HZ/HN))
END IF
IF(PT.GT..1) THEN
    SHPT=(SINH(PT)-T*SINHP)/(SINHP-PP)
ELSE
    PT2=PT*PT
    HZ=H1+PT2*(H2+PT2*(H3+PT2*H4))
    SHPT=T*(-1.+T*T*(HZ/HN))
END IF
EXPVAL=A(I)*U+B(I)*T+C(I)*SHPU+D(I)*SHPT
RETURN
END
```

Figure 7.5. Program listing of EXPVAL.

FUNCTION EXPVAL(N,X,P,A,B,C,D,V,IFLAG)

Purpose:
EXPVAL is a FUNCTION subroutine for the calculation of a function value of an exponential spline interpolant at a point $V \in [X(1),X(N)]$.

Description of the parameters:

N	Number of given points.
X	ARRAY(N): x-values.
P	ARRAY(N): Vector of factors $p_k \geq 0$, $k = 1, \cdots, n - 1$.
A,B,C,D	ARRAY(N): Vectors of the spline coefficients.
V	Value of the point at which the spline function is to be evaluated.
IFLAG	=0: Normal execution.
	=1: $N \geq 2$ required.
	=3: Error in interval determination (INTONE).

Required subroutine: INTONE.

Remark: The statement 'DATA I/1/' has the effect that I is set to 1 at the first call to EXPVAL.

Figure 7.6. Description of EXPVAL.

if $p_k < .1$, it uses the power series expansions (7.33) and (7.34) of the order shown, while if $p_k \geq .1$, it calls the Fortran built-in functions SINH and COSH. The subroutine EXPSP1 (Figs. 7.3 and 7.4) then calculates the coefficients of the for $p_k \geq 0$ uniquely determined C^2 exponential spline interpolant with boundary values y_1' and y_n'. Attention to numerical instabilities must also be paid when evaluating the spline. Hence, in EXPVAL (Figs. 7.5 and 7.6), the hyperbolic sine is evaluated using the power series expansion (7.5) of the order shown when one of its (positive) arguments, $p_k, p_k h$, or $p_k t$, is smaller than .1. This expansion guarantees at least 14 digits of accuracy. There is a somewhat more efficient approximation with this same accuracy. (For an example, see [118].)

The pairs of examples in Figs. 7.7 and 7.8 show the different results for $p_k = 5$ and $p_k = 10$. Figure 7.8c shows the results for the second example but with the parameters chosen to be p=(1,20,10,1,1,10,5,1,20,5,1,5,1) (boundary values y_1' and y_n' determined with GRAD2B).

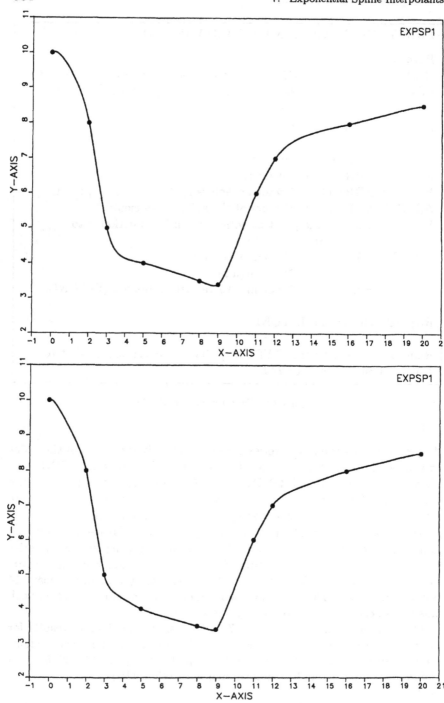

Figure 7.7. a, b.

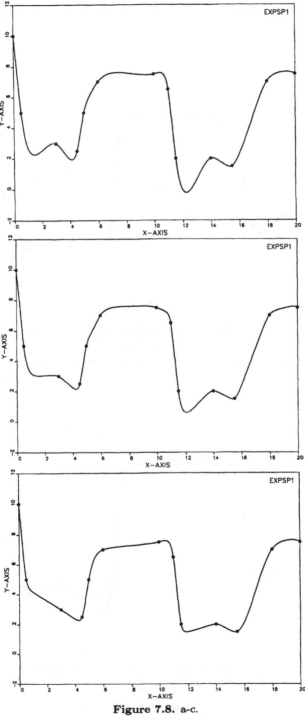

Figure 7.8. a-c.

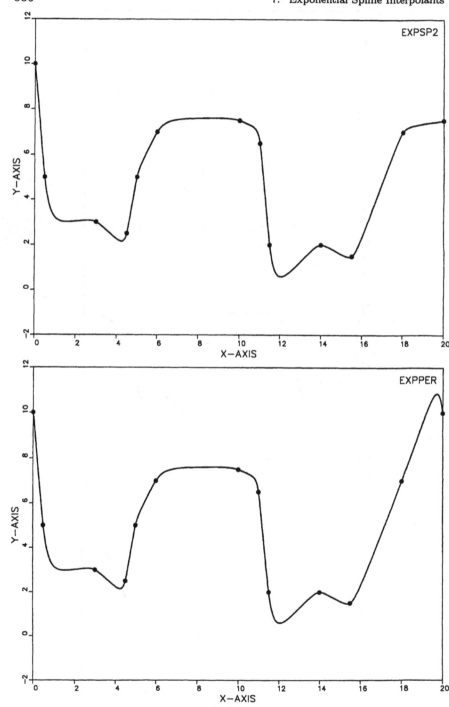

Figure 7.9. a, b.

One has the impression that it is easier to find suitable tension parameters than it is to set the location of the poles for the rational spline interpolants. The former are more directly related to the "tightening" of the curve towards a piecewise linear shape.

7.2. Second Derivatives as Unknowns

If the y_k'' are used as unknowns, then besides the interpolation conditions from which (7.19) follows, we obtain, for general ϕ_k, the requirements that

$$\phi_1'' C_k + \phi_0'' D_k = h_k^2 y_k'',$$

$$\phi_0'' C_k + \phi_1'' D_k = h_k^2 y_{k+1}''. \tag{7.40}$$

Provided that

$$\phi_0'' \neq \pm \phi_1'', \tag{7.41}$$

these may be solved for the remaining coefficients,

$$C_k = \frac{h_k^2}{(\phi_1'')^2 - (\phi_0'')^2}(\phi_1'' y_k'' - \phi_0'' y_{k+1}''),$$

$$D_k = \frac{h_k^2}{(\phi_1'')^2 - (\phi_0'')^2}(\phi_1'' y_{k+1}'' - \phi_0'' y_k''). \tag{7.42}$$

Then from the C^1 continuity conditions, $s_{k-1}'(x_k) = s_k'(x_k)$, $k = 2, \cdots, n - 1$, we obtain the linear system,

$$\tilde{\alpha}_{k-1} h_{k-1} y_{k-1}'' + (\tilde{\beta}_{k-1} h_{k-1} + \tilde{\beta}_k h_k) y_k'' + \tilde{\alpha}_k h_k y_{k+1}'' = d_k - d_{k-1}, \tag{7.43}$$

$$k = 2, \cdots, n - 1,$$

where

$$\tilde{\alpha}_k = -\frac{\phi_0' \phi_1'' + \phi_1' \phi_0''}{(\phi_1'')^2 - (\phi_0'')^2}$$

and

$$\tilde{\beta}_k = \frac{\phi_0' \phi_0'' + \phi_1' \phi_1''}{(\phi_1'')^2 - (\phi_0'')^2}. \tag{7.44}$$

In the special case of (7.4),

$$\tilde{\alpha}_k = \frac{1}{a_k} \quad \text{and} \quad \tilde{\beta}_k = b_k \tilde{\alpha}_k = \frac{b_k}{a_k}, \tag{7.45}$$

```
      SUBROUTINE EXPSP2(N,X,Y,P,A,B,C,D,EPS,BIG,IFLAG)
      DIMENSION X(N),Y(N),P(N),A(N),B(N),C(N),D(N)
      IFLAG=0
      IF(N.LT.3) THEN
          IFLAG=1
          RETURN
      END IF
      N1=N-1
      N2=N-2
      Y21=C(1)
      Y2N=C(N)
      DO 20 K=1,N1
          KP1=K+1
          KM1=K-1
          H=X(KP1)-X(K)
          CALL ABKVAL(P(K),A(K),B(K),BIG,IFLAG)
          IF(IFLAG.NE.0) RETURN
          D2=(Y(KP1)-Y(K))/H
          C2=B(K)/A(K)*H
          B(K)=H/A(K)
          IF(K.EQ.1) GOTO 10
          C(KM1)=C1+C2
          D(KM1)=D2-D1
          IF(K.EQ.2) D(KM1)=D(KM1)-B(KM1)*Y21
          IF(K.EQ.N1) D(KM1)=D(KM1)-B(K)*Y2N
          B(KM1)=B(K)
10        C1=C2
          D1=D2
20    CONTINUE
      CALL TRIDIS(N2,C,B,D,EPS,IFLAG)
      IF(IFLAG.NE.0) RETURN
      DO 30 K=1,N2
          B(K+1)=D(K)
30    CONTINUE
      B(1)=Y21
      B(N)=Y2N
      DO 40 K=1,N1
          K1=K+1
          H=X(K1)-X(K)
          H1=H*H/A(K)
          D(K)=H1*B(K1)
          C(K)=H1*B(K)
          A(K)=B(K)
          B(K)=Y(K1)
40    CONTINUE
      RETURN
      END
```

Figure 7.10. Program listing of EXPSP2.

Calling sequence:

CALL EXPSP2(N,X,Y,P,A,B,C,D,EPS,BIG,IFLAG)

Purpose:
EXPSP2 determines coefficients Y_k, B_k, C_k and D_k, $k = 1, \cdots, n-1$ of the same type of exponential spline interpolant as does EXPSP1. However, values for y_1'' and y_n'' must be given in C(1) and C(N), respectively.

Description of the parameters:

N,X,Y,P,B,C,D,EPS,BIG,IFLAG as in EXPSP1 (N$\geq$ 3 required).
A   ARRAY(N): Output: Upon execution contains the values
$$y_k'', \quad k = 1, \cdots, n-1, \text{ of the second derivatives.}$$

Required subroutines: ABKVAL, TRIDIS.

Figure 7.11. Description of EXPSP2.

with the a_k and b_k as in (7.31). From (7.35), it follows that

$$\tilde{\alpha}_k > 0, \qquad \tilde{\beta}_k > 2\tilde{\alpha}_k, \tag{7.46}$$

and hence (for given values of y_1'' and y_n'') the (symmetric) coefficient matrix of the system (7.43) is strictly diagonally dominant. Since

$$\lim_{p_k \to 0} \tilde{\alpha}_k = \frac{1}{6} \quad \text{and} \quad \lim_{p_k \to 0} \tilde{\beta}_k = \frac{1}{3}, \tag{7.47}$$

(7.43) reduces to the system (4.21) for cubic spline interpolants as $p_k \to 0$, $k = 1, \cdots, n-1$.

The other two coefficients, (7.42), simplify to

$$C_k = \frac{1}{a_k} h_k^2 y_k'', \qquad D_k = \frac{1}{a_k} h_k^2 y_{k+1}''. \tag{7.48}$$

Because of (7.39), (7.14) again holds when the y_k'' remain bounded. If the p_k, $k = 1, \cdots, n-1$, go uniformly to infinity, then by (7.39) the system (7.43) simplifies to

$$y_k'' = \frac{1}{2} \frac{d_k - d_{k-1}}{h_k + h_{k+1}}, \tag{7.49}$$

i. e., the second derivative of the quadratic passing through points $k-1, k$, and $k+1$. For the general situation of the p_k, we refer the reader to the investigations of [111].

The subroutine EXPSP2 (Figs. 7.10 and 7.11) computes the exponential spline interpolant in the case of given values for y_1'' and y_n''. At the price of two additional equations, this could also have been accomplished by making the appropriate modifications to the system (7.24) and EXPSP1.

Figure 7.9a shows an example comparable to that of Fig. 7.8b but with $p_k = 10$ and $y_1'' = y_n'' = 0$. The rightmost interval shows the effect of the different end conditions.

7.3.   Periodic Exponential Spline Interpolants

As with all the constructions of periodic splines, we suppose that

$$y_n = y_1, \qquad (7.50)$$

and then in addition ask that

$$s_1'(x_1) = s_{n-1}'(x_n), \qquad (7.51)$$
$$s_1''(x_1) = s_{n-1}''(x_n). \qquad (7.52)$$

Of the two systems, (7.25) and (7.43), we choose to make the appropriate modifications to (7.43), since this one will not involve any complicated expressions on the right-hand side. Condition (7.52) is satisfied by setting $y_n'' = y_1''$ in the last equation of (7.43) while observing (7.50). This results in

$$\tilde{\alpha}_{n-1}h_{n-1}y_1'' + \tilde{\alpha}_{n-2}h_{n-2}y_{n-2}'' + (\tilde{\beta}_{n-2}h_{n-2} + \tilde{\beta}_{n-1}h_{n-1})y_{n-1}''$$
$$\qquad (7.53)$$
$$= d_{n-1} - d_{n-2}.$$

To this is added (7.51), which by (7.20) results in

$$(\tilde{\beta}_1 h_1 + \tilde{\beta}_{n-1}h_{n-1})y_1'' + \tilde{\alpha}_1 h_1 y_2'' + \tilde{\alpha}_{n-1}h_{n-1}y_{n-1}''$$
$$\qquad (7.54)$$
$$= d_1 - d_{n-1}.$$

Altogether, we have $n-1$ equations in $n-1$ unknowns, $y_1'', \cdots, y_{n-1}''$. The coefficient matrix is symmetric, cyclically tridiagonal, and strictly diagonally dominant. This system is solved by the subroutine EXPPER (Figs. 7.12 and 7.13). The coefficients of (7.3) with (7.4) are calculated according to (7.19) and (7.38). Again, EXPVAL may be used for evaluation.

Figure 7.9b is the example of 7.9a, again for $p_k = 10$.

```
      SUBROUTINE EXPPER(N,X,Y,P,A,B,C,D,EPS,BIG,IFLAG)
      DIMENSION X(N),Y(N),P(N),A(N),B(N),C(N),D(N)
      IFLAG=0
      IF(N.LT.4) THEN
         IFLAG=1
         RETURN
      END IF
      Y(N)=Y(1)
      N1=N-1
      C1=0.
      D1=0.
      DO 10 K=1,N1
         KP1=K+1
         KM1=K-1
         H=X(KP1)-X(K)
         CALL ABKVAL(P(K),A(K),B(K),BIG,IFLAG)
         IF(IFLAG.NE.0) RETURN
         D2=(Y(KP1)-Y(K))/H
         C2=B(K)/A(K)*H
         B(K)=H/A(K)
         C(K)=C1+C2
         D(K)=D2-D1
         C1=C2
         D1=D2
   10 CONTINUE
      C(1)=C(1)+C2
      D(1)=D(1)-D2
      CALL TRIPES(N1,C,B,D,EPS,IFLAG)
      IF(IFLAG.NE.0) RETURN
      D(N)=D(1)
      DO 20 K=1,N1
         K1=K+1
         H=X(K1)-X(K)
         H1=H*H/A(K)
         A(K)=D(K)
         B(K)=Y(K1)
         C(K)=H1*D(K)
         D(K)=H1*D(K1)
   20 CONTINUE
      RETURN
      END
```

Figure 7.12. Program listing of EXPPER.

Calling sequence:

CALL EXPPER(N,X,Y,P,A,B,C,D,EPS,BIG,IFLAG)

Purpose:
EXPPER determines coefficients Y_k, B_k, C_k, and D_k, $k = 1, \cdots, n - 1$, of a periodic exponential spline interpolant. $y_n = y_1$ is enforced by the program.

Description of the parameters:

N,X,Y,P,A,B,C,D,EPS,BIG,IFLAG as in EXPSP2 (N$\geq$ 4 required).

Required subroutines: ABKVAL, TRIPES.

Figure 7.13. Description of EXPPER.

7.4. Shape Preservation and Other Considerations

To begin with, we should explain why in (7.3) and in the later derivations of the coefficient formulas (7.22) and (7.42), and even of the linear systems (7.24) and (7.43), we allowed ϕ_k, instead of just (7.4), to be a general function having property (7.16). The reason is that in this way, the rational spline interpolants of (6.63) for $p_k = q_k$ with

$$\phi_k(t) = \frac{t^3}{1 + p_k u} - t \tag{7.55}$$

and other functions, e. g. ([154]),

$$\phi_k(t) = t^3 e^{-p_k u} - t, \tag{7.56}$$

for which $\phi_k''(t) \geq 0$ and have other properties ([154]) in common, can all be treated together. The systems of equations (7.25) and (7.43) still hold with the appropriately adjusted values of ϕ_0', ϕ_1', ϕ_0'', and ϕ_1'', provided that the restrictions (7.21) and (7.41) also still hold. Further possibilities are discussed in [110]. It is our conviction that because of the simplicity of (7.55), rational spline interpolants cannot be surpassed in their effectiveness and efficiency.

Shape perservation, i. e., the removal of undesirable inflection points ([112,148,162]), monotonicity, convexity/concavity, and positivity of exponential spline interpolants when the data *locally* indicates these, can in principle be attained by taking sufficiently large values of the p_k. It is not always clear whether only some of the p_k have to be large enough or whether all of them must have a certain minimum value. Moreover, all the values should be chosen to be as small as possible so that the resulting interpolant does not too closely resemble a piecewise linear.

Several methods, some heuristic, others very expensive (quadratic optimization), for the determination of good values for the p_k are discussed in the literature. A number of convincing examples have been presented, for C^1 exponential Hermite splines ([56,117]) as well as C^2 exponential splines ([82,56,118,119,120,172]). (Reference [117] has a very good suggestion for another GRAD-type routine.) Reference [65] discusses how to determine the p_k so that one of C_k and D_k become zero. This is similar to what we did in the section on rational splines with two prescribable real poles. However, the procedure of [65] is far more expensive, since a complicated nonlinear equation must be solved for each interval. Also, the choice of tension parameter for exponential smoothing splines is discussed in [112]. In our opinion, the simplest and most efficient way is to interactively adjust the p_k on a computer screen until the user arrives at a visually pleasing curve. Being able to freely choose the parameters gives a flexibility that only at great expense can be replaced by a mathematical analysis. This perhaps applies equally well to the case of rational spline interpolants.

In [53], unknown ordinates and values for the second derivatives of an exponential spline are determined so as to minimize the sum of the squares of the distances to given ordinates. An unusual feature is that this involves an *underdetermined* linear system of equations.

The problem underlying subroutines INCUB1 and INCUB2, i.e., given values of y'_k (or y''_k) solve for the y_k, is discussed in [74].

Quadratic exponential splines, i.e., those that reduce to quadratics instead of cubics as $p_k \to 0$, are treated in [124,141]. In [141], the knots and nodes are the same. Conditions for monotonicity and convexity are also given. In [124], the knots are between the nodes.

Naturally, we could also, by the same method as in the last chapter, have constructed exponential histosplines.

Finally, we mention that the theory of polynomial B-splines has been generalized to the case of exponential splines in [125,146]. To date, this has not been done for our rational splines with two free poles.

8

Spline Interpolation and Smoothing in the Plane

8.1. Interpolation with Arbitrary Splines

In this chapter, we will no longer suppose, as in (2.1), that the abscissas are strictly increasing. Instead, we assume that every two consecutive points (x_k, y_k) are distinct, i. e.,

$$x_k \neq x_{k+1} \quad \text{or} \quad y_k \neq y_{k+1}, \qquad k = 1, \cdots, n-1. \tag{8.1}$$

This condition allows for the possibility of points being repeated.

Now, if it is desired to construct a smooth interpolant through these points in the plane — space curves are analogous — we can choose suitable parameters v_k, $k = 1, \cdots, n$, with

$$v_1 < v_2 < \cdots < v_n, \tag{8.2}$$

use one of our previous methods to construct spline interpolants $\xi = \xi(v)$ and $\eta = \eta(v)$ through the points (v_k, x_k) and (v_k, y_k), respectively, and then define the curve as

$$s(v) = \begin{pmatrix} \xi(v) \\ \eta(v) \end{pmatrix} \tag{8.3}$$

345

for $v \in [v_1, v_n]$. A common choice ([26,52,163]) is to set

$$v_1 = 0, \quad v_{k+1} = v_k + \sqrt{(\Delta x_k)^2 + (\Delta y_k)^2}, \qquad k = 1, \cdots, n-1, \quad (8.4)$$

so that the v_k correspond to the cumulative average arclength of the piece-wise linear going through the given points. But what is wrong with choosing

$$v_k = k - 1, \qquad k = 1, \cdots, n? \tag{8.5}$$

In fact, the values of the v_k have a very pronounced effect on the shape of the resulting curve and it is not at all clear how to choose suitable values. (This is in fact one of the reasons why some very different parameter choices have been suggested ([24,51,83]).) Hence, we first want to emperically compare the curves resulting from using (8.4) and (8.5) for several different spline methods.

We first consider (8.4). Figures 8.1a,b,c show the results of using CUB1R5 with IR=2, i. e., ξ and η are cubic splines with the end conditions,

$$\left.\frac{d\xi}{dv}\right|_{v=v_1} = \frac{\Delta x_1}{\Delta v_1}, \quad \left.\frac{d\xi}{dv}\right|_{v=v_n} = \frac{\Delta x_{n-1}}{\Delta v_{n-1}},$$

$$\left.\frac{d\eta}{dv}\right|_{v=v_1} = \frac{\Delta y_1}{\Delta v_1}, \quad \left.\frac{d\eta}{dv}\right|_{v=v_n} = \frac{\Delta y_{n-1}}{\Delta v_{n-1}}. \tag{8.6}$$

(In Fig. 8.1c, $(x_1, y_1) = (x_n, y_n)$ without, as we will do later, requiring that ξ and η are C^2 there.) The subroutine CUB2R7 with IR=12, i. e., with "natural" end conditions,

$$\left.\frac{d^2\xi}{dv^2}\right|_{v=v_1} = \left.\frac{d^2\xi}{dv^2}\right|_{v=v_n} = \left.\frac{d^2\eta}{dv^2}\right|_{v=v_1} = \left.\frac{d^2\eta}{dv^2}\right|_{v=v_n} = 0, \tag{8.7}$$

was used on the same three examples. The results are shown in Figs. 8.2a,b,c.

Figures 8.3a,b,c show the results of RATSP1 for end conditions (8.6) and $p_k = q_k = 5$ on both ξ and η. EXPSP1 gives, for $p_k = 10$, very similar curves. The selection of control parameters (poles or tension parameters) for this type of problem is complicated by the fact that they must be chosen to be simultaneously good for two curves, and their influence on the shape of the curve depends on this as well as on the values of the v_k.

Figures 8.4a,b,c correspond to 8.2a,b,c, with the difference that the v_k are chosen according to (8.5) instead of (8.4). By comparing these, it is clear that the choice of the v_k has very strong influence on the resulting curve. We will directly return to some even more drastic examples as well

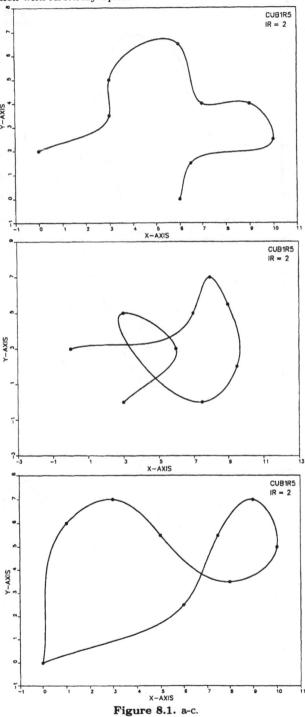

Figure 8.1. a-c.

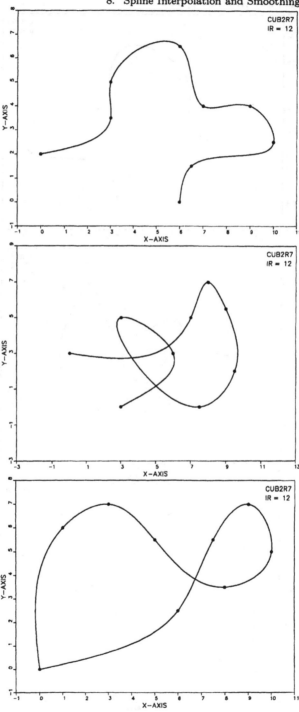

Figure 8.2. a-c.

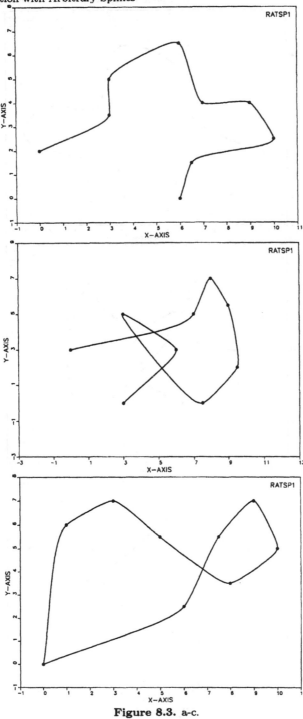

Figure 8.3. a-c.

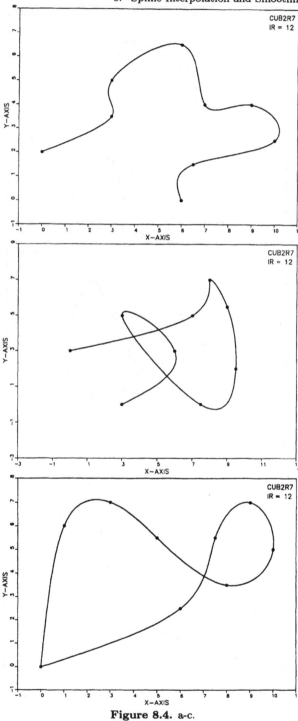

Figure 8.4. a-c.

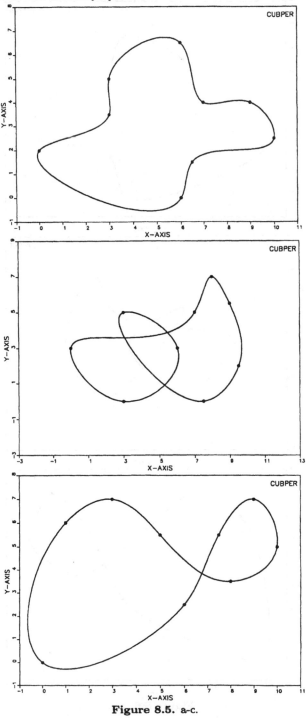

Figure 8.5. a-c.

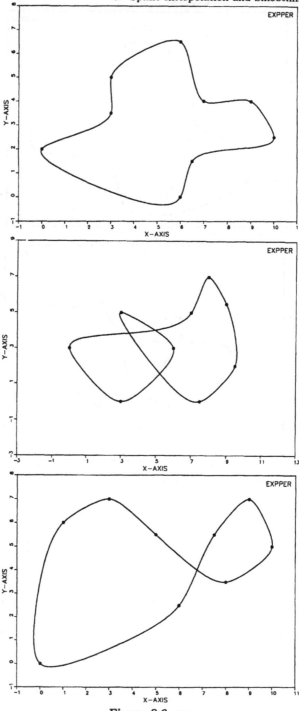

Figure 8.6. a-c.

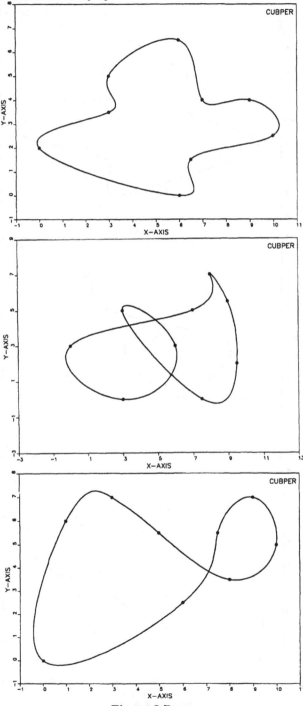

Figure 8.7. a-c.

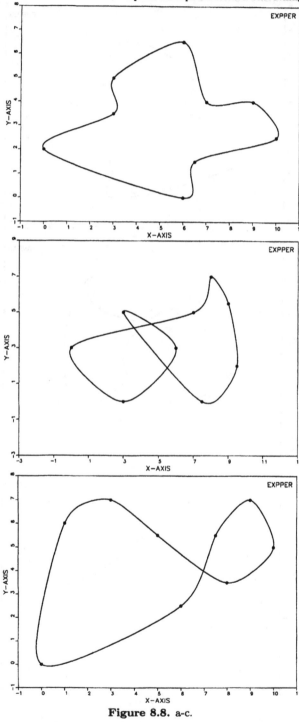

Figure 8.8. a-c.

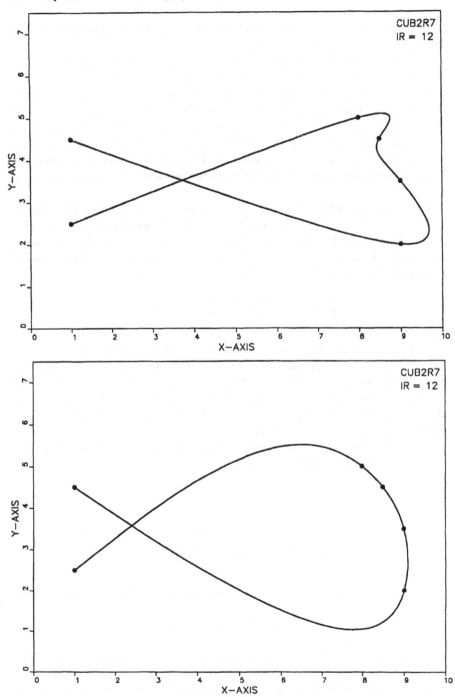

Figure 8.9. a, b.

as a suggestion for an "optimal" choice of the v_k, but first we wish to give some examples of periodic curves.

For parametric periodic spline interpolants, we assume that $(x_1, y_1) = (x_n, y_n)$ and construct periodic spline interpolants ξ and η through the points (v_k, x_k) with $x_1 = x_n$ and (v_k, y_k) with $y_1 = y_n$, respectively. Figures 8.5a,b,c show the results of CUBPER and Figs. 8.6a,b,c those of EXPPER ($p_k = 10$), each with parameters chosen according to (8.4). Figures 8.7a,b,c and 8.8a,b,c give the corresponding curves for v_k chosen according to (8.5). Again with respect to the question of the suitable choice of parameters, Figs. 8.9a,b show rather drastic differences between (8.4) and (8.5) for an open curve.

From the preceding considerations, it is quite reasonable ([1]) to try to choose the v_k so that the resulting parameterization gives a *canonical representation* of the curve, i.e., that the v_k correspond to the cumulative arclength of the desired curve. Equivalently (assuming without loss of generality that $v_1 = 0$), we ask that

$$v_{k+1} - v_k = \int_{v_k}^{v_{k+1}} \sqrt{\dot{\xi}_k^2(v) + \dot{\eta}_k^2(v)}\, dv, \qquad k = 1, \cdots, n-1. \qquad (8.8)$$

This is a rather highly nonlinear system of equations in the v_k, as the segments ξ_k and η_k, at least in the C^2 case, depend on *all* of the v_k. But even when, for a given spline method, the v_k can be determined by the optimality conditions (8.8), it remains an open question whether and why the corresponding parametric curve is always visually preferable to that corresponding to the use of (8.4).

An obvious iteration (with index t) that could be used to solve the system (8.8) of $n - 1$ equations in $v_2, \cdots, v_n$ is

$$v_1^{(t)} = 0,$$

$$v_{k+1}^{(t)} = v_k^{(t)} + \int_{v_k^{(t-1)}}^{v_{k+1}^{(t-1)}} \sqrt{\left[\dot{\xi}_k^{(t-1)}(v)\right]^2 + \left[\dot{\eta}_k^{(t-1)}(v)\right]^2}\, dv. \qquad (8.9)$$

The starting values $v_2^{(0)}, \cdots, v_n^{(0)}$ are arbitrary except that they must satisfy

$$0 < v_2^{(0)} < \cdots < v_n^{(0)}.$$

For example, the $v_k^{(0)}$ could be chosen according to either (8.4) or (8.5). As the arclength of the piecewise linear interpolant is quite a good first approximation to the arclength of the curve, (8.4) is to be especially recommended. In practice, (8.9) means that starting with values for the $v_k^{(t-1)}$, the corresponding spline interpolants $\xi_k^{(t-1)}$ and $\eta_k^{(t-1)}$ must be computed, and then finally the new parameter values $v_k^{(t)}$ are computed by (numerically) evaluating the appropriate integrals. In principle, this algorithm can be carried out for any type of spline interpolation. We only implemented it, however, for the case of natural cubic splines. (SPBOGL has about 70 statements.) It is not clear, although in practice it always does, that the method always converges to a unique solution. It does not appear that we are dealing here with a fixed point method that has a contraction constant less than one on all of $\mathbb{R}^{n-1}$. Because of the lack of a convergence proof, we will not give the subroutine SPBGOL but only report on the results of some experiments.

Figures 8.10a,b show the resulting curves for the example of Figs. 8.9a,b with starting values (8.5) for the zeroth (dotted), first (dashed), and eighth (heavy), iterations and then with starting values (8.4) for the zeroth, first, and sixth iterations. The circumstances in the next example are even more extreme. Figure 8.11a shows the result for (8.4), Fig. 8.11b for (8.5) and Fig. 8.11c for starting value (8.5) and the first and 11th iterations. In particular, the method seems empirically to converge relatively fast. The resulting curves for the canonical representation are visually preferable to those for other choices of the parameters v_k. Why this is so and why (at least in the case of cubic splines) the iteration starting with (8.4) presumably always converges remain for the time being open questions. Also, one cannot imagine that, in the case of interpolation of planar data, other splines would produce nicer curves than do cubic splines.

8.2.   Smoothing with Cubic Spline Interpolants

The cubic smoothing splines of Chapter 4 can also be used for the smoothing of points in the plane. Given v_k satisfying (8.2), we simply calculate two smoothing splines ξ and η with control parameters p_k and q_k, $k = 1, \cdots, n$, respectively. This is possible for both open (CUBSM1) and closed (CUBSM2) curves.

For a first example, Figs. 8.12a–8.13b show the sequence of results for CUBSM1 with parameters (8.4) and $p_k = q_k = .1, .5, 1, 5, 10$.

The next example is taken from [163]. Figures 8.14a–8.15b show the results of CUBSM2 for the four cases of $p_k = q_k = .1$, $p_k = q_k = 1$, $p_k = q_k = 10$, and $p_k = q_k = 1$ except for $p_{10} = q_{16} = 10$. The last figure, 8.15b, demonstrates how the p_k can be chosen to influence the shape of the curve. In comparison to Fig. 8.14a, the value $p_{10} = 10$ has caused the

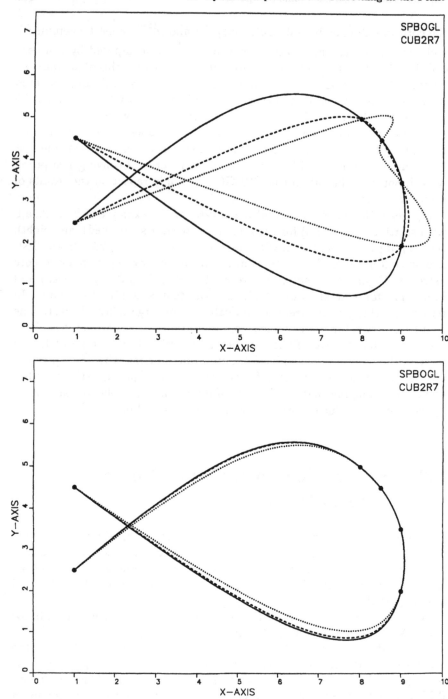

Figure 8.10. a, b.

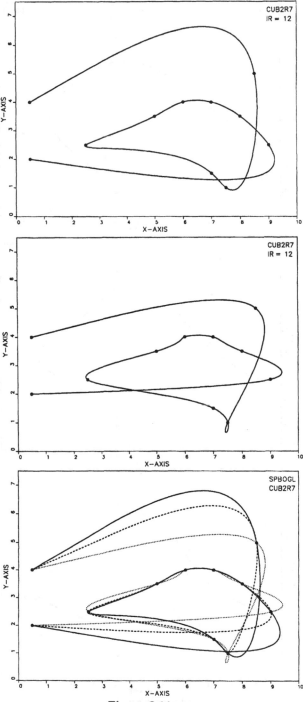

Figure 8.11. a-c.

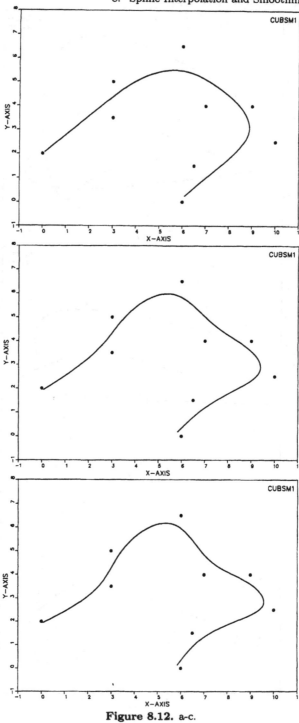

Figure 8.12. a–c.

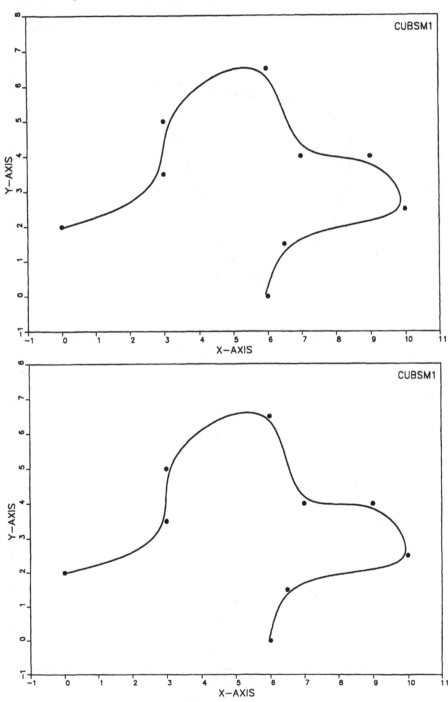

Figure 8.13. a, b.

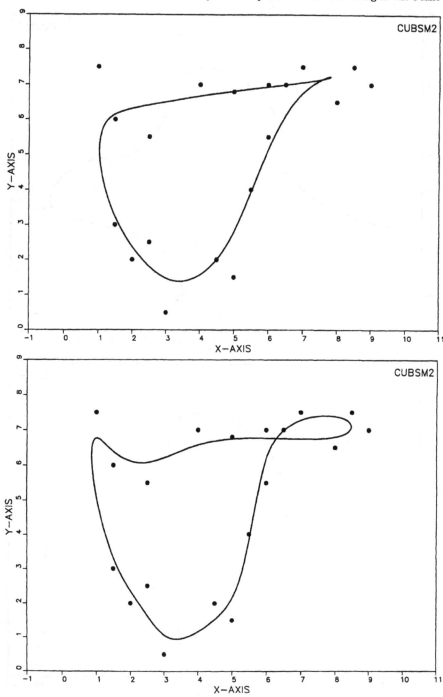

Figure 8.14. a, b.

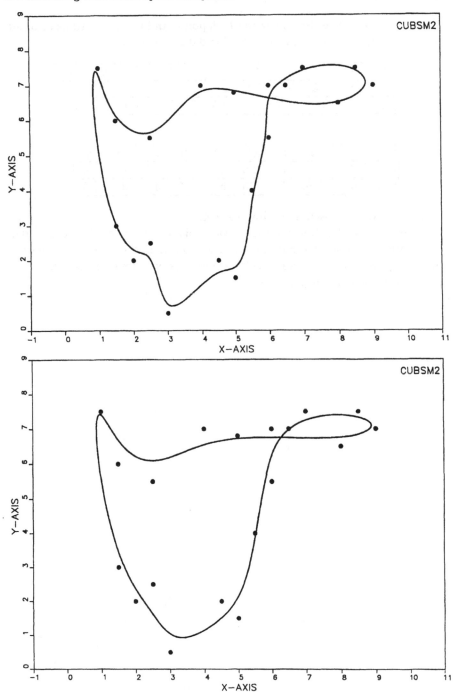

Figure 8.15. a, b.

curve to move to the right near the 10th point, and the $q_{16} = 10$ has caused it to move up near the 16th point. The data used were

k	1	2	3	4	5	6	7	8	9	10
x_k	3	4.5	5	5.5	6	6	6.5	7	8.5	9
y_k	.5	2	1.5	4	5.5	7	7	7.5	7.5	7

k	11	12	13	14	15	16	17	18	19
x_k	8	5	4	2.5	1.5	1	1.5	2	2.5
y_k	6.5	6.8	7	5.5	6	7.5	3	2	2.5

and the v_k were chosen according to (8.4).

In both cases, the choice of the v_k was arbitrary. It is possible to consider an iteration method like that of the last section when the control parameters p_k and q_k are fixed. Or, conversely, one can, as was done before, fix the v_k and adjust the control parameters.

Postscript

As one can never in one book, be it for reasons of space or for having deadlines to meet, accommodate all that was planned or that might have been included in the detail that it deserves, we would at least like to call attention to some important literature references as well as to the bibliography [91], which became known to us only after the completion of this book.

Nonlinear splines are those that minimize, for example, the total curvature,

$$E(s) = \int_{x_1}^{x_n} \frac{[s''(x)]^2}{(1 + [s'(x)]^2)^{5/2}} dx.$$

(For slowly varying s', and only in this case, these are close to cubic splines because of (4.29).) There are several references to these in the literature [45,64,68,76,85,107,174]. The method of [76] appeared to be the most promsing, but we were somewhat disappointed in it. Depending on the example, it converged to stationary points of the above functional that did not correspond to a minimum.

As a practitioner informs us that his interpolants were restricted to be composed of line or circle segments, the publications [61,84,165] may be of special importance.

Naturally, it would be nice, but also more difficult, to continue this book with two-dimensional spline interpolation methods. Tensor product splines over rectangles are relatively simple. If the points were irregularly distributed in the plane, then it would first be necessary to triangulate the points, then to define spline elements on the triangles, and finally to join them smoothly together. For practical purposes, C^1 should suffice.

A

Appendix

Before we can describe the workings of the subroutines TRIDIS, TRIDIU, TRIPES, TRIPEU, PENTAS, and PENPES, which have been used throughout this book, we will need some definitions and theorems from numerical linear algebra.

A square $n \times n$, real or complex matrix, $A = [a_{ik}]$, is said to be *diagonally dominant* ([166]) when

$$\sum_{\substack{k=1 \\ k \neq i}} |a_{ik}| \leq |a_{ii}| \quad i = 1, \cdots, n, \tag{A.1}$$

i.e., when for each row the sum of the absolute values of the off-diagonal elements is less than or equal to the absolute value of the diagonal element. To be more precise, a matrix A with property (A.1) should be called *row diagonally dominant*, and then those with the property that

$$\sum_{\substack{k=1 \\ k \neq i}} |a_{ki}| \leq |a_{ii}| \quad i = 1, \cdots, n \tag{A.2}$$

should be called *column diagonally dominant*.

For symmetric matrices, $A = A^t$, the two notions coincide. Since A and A^t are either both singular or both nonsingular, by diagonal dominance we may in general understand either row or column diagonal dominance.

A matrix A is said to be *strictly diagonally dominant* ([166]) if in (A.1) or (A.2) there is strict inequality for all $i = 1, \cdots, n$. Further, A is said to be *irreducibly diagonally dominant* ([166]) when A is irreducible, diagonally dominant, and at least one of the inequalities is strict. We will not explain here what irreducible means. For the matrices that are of special interest to us, we will say in which cases they are irreducible.

Now the following theorem is true ([166]): if A is strictly or irreducibly diagonally dominant, then A is nonsingular. If in addition, $A = A^t$ and $a_{ii} > 0$, $i = 1, \cdots, n$, then A is positive definite.

In this book, only very special matrices arise, namely: symmetric and nonsymmetric tridiagonal, symmetric five-diagonal, symmetric and non-symmetric cyclically tri-diagonal, and symmetric cyclically five-diagonal martrices. A matrix A is said to be tridiagonal if $a_{ik} = 0$ for $|i - k| \geq 2$ or, equivalently, has the form,

$$\begin{pmatrix} b_1 & c_1 & & & & \\ a_2 & b_2 & c_2 & & & \\ & a_3 & b_3 & c_3 & & \\ & & \cdot & \cdot & \cdot & \\ & & & \cdot & \cdot & \cdot \\ & & & a_{n-1} & b_{n-1} & c_{n-1} \\ & & & & a_n & b_n \end{pmatrix}. \tag{A.3}$$

Similarly, a matrix is said to be five-diagonal if $a_{ik} = 0$ for $|i - k| \geq 3$, i.e., if it has the form,

$$\begin{pmatrix} b_1 & c_1 & d_1 & & & & & \\ a_2 & b_2 & c_2 & d_2 & & & & \\ e_3 & a_3 & b_3 & c_3 & d_3 & & & \\ & e_4 & a_4 & b_4 & c_4 & d_4 & & \\ & & \cdot & \cdot & \cdot & \cdot & \cdot & \\ & & & \cdot & \cdot & \cdot & \cdot & \cdot \\ & & & e_{n-2} & a_{n-2} & b_{n-2} & c_{n-2} & d_{n-2} \\ & & & & e_{n-1} & a_{n-1} & b_{n-1} & c_{n-1} \\ & & & & & e_n & a_n & b_n \end{pmatrix}. \tag{A.4}$$

Further, A is cyclically tridiagonal if it has the form,

$$\begin{pmatrix} b_1 & c_1 & & & & a_1 \\ a_2 & b_2 & c_2 & & & \\ & a_3 & b_3 & c_3 & & \\ & & \cdot & \cdot & \cdot & \\ & & & \cdot & \cdot & \\ & & & a_{n-1} & b_{n-1} & c_{n-1} \\ c_n & & & & a_n & b_n \end{pmatrix}, \tag{A.5}$$

and cyclically five-diagonal if it has the form,

$$
\begin{pmatrix}
b_1 & c_1 & d_1 & & & & & e_1 & a_1 \\
a_2 & b_2 & c_2 & d_2 & & & & & e_2 \\
e_3 & a_3 & b_3 & c_3 & d_3 & & & & \\
 & e_4 & a_4 & b_4 & c_4 & d_4 & & & \\
 & & \cdot & \cdot & \cdot & \cdot & \cdot & & \\
 & & & \cdot & \cdot & \cdot & \cdot & \cdot & \\
 & & & e_{n-2} & a_{n-2} & b_{n-2} & c_{n-2} & d_{n-2} & \\
d_{n-1} & & & & e_{n-1} & a_{n-1} & b_{n-1} & c_{n-1} \\
c_n & d_n & & & & e_n & a_n & b_n
\end{pmatrix} . \tag{A.6}
$$

In all of these, the matrix elements not explicitly indicated are zero. In our context, it is important that in *all the cases arising*, the indicated elements, i.e., a_i, b_i, c_i, d_i, e_i, $i = 1, \cdots, n$, are nonzero. The irreducibility of each of the matrices (A.3) through (A.6) is thus guaranteed, as the *associated directed graphs* are strongly connnected ([166]).

We used the first part of the previous theorem repeatedly to show the existence and uniqueness of the spline interpolants under consideration.

The second part of the theorem is useful for the solution of linear systems of equations, as it is known that Gaussian elimination when applied to positive definite matrices does not require pivoting ([164]). But also pivoting is not required for nonsymmetric irreducibly or strictly diagonally dominant tridiagonal matrices ([60, p. 59] and [8, p. 234]). (We did not encounter any nonsymmetric five-diagonal matrices.) This is important, since elimination with pivoting destroys the banded structure of the matrix, thus making it more expensive in both execution time and storage space. See [62, p. 225] and [147, p. 47] for a treatment of tridiagonal matrices.

For the sake of clarity and to avoid special cases of no practical importance, in what follows we will ask that the tridiagonal matrices of the forms (A.3) and (A.5) not only be irreducible but also that $n \geq 1$ and $n \geq 3$, respectively, and that $n \geq 3$ for (A.4) as well as that $n \geq 5$ for (A.6).

We now consider Gaussian elimination without pivoting applied to a linear system of equations,

$$
Ax = r. \tag{A.7}
$$

We will treat the special cases of A being of the type (A.3), (A.5), (A.4), and (A.6), one at a time, in that order. We will only consider the general case of nonsymmetric matrices. The symmetric case is easily handled by setting $a_{k+1} = c_k$, $k = 1, \cdots, n$ (with $a_{n+1} := a_1$), and $e_{k+2} = d_k$, $k = 1, \cdots, n$ (with $e_{n+1} := e_1$ and $e_{n+2} := e_2$).

Gaussian elimination with no pivoting may be interpreted ([62,147,164]) as the decomposition of A into the product of a lower triangular matrix L

and an upper triangular matrix U and then solving, in sequence,

$$Ly = r \quad \text{(forward substitution)},$$
$$Ux = y \quad \text{(back substitution)}. \tag{A.8}$$

For our very special matrices, L and U have special forms. For (A.3),

$$L = \begin{pmatrix} \beta_1 & & & & & & \\ \alpha_2 & \beta_2 & & & & & \\ & \alpha_3 & \beta_3 & & & & \\ & & & \cdot & & & \\ & & & & \cdot & \cdot & \\ & & & & & \cdot & \\ & & & & & \alpha_{n-1} & \beta_{n-1} \\ & & & & & & \alpha_n & \beta_n \end{pmatrix},$$

$$U = \begin{pmatrix} 1 & \gamma_1 & & & & & \\ & 1 & \gamma_2 & & & & \\ & & 1 & \gamma_3 & & & \\ & & & \cdot & \cdot & & \\ & & & & \cdot & \cdot & \\ & & & & 1 & \gamma_{n-2} & \\ & & & & & 1 & \gamma_{n-1} \\ & & & & & & 1 \end{pmatrix};$$

for (A.5),

$$L = \begin{pmatrix} \beta_1 & & & & & & \\ \alpha_2 & \beta_2 & & & & & \\ & \alpha_3 & \beta_3 & & & & \\ & & & \cdot & \cdot & & \\ & & & & \cdot & \cdot & \\ & & & & & \cdot & \cdot & \\ & & & & & \alpha_{n-1} & \beta_{n-1} \\ v_1 & v_2 & v_3 & \cdot & \cdot & v_{n-2} & \alpha_n & \beta_n \end{pmatrix},$$

$$U = \begin{pmatrix} 1 & \gamma_1 & & & & & w_1 \\ & 1 & \gamma_2 & & & & w_2 \\ & & 1 & \gamma_3 & & & w_3 \\ & & & \cdot & \cdot & & \cdot \\ & & & & \cdot & \cdot & \cdot \\ & & & & 1 & \gamma_{n-2} & w_{n-2} \\ & & & & & 1 & \gamma_{n-1} \\ & & & & & & 1 \end{pmatrix};$$

for (A.4),

$$
L = \begin{pmatrix}
\gamma_1 & & & & & & & \\
\beta_2 & \gamma_2 & & & & & & \\
\alpha_3 & \beta_3 & \gamma_3 & & & & & \\
& & & \cdot & \cdot & \cdot & & \\
& & & & \cdot & \cdot & \cdot & \\
& & & & & \cdot & \cdot & \cdot \\
& & & & \alpha_{n-1} & \beta_{n-1} & \gamma_{n-1} & \\
& & & & & \alpha_n & \beta_n & \gamma_n
\end{pmatrix},
$$

$$
U = \begin{pmatrix}
1 & \delta_1 & \epsilon_1 & & & & \\
& 1 & \delta_2 & \epsilon_2 & & & \\
& & \cdot & \cdot & \cdot & & \\
& & & \cdot & \cdot & \cdot & \\
& & & & \cdot & \cdot & \cdot \\
& & & & 1 & \delta_{n-2} & \epsilon_{n-2} \\
& & & & & 1 & \delta_{n-1} \\
& & & & & & 1
\end{pmatrix};
$$

and finally for (A.6),

$$
L = \begin{pmatrix}
\gamma_1 & & & & & & & \\
\beta_2 & \gamma_2 & & & & & & \\
\alpha_3 & \beta_3 & \gamma_3 & & & & & \\
& & \cdot & \cdot & \cdot & & & \\
& & & \cdot & \cdot & \cdot & & \\
& & & & \cdot & \cdot & \cdot & \\
v_1 & v_2 & v_3 & \cdot & \alpha_{n-1} & \beta_{n-1} & \gamma_{n-1} & \\
w_1 & w_2 & w_3 & \cdot & w_{n-3} & \alpha_n & \beta_n & \gamma_n
\end{pmatrix},
$$

$$
U = \begin{pmatrix}
1 & \delta_1 & \epsilon_1 & & & & p_1 & q_1 \\
& 1 & \delta_2 & \epsilon_2 & & & p_2 & q_2 \\
& & \cdot & \cdot & \cdot & & \cdot & \cdot \\
& & & \cdot & \cdot & \cdot & p_{n-4} & q_{n-4} \\
& & & & 1 & \delta_{n-3} & \epsilon_{n-3} & q_{n-3} \\
& & & & & 1 & \delta_{n-2} & \epsilon_{n-2} \\
& & & & & & 1 & \delta_{n-1} \\
& & & & & & & 1
\end{pmatrix}.
$$

These are easily verified. By multiplying L and U together and equating this product with A, we obtain simple linear recursion formulas for the unknown elements of L and U. Once these are known, (A.7) can be solved by (A.8). We will give only the details for the case of (A.5), seeing that for this example there is a typographical error in [154]. We calculate LU to be

$$
\begin{bmatrix}
\beta_1 & \beta_1\gamma_1 & & & & \beta_1 w_1 \\
\alpha_2 & \alpha_2\gamma_1 + \beta_2 & \beta_2\gamma_2 & & & \alpha_2 w_1 + \beta_2 w_2 \\
& & \cdot & & & \cdot \\
& & & \cdot & & \alpha_{n-2} w_{n-3} + \beta_{n-2} w_{n-2} \\
& & & \alpha_{n-1}\gamma_{n-2} + \beta_{n-1} & \alpha_{n-1} w_{n-2} + \beta_{n-1} w_{n-1} \\
v_1 & v_1\gamma_1 + v_2 & v_2\gamma_2 + v_3 & \cdot & v_{n-2}\gamma_{n-2} + \alpha_n & \sum_{k=1}^{n-2} v_k w_k + \alpha_n \gamma_{n-1} + \beta_n
\end{bmatrix}.
$$

By equating entries, we see that

$$
\begin{aligned}
\alpha_k &= a_k \quad k = 2, \cdots, n-1, \\
\beta_1 &= b_1, \\
\gamma_1 &= \frac{c_1}{\beta_1}, \\
w_1 &= \frac{a_1}{\beta_1}, \\
\beta_k &= b_k - \alpha_k \gamma_{k-1}, \quad k = 2, \cdots, n-1, \\
\gamma_k &= \frac{c_k}{\beta_k}, \quad k = 2, \cdots, n-2, \\
w_k &= -\frac{\alpha_k w_{k-1}}{\beta_k}, \quad k = 2, \cdots, n-2, \\
\gamma_{n-1} &= \frac{c_{n-1} - \alpha_{n-1} w_{n-2}}{\beta_{n-1}}, \\
v_1 &= c_n, \\
v_k &= -\frac{v_{k-1}}{\gamma_{k-1}}, \quad k = 2, \cdots, n-2, \\
\alpha_n &= a_n - v_{n-2}\gamma_{n-2}, \\
\beta_n &= b_n - \sum_{k=1}^{n-2} v_k w_k - \alpha_n \gamma_{n-1}.
\end{aligned}
$$

From $Ly = r$, i.e., from

$$
\begin{aligned}
\beta_1 y_1 &= r_1, \\
\alpha_k y_{k-1} + \beta_k y_k &= r_k, \quad k = 2, \cdots, n-1, \\
\sum_{k=1}^{n-2} v_k y_k + \alpha_n y_{n-1} &= r_n,
\end{aligned}
$$

we obtain

$$
\begin{aligned}
y_1 &= \frac{r_1}{\beta_1}, \\
y_k &= \frac{r_k - \alpha_k y_{k-1}}{\beta_k}, \quad k = 2, \cdots, n-1, \\
y_n &= \frac{r_n - \sum_{k=1}^{n-2} v_k y_k - \alpha_n y_{n-1}}{\beta_n},
\end{aligned}
$$

and then from $Ux = y$, i.e., from

$$
\begin{aligned}
x_k + \gamma_k x_{k+1} + w_k x_n &= y_k, \quad k = 1, \cdots, n-2, \\
x_{n-1} + \gamma_{n-1} x_n &= y_{n-1}, \\
x_n &= y_n,
\end{aligned}
$$

we have

$$
\begin{aligned}
x_n &= y_n, \\
x_{n-1} &= y_{n-1} - \gamma_{n-1} x_n, \\
x_k &= y_k - \gamma_k x_{k+1} - w_k x_n, \quad k = n-2, \cdots, 1.
\end{aligned}
$$

Since we assume that A is either irreducibly or strictly diagonally dominant, the β_k are all nonzero and hence the algorithm will run to completion.

In each of the subroutines,

- TRIDIS (symmetric tridiagonal matrix, $n \geq 1$),

- TRIDIU (nonsymmetric tridiagonal matrix, $n \geq 1$),

- TRIPES (symmetric cyclically tridiagonal matrix, $n \geq 3$),

- TRIPEU (nonsymmetric cyclically tridiagonal matrix, $n \geq 3$),

- PENTAS (symmetric five-diagonal matrix, $n \geq 3$),

- PENPES (symmetric cyclically five-diagonal matrix, $n \geq 5$),

the denominators of the quotients corresponding to that just mentioned are tested to see whether or not they are larger or equal in magnitude than a given small positive number EPS. If this is not numerically the case, then each of these routines is terminated with IFLAG=2 (instead of IFLAG=0). Moreover, the algorithms are implemented so as to minimize storage requirements, i.e., when setting up and solving (A.8), storage locations are overwritten whenever possible. The details just given are actually for TRIPEU. There are no built-in checks on the allowability of the values of n, as this is done in the calling subroutines.

```
      SUBROUTINE TRIDIS(N,B,C,D,EPS,IFLAG)
      DIMENSION B(N),C(N),D(N)
      IFLAG=0.
      H1=0.
      H2=0.
      H3=0.
      DO 10 K=1,N
         Z=B(K)-H3*H1
         IF(ABS(Z).LT.EPS) THEN
            IFLAG=2
            RETURN
         END IF
         H1=C(K)/Z
         H2=(D(K)-H3*H2)/Z
         H3=C(K)
         C(K)=H1
         B(K)=H2
   10 CONTINUE
      D(N)=B(N)
      DO 20 K=N-1,1,-1
         D(K)=B(K)-C(K)*D(K+1)
   20 CONTINUE
      RETURN
      END
```

Calling sequence:

CALL TRIDIS(N,B,C,D,EPS,IFLAG)

Purpose:
Solution of a linear system of equations having symmetric tridiagonal coefficient matrix.

Description of the parameters:

N	Dimension of the matrix.
B	ARRAY(N): Diagonal $(1, \cdots, N)$ (is overwritten).
C	ARRAY(N): Superdiagonal $(1, \cdots, N-1)$ (is overwritten).
D	ARRAY(N): Input: Right-hand side.
	Output: Solution vector (if IFLAG=0).
EPS	Value used for accuracy test. If the absolute value of a transformed element on the main diagonal is smaller than EPS, then execution is terminated with IFLAG=2. Recommendation: EPS$= 10^{-t+2}$, where t is the number of available decimal digits.
IFLAG	=0: Normal execution.
	=2: The matrix is numerically singular.

Figure A.1. Program listing of TRIDIS and its description.

```
    SUBROUTINE TRIDIU(N,A,B,C,D,EPS,IFLAG)
    DIMENSION A(N),B(N),C(N),D(N)
    IFLAG=0
    H1=0.
    H2=0.
    H3=0.
    DO 10 K=1,N
        Z=B(K)-H3*H1
        IF(ABS(Z).LT.EPS) THEN
            IFLAG=2
            RETURN
        END IF
        H1=C(K)/Z
        C(K)=H1
        H2=(D(K)-H3*H2)/Z
        B(K)=H2
        H3=A(K)
 10 CONTINUE
    D(N)=B(N)
    DO 20 K=N-1,1,-1
        D(K)=B(K)-C(K)*D(K+1)
 20 CONTINUE
    RETURN
    END
```

Calling sequence:

CALL TRIDIU(N,A,B,C,D,EPS,IFLAG)

Purpose:
Solution of a linear system of equations having nonsymmetric tridiagonal
coefficient matrix.

Description of the parameters:

N,B,C,D,EPS,IFLAG as in TRIDIS.
A   ARRAY(N): Subdiagonal $(1, \cdots, N-1)$.

Figure A.2. Program listing of TRIDIU and its description.

```
      SUBROUTINE TRIPES(N,B,C,D,EPS,IFLAG)
      DIMENSION B(N),C(N),D(N)
      IFLAG=0
      N1=N-1
      Z=B(1)
      IF(ABS(Z).LT.EPS) GOTO 30
      Z=1./Z
      BH=B(2)
      B(2)=Z*C(1)
      CH1=C(1)
      CH2=C(2)
      C(2)=Z*C(N)
      D(1)=Z*D(1)
      S1=B(N)
      S2=D(N)
      S3=C(N)
      DO 10 K=2,N1
         KM1=K-1
         KP1=K+1
         Z=BH-CH1*B(K)
         IF(ABS(Z).LT.EPS) GOTO 30
         Z=1./Z
         S1=S1-S3*C(K)
         S2=S2-S3*D(KM1)
         S3=-B(K)*S3
         BH=B(KP1)
         B(KP1)=Z*CH2
         D(K)=Z*(D(K)-CH1*D(KM1))
         CH=CH2
         CH2=C(KP1)
         C(KP1)=-Z*CH1*C(K)
         CH1=CH
10    CONTINUE
      S1=S1-(S3+CH1)*(B(N)+C(N))
      D(N)=S2-(S3+CH1)*D(N1)
      IF(ABS(S1).LT.EPS) GOTO 30
      H=D(N)/S1
      D(N)=H
      D(N1)=D(N1)-(B(N)+C(N))*H
      DO 20 K=N-2,1,-1
         K1=K+1
         D(K)=D(K)-B(K1)*D(K1)-C(K1)*H
20    CONTINUE
      RETURN
30    IFLAG=2
      RETURN
      END
```

Figure A.3. Program listing of TRIPES.

Calling sequence:

CALL TRIPES(N,B,C,D,EPS,IFLAG)

Purpose:
Solution of a linear system of equations having symmetric, cyclically tridiagonal coefficient matrix.

Description of the parameters:

N,B,D,EPS,IFLAG as in TRIDIS.
C   ARRAY(N): Superdiagonal $(1, \cdots, N-1)$ (is overwritten).
             $C(N)$ must contain that element of the
             coefficient matrix with indices (n,1) or (1,n).

Figure A.4. Description of TRIPES.

```
SUBROUTINE TRIPEU(N,A,B,C,D,EPS,IFLAG)
DIMENSION A(N),B(N),C(N),D(N)
IFLAG=0
N1=N-1
Z=B(1)
IF(ABS(Z).LT.EPS) GOTO 30
Z=1./Z
BH=B(2)
B(2)=Z*C(1)
CH=C(2)
C(2)=Z*A(1)
D(1)=Z*D(1)
S1=B(N)
S2=D(N)
S3=C(N)
DO 10 K=2,N1
   KM1=K-1
   KP1=K+1
   Z=BH-A(K)*B(K)
   IF(ABS(Z).LT.EPS) GOTO 30
   Z=1./Z
   S1=S1-S3*C(K)
   S2=S2-S3*D(KM1)
   S3=-B(K)*S3
   BH=B(KP1)
   B(KP1)=Z*CH
   D(K)=Z*(D(K)-A(K)*D(KM1))
   CH=C(KP1)
   C(KP1)=-Z*A(K)*C(K)
```

(cont.)

```
10 CONTINUE
   S1=S1-(S3+A(N))*(B(N)+C(N))
   D(N)=S2-(S3+A(N))*D(N1)
   IF(ABS(S1).LT.EPS) GOTO 30
   H=D(N)/S1
   D(N)=H
   D(N1)=D(N1)-(B(N)+C(N))*H
   DO 20 K=N-2,1,-1
      K1=K+1
      D(K)=D(K)-B(K1)*D(K1)-C(K1)*H
20 CONTINUE
   RETURN
30 IFLAG=2
   RETURN
   END
```

Calling sequence:

CALL TRIPEU(N,A,B,C,D,EPS,IFLAG)

Purpose:
Solution of a linear system of equations with nonsymmetric, cyclically tridiagonal coefficient matrix.

Description of the parameters:

N,B,D,EPS,IFLAG as in TRIDIS.

A   ARRAY(N): Subdiagonal $(2, \cdots, N)$. A(1) must contain the upper rightmost element of the matrix.
C   ARRAY(N): Superdiagonal $(1, \cdots, N - 1)$ (is overwritten). C(N) must contain the lower leftmost element of the matrix.

Figure A.5. Program listing of TRIPEU and its description.

```
      SUBROUTINE PENTAS(N,A,B,C,F,EPS,IFLAG)
      DIMENSION A(N),B(N),C(N),F(N)
      IFLAG=0
      H1=0.0
      H2=0.0
      H3=0.0
      H4=0.0
      H5=0.0
      HH1=0.0
      HH2=0.0
      HH3=0.0
      HH4=0.0
      HH5=0.0
      DO 10 K=1,N
         Z=A(K)-H4*H1-HH5*HH2
         IF(ABS(Z).LT.EPS) THEN
            IFLAG=2
            RETURN
         END IF
         HB=B(K)
         HH1=H1
         H1=(HB-H4*H2)/Z
         B(K)=H1
         HC=C(K)
         HH2=H2
         H2=HC/Z
         C(K)=H2
         A(K)=(F(K)-HH3*HH5-H3*H4)/Z
         HH3=H3
         H3=A(K)
         H4=HB-H5*HH1
         HH5=H5
         H5=HC
   10 CONTINUE
      H2=0.
      H1=A(N)
      F(N)=H1
      DO 20 K=N-1,1,-1
         F(K)=A(K)-B(K)*H1-C(K)*H2
         H2=H1
         H1=F(K)
   20 CONTINUE
      RETURN
      END
```

Figure A.6. Program listing of PENTAS.

Calling sequence:

CALL PENTAS(N,A,B,C,F,EPS,IFLAG)

Purpose:
Solution of a linear system of equations having symmetric five-diagonal coefficient matrix.

Description of the parameters:

N	Dimension of the matrix.
A	ARRAY(N): Diagonal $(1, \cdots, N)$ (is overwritten).
B	ARRAY(N): First subdiagonal $(1, \cdots, N-1)$ (is overwritten).
C	ARRAY(N): Second subdiagonal $(1, \cdots, N-2)$ (is overwritten).
F	ARRAY(N): Input: Right-hand side. Output: Solution vector (if IFLAG=0).
EPS	Value used for accuracy test. If the absolute value of a transformed element on the main diagonal is smaller than EPS, then execution is terminated with IFLAG=2. Recommendation: EPS$= 10^{-t+2}$, where t is the number of available decimal digits.
IFLAG	=0: Normal execution.
	=2: The matrix is numerically singular.

Figure A.7. Description of PENTAS.

```
SUBROUTINE PENPES(N,A,B,C,D,E,F,H,EPS,IFLAG)
DIMENSION A(N),B(N),C(N),D(N),E(N),F(N),H(N)
IFLAG=0
NM1=N-1
NM2=N-2
NM3=N-3
NM4=N-4
Z=A(1)
IF(ABS(Z).LT.EPS) GOTO 90
A(1)=F(1)/Z
D(1)=C(NM1)/Z
E(1)=B(N)/Z
F(1)=C(NM1)
H(1)=B(N)
H3=B(1)
H1=C(1)
H4=B(1)/Z
B(1)=H4
C(1)=C(1)/Z
Z=A(2)-H3*H4
IF(ABS(Z).LT.EPS) GOTO 90
A(2)=(F(2)-H3*A(1))/Z
D(2)=-(H3*D(1))/Z
E(2)=(C(N)-H3*E(1))/Z
F(2)=-F(1)*H4
H(2)=C(N)-H(1)*H4
HH3=H3
H3=B(2)-H1*H4
H4=(B(2)-HH3*C(1))/Z
B(2)=H4
IF(N.GT.5) THEN
    NM5=N-5
    DO 10 K=3,NM3
        KM1=K-1
        KM2=K-2
        KP1=K+1
        HH1=H1
        H1=C(KM1)
        C(KM1)=C(KM1)/Z
        H2=C(KM2)
        Z=A(K)-HH1*H2-H3*H4
        IF(ABS(Z).LT.EPS) GOTO 90
        A(K)=(F(K)-HH1*A(KM2)-H3*A(KM1))/Z
        D(K)=-(HH1*D(KM2)+H3*D(KM1))/Z
        E(K)=-(HH1*E(KM2)+H3*E(KM1))/Z
        F(K)=-F(KM2)*H2-F(KM1)*H4
        H(K)=-H(KM2)*H2-H(KM1)*H4
        HH3=H3
        H3=B(K)-H1*H4
        H4=(B(K)-HH3*C(KM1))/Z
        B(K)=H4
10  CONTINUE
    HHH1=HH1
    HH1=H1
    H1=C(NM3)-F(NM5)*C(NM5)-F(NM4)*B(NM4)
    C(NM3)=(C(NM3)-HHH1*D(NM5)-HH3*D(NM4))/Z
```

(cont.)

```
    ELSE
        HH1=H1
        H1=C(NM3)-F(NM4)*B(NM4)
        C(NM3)=(C(NM3)-HH3*D(NM4))/Z
    END IF
    H2=C(NM4)
    Z=A(NM2)-HH1*H2-H3*H4
    IF(ABS(Z).LT.EPS) GOTO 90
    A(NM2)=(F(NM2)-HH1*A(NM4)-H3*A(NM3))/Z
    HH3=H3
    H3=B(NM2)-F(NM4)*H2-H1*H4
    H4=(B(NM2)-D(NM4)*HH1-HH3*C(NM3))/Z
    B(NM2)=H4
    HHH1=HH1
    HH1=H1
    H1=C(NM2)-H(NM4)*H2-H(NM3)*B(NM3)
    C(NM2)=(C(NM2)-HHH1*E(NM4)-HH3*E(NM3))/Z
    SUM=0.0
    DO 20 I=1,NM4
        SUM=SUM+F(I)*D(I)
20  CONTINUE
    Z=A(NM1)-HH1*C(NM3)-H3*H4-SUM
    IF(ABS(Z).LT.EPS) GOTO 90
    SUM=0.0
    DO 30 I=1,NM4
        SUM=SUM+F(I)*A(I)
30  CONTINUE
    A(NM1)=(F(NM1)-SUM-HH1*A(NM3)-H3*A(NM2))/Z
    SUM=0.0
    DO 40 I=1,NM4
        SUM=SUM+F(I)*E(I)
40  CONTINUE
    H5=B(NM1)
    H4=(H5-H3*C(NM2)-SUM-HH1*E(NM3))/Z
    B(NM1)=H4
    SUM=0.0
    DO 50 I=1,NM4
        SUM=SUM+H(I)*D(I)
50  CONTINUE
    HH3=H3
    H3=H5-H1*B(NM2)-SUM-H(NM3)*C(NM3)
    SUM=0.0
    DO 60 I=1,NM3
        SUM=SUM+H(I)*E(I)
60  CONTINUE
    Z=A(N)-H1*C(NM2)-H3*H4-SUM
    IF(ABS(Z).LT.EPS) GOTO 90
    SUM=0.0
    DO 70 I=1,NM3
        SUM=SUM+H(I)*A(I)
70  CONTINUE
    A(N)=(F(N)-SUM-H1*A(NM2)-H3*A(NM1))/Z
    F(N)=A(N)
    F(NM1)=A(NM1)-B(NM1)*F(N)
    F(NM2)=A(NM2)-B(NM2)*F(NM1)-C(NM2)*F(N)
    F(NM3)=A(NM3)-B(NM3)*F(NM2)-C(NM3)*F(NM1)-E(NM3)*F(N)
```

(cont.)

```
   DO 80 I=NM4,1,-1
      F(I)=A(I)-B(I)*F(I+1)-C(I)*F(I+2)-D(I)*F(NM1)-E(I)*F(N)
80 CONTINUE
   RETURN
90 IFLAG=2
   RETURN
   END
```

Figure A.8. Program listing of PENPES.

Calling sequence:

CALL PENPES(N,A,B,C,D,E,F,H,EPS,IFLAG)

Purpose:
Solution of a linear system of equations having symmetric, cyclically five-diagonal coefficient matrix.

Description of the parameters:

N,A,F,EPS,IFLAG as in PENTAS.

B	ARRAY(N):	First subdiagonal $(1, \cdots, N-1)$ (is overwritten). B(N) must contain the matrix element with indices (n,1) or (1,n).
C	ARRAY(N):	Second subdiagonal $(1, \cdots, N-2)$ (is overwritten). C(N − 1) must contain the matrix element with indices $(n-1,1)$ or $(1, n-1)$. C(N) must contain the element of the matrix with indices (n,2) or (2,n).
D,E,H	ARRAY(N):	Work space.

Figure A.9. Description of PENPES.

B

List of Subroutines

No.	Name	Required Subroutines	Listing on Page	Description on Page
		List of Subroutines		
1	NEWDIA		3	3
2	NEWSOL		4	4
3	INTONE		12	12
4	POLVAL	INTONE	13	13
5	POLSYM		17	17
6	POLSM1	TRIDIS	21	22
7	POLSM2	TRIDIS	25	26
8	QUAOPT		33	34
9	QUAVAL	INTONE	35	35
10	QUASKF	TRIDIS	42	43
11	QUAVAM	INTONE	44	44
12	QUASKV	TRIDIU	48	49
13	QUAPKV	TRIPEU	53	54
14	QUHIST	TRIDIS	56	57
15	HEQUA		63	64
16	GRAD1		66	66

No.	Name	Required Subroutines	Listing on Page	Description on Page
17	GRAD2		67	67
18	GRAD3	NEWDIA	68	69
19	GRAD4		70	71
20	QUAFZ		76	76
21	INQUA1	TRIDIU	79	80
22	CUB1R5	TRIDIS	88	90
23	CUB2R7	TRIDIS	93	95
24	CUBVAL	INTONE	97	97
25	CUBPER	TRIPES	109	110
26	INCUB1	TRIDIU	113	114
27	INCUB2	TRIDIS	116	117
28	CUBSM1	PENTAS	121	122
29	CUBSM2	PENPES	127	128
30	CUBXSP	TRIDIS, TRIDIU	131	132
31	HECUB		137	137
32	GRAD2B		140	141
33	GRAD2R		142	142
34	GRAD5		146	147
35	GRAD6		150	151
36	GRAD7		154	155
37	GRAD8	COMPB	157	158
38	COMPB		159	159
39	QUINAT		185	188
40	QUIVAL	INTONE	189	190
41	HEQUI1	TRIDIS	196	197
42	HEQUIP	TRIPES	203	204
43	HEQUI2	TRIDIU	205	206
44	HIST42	HEQUI1, TRIDIS	208	209
45	QATVAL	INTONE	210	210
46	RATSCH	TRIDIU	224	227
47	RATVAN	INTONE	229	229
48	RATSAK	TRIDIU	240	241
49	RSKVAL	INTONE	242	242
50	RATSP1	TRIDIU	254	255
51	RATPS1	TRIDIS	256	257
52	RATSP2	TRIDIU	258	259
53	RATPS2	TRIDIS	260	261
54	RATVAL	INTONE	262	262
55	RATPER	TRIPEU	273	274

No.	Name	Required Subroutines	Listing on Page	Description on Page
56	RATPPR	TRIPES	274	275
57	RATGRE	TRIDIS	279	280
58	GREVAL	INTONE	281	281
59	RMONC1	GRAD2B, GRAD2R	289	290
60	GREVAM	INTONE	290	291
61	RMONC2		299	300
62	RCONC2		306	307
63	RHIST1	RATSAK, TRIDIU	312	313
64	RH1VAL	INTONE	314	314
65	RHIST2	RATSP1, TRIDIU	317	318
66	RH2VAL	INTONE	319	319
67	ABKVAL		329	329
68	EXPSP1	ABKVAL, TRIDIS	330	331
69	EXPVAL	INTONE	332	333
70	EXPSP2	ABKVAL, TRIDES	338	339
71	EXPPER	ABKVAL, TRIPES	341	342
72	TRIDIS		374	374
73	TRIDIU		375	375
74	TRIPES		376	377
75	TRIPEU		377	378
76	PENTAS		379	380
77	PENPES		381	383

Remark: The 77 subroutines can be obtained on diskette from the author for a cover charge.

Bibliography

[1] Ahlberg, J. H., Nilson, E. N., and Walsh, J. L. *The Theory of Splines and their Applications*, Academic Press, 1967.

[2] Ahlin, A. C. "Computer algorithms and theorems for generalized spline interpolation," presented at SIAM National Meeting, New York, 1965.

[3] Akima, H. "A new method of interpolation and smooth curve fitting based on local procedures," *J. ACM* 17, 589–602 (1970).

[4] Albasiny, E. L., and Hoskins, W. D. "The numerical calculation of odd-degree polynomial splines with equi-spaced knots," *J. Inst. Maths Applics* 7, 384–397 (1971).

[5] Behforooz, G. H., Papamichael, N., and Worsey, A. J. "A class of piecewise-cubic interpolatory polynomials," *J. Inst. Maths Applics* 25, 53–65 (1980).

[6] Boneva, L. I., Kendall, D., and Stefanov, I. "Spline transformations: Three new diagnostic aids for the statistical data-analyst," *J. Royal Stat. Soc. B* 33, 1–71 (1971).

[7] Burmeister, W., Hess, W., and Schmidt, J. W. "Convex spline interpolants with minimal curvature," *Computing* 35, 219–229 (1985).

[8] Cheney, W., and Kincaid, D. *Numerical Mathematics and Computing*, 2nd Edition, Brooks/Cole Publishing Company, California, 1985.

[9] Clenshaw, C. W., and Negus, B. "The cubic X-spline and its application to interpolation," *J. Inst. Maths Applics 22*, 109–119 (1978).

[10] Cline, A. K. "Scalar- and planar-valued curve fitting using splines under tension," *Comm. ACM* 17, 218–220 (1974).

[11] Cline, A. K. "Algorithm 476: Six subprograms for curve fitting using splines under tension," *Collected Algorithms from CACM*.

[12] Costantini, P. "On monotone and convex spline interpolation," *Math. Comput.* 46, 203–214 (1986).

[13] Costantini, P. "Co-monotone interpolating splines of arbitrary degree — a local approach," *SIAM J. Sci. Stat. Comput.* 8, 1026–1034 (1987).

[14] Costantini, P. "An algorithm for computing shape-preserving interpolating splines of arbitrary degree," *J. Comput. Appl. Math.* 22, 89–136 (1988).

[15] Costantini, P., and Morandi, R. "Monotone and convex cubic spline interpolation," *Calcolo* 21, 281–294 (1984).

[16] Costantini, P., and Morandi, R. "An algorithm for computing shape-preserving cubic spline interpolation to data," *Calcolo* 21, 295–305 (1984).

[17] Culpin D. "Calculation of cubic smoothing splines for equally spaced data," *Numer. Math.* 48, 627–638 (1986).

[18] de Boor, C. "Package for calculating with B-Splines," *SIAM J. Numer. Anal.* 14, 441–472 (1977).

[19] de Boor, C. *A Practical Guide to Splines*, Springer-Verlag, 1978.

[20] de Boor, C. and Höllig, K. "B-splines without divided differences," In Farin, G. E. (Ed.), *Geometric Modelling: Algorithms and New Trends*, SIAM 1987.

[21] Delbourgo, R. "Shape preserving interpolation to convex data by rational functions with quadratic numerator and linear denominator," *IMA J. Numer. Anal.* 9, 123–136 (1989).

[22] Delbourgo, R., and Gregory, J. A. "C^2 rational quadratic spline interpolation to monotonic data," *IMA J. Numer. Anal.* 3, 141–152 (1983).

[23] Delbourgo, R., and Gregory, J. A. "Shape preserving piecewise rational interpolation," *SIAM J. Sci. Stat. Comput.* 6, 967–976 (1985).

[24] Dettori, G. "Interpolation for CAD using curves with tension," *Méthodes numériques dans les sciences de l'ingénieur, 2e Congr. int. G. A. M. N. I., Paris,* 2, 963–973 (1980).

[25] Dierckx, P. "Algorithm 42: An algorithm for cubic spline fitting with convexity constraints," *Computing* 24, 349–371 (1980).

[26] Dierckx, P. "Algorithms for smoothing data with periodic and parametric splines," *Computer Graphics and Image Processing* 20, 171–184 (1982).

[27] Dierckx, P. "An algorithm for smoothing, differentiation and integration of experimental data using spline functions," *J. Comput. Appl. Math.* 1, 165–184 (1975).

[28] Dikshit, H. P., and Powar, P. "Discrete cubic spline interpolation," *Numer. Math.* 40, 71–78 (1982).

[29] Dikshit, H. P., and Powar, P. "Area matching interpolation by discrete cubic splines," *Research Notes in Math.* 133, 35–45 (1985).

[30] Dimsdale, B. "Convex cubic splines," *IBM J. Res. Develop.* 22, 168–178 (1978).

[31] Dodd, S. L., and McAllister, D. F. "Algorithms for computing shape preserving spline approximations to data," *Numer. Math.* 46, 159–174 (1985).

[32] Dubeau, F., Savoie, J. "Periodic quadratic spline interpolation," *J. Approx. Th.* 39, 77–88 (1983).

[33] Duris, C. S. "Discrete interpolating and smoothing spline functions," *SIAM J. Numer. Anal.* 14, 686–698 (1977).

[34] Duris, C. S. "Algorithm 547: Fortran routines for discrete cubic spline interpolation and smoothing," *ACM Trans. Math. Softw.* 6, 92–103 (1980).

[35] Elfving, T., and Andersson, L.-E. "An algorithm for computing constrained smoothing spline functions," *Numer. Math.* 52, 583–595 (1988).

[36] Ellis, T. M. R., and McLain, D. H. "Algorithm 514: A new method of cubic curve fitting using local data," *ACM Trans. Math. Softw.* 3, 175–178 (1977).

[37] Encarnacao, J., and Strasser, W. *Computer Graphics*, R. Oldenbourg, 3rd Edition.

[38] Engeln-Müllges, G., and Reutter, F. *Formelsammlung zur Numerischen Mathematik mit Standard-FORTRAN-Programmen*, Bibliographisches Institut, Zürich, 4th Edition, 1984.

[39] Ertel, J. E., and Fowlkes, E. B. "Some algorithms for linear spline and piecewise multiple linear regression," *J. Am. Stat. Assoc.* 71, 640–648 (1976).

[40] Fischer, B., Opfer, G., and Puri, M. L. "A local algorithm for constructing non-negative cubic splines," Hamburger Beiträäge zur Agnew. Math., Reihe A, Preprint 7, Universität Hamburg, 1987.

[41] Fritsch, F. N., and Carlson, R. E. "Monotone piecewise cubic interpolation," *SIAM J. Numer. Anal.* 17, 238–246 (1980).

[42] Fritsch, F. N., and Carlson, R. E. "Piecewise cubic Hermite interpolation package," UCRL-87285 preprint, Stanford University, 1982.

[43] Fritsch, F. N., and Butland, J. "A method for constructing local monotone piecewise cubic interpolants," *SIAM J. Sci. Stat. Comput.* 5, 300–304 (1984).

[44] Forsythe, G. E., Malcolm, M. A., and Moler C. B. *Computer Methods for Mathematical Computations*, Prentice Hall, 1977.

[45] Glass, J. M. "Smooth-curve interpolation: a generalized spline-fit procedure," *BIT* 6, 277–293 (1966).

[46] Graham, N. Y. "Smoothing with periodic cubic splines," *The Bell System Techn. J.* 62, 101–110 (1983).

[47] Gregory, J. A. "Shape preserving rational spline interpolation." In Graves-Morris, R. R., Saff, E. B., Varga, R. S. (Eds.), *Rational Approximation and Interpolation*, 431–441, Springer-Verlag, 1984.

[48] Gregory, J. A. "Shape preserving spline interpolation," *Computer-Aided Design* 18, 53–57 (1986).

[49] Gregory, J. A., and Delbourgo, R. "Piecewise rational quadratic interpolation to monotonic data," *IMA J. Numer. Anal.* 2, 123–130 (1982).

[50] Gyorvari, J. "Lakunäre Interpolation mit Spline-Funktionen: Die Fälle (0,2,3) und (0,2,4)," *Acta. Math. Hung.* 42, 25–33 (1983).

[51] Hagen, H. "Geometric spline curves," *Comp. Aided Geom. Design* 2, 223–227 (1985).

[52] Hanna, M. S., Evans, D. G., and Schweitzer, P. N. "On the approximation of plane curves by parametric cubic splines," *BIT* 26, 217–232 (1986).

[53] Heidemann, U. "Linearer Ausgleich mit Exponentialsplines bei automatischer Bestimmung der Intervallteilungspunkte," *Computing* 36, 217–227 (1986).

[54] Herriot, J. G., and Reinsch, C. H. "Algorithm 507: Procedures for quintic natural spline interpolation," *ACM Trans. Math. Softw.* 2, 281–289 (1976).

[55] Herriot, J. G., and Reinsch, C. H. "Algorithm 600: Translation of algorithm 507: Procedures for quintic natural spline interpolation," *ACM Trans. Math. Softw.* 9, 258–259 (1983).

[56] Hess, W., and Schmidt J. W. "Convexity preserving interpolation with exponential splines," *Computing* 36, 335–342 (1986).

[57] Holladay, J. A. "Smoothest curve approximation," *Math. Tables Aids Comput.* 11, 233–243 (1957).

[58] Ichida, K., Voshimoto, F., and Kiyoko, T. "Curve fitting by a piecewise cubic polynomial," *Computing* 16, 329–338 (1976).

[59] Irvine, L. D., Marin, S. P., and Smith, P. W. "Constrained interpolation and smoothing," *Constr. Approx.* 2, 129–151 (1986).

[60] Isaacson, E., and Keller, H. B. *Analyse numerischer Verfahren*, Verlag Harri Deutsch, Frankfurt, 1973.

[61] Jakubczyk, K. "Approximation by circular splines for solutions of ordinary differential equations," *Zastosowania Matematyki* 16, 283–292 (1978).

[62] Kielbasinski, A., and Schwetlick, H. *Numerische lineare Algebra*, Verlag Harri Deutsch, Frankfurt, 1988.

[63] Knuth, D. E. *Sorting and Searching, Vol. 3 of The Art of Computer Programming*, Addison-Wesley, Reading, Massachusetts, 1973.

[64] Kosma, Z. "On a special type of cubic splines," *Bull. Acad. Pol. Sci.* 26, 1–10 (1978).

[65] Kozak, J. "Shape preserving approximation," *Computers in Industry* 7, 435–440 (1986).

[66] Kvasov, B. I. "Parabolic B-Splines in interpolation problems," *U.S.S.R. Comput. Maths. Math. Phys.* 23, 13–19 (1983).

[67] Lancaster, P., and Salkauskas, K. *Curve and Surface Fitting*, Academic Press 1986.

[68] Lee, E. H., and Forsythe, G. E. "Variational study of nonlinear spline curves," *SIAM Rev.* 15, 120–133 (1973).

[69] Lerman, P. M. "Fitting segmented regression models by grid search," *Appl. Stat.* 29, 77–84 (1980).

[70] Lyche, T., and Schumaker, L. L. "Computation of smoothing and interpolating natural splines via local bases," *SIAM J. Numer. Anal.* 10, 1027–1038 (1973).

[71] Lyche, T., and Schumaker, L. L. "Algorithm 480: Procedures for computing smoothing and interpolating natural splines," *Commun. ACM* 17, 463–467 (1974).

[72] Lyche, T. "Discrete cubic spline interpolation," *BIT* 16, 281–290 (1976).

[73] Lyche, T., Schumaker, L. L., and Sepehrnoori, K. "Fortran subroutines for computing smoothing and interpolating natural splines," *Adv. Eng. Software* 5, 2–5 (1983).

[74] Maciejewski, M. "Hyperbolic splines with given derivatives at the knots," *Zastos. Matematyki* 18, 319–336 (1984).

[75] Maess, B., and Maess, G. "Interpolating quadratic splines with norm-minimal curvature," *Rostock. Math. Kolloq.* 26, 83–88 (1984).

[76] Malcolm, M. A. "On the computation of nonlinear spline functions," *SIAM J. Numer. Anal.* 14, 254–282 (1977).

[77] Maude, A. D. "Interpolation — mainly for graph plotters," *Comp. J.* 16, 64–65 (1973).

[78] McAllister, D. F., and Roulier, J. A. "Interpolation by convex quadratic splines," *Math. Comput.* 32, 1154–1162 (1978).

[79] McAllister, D. F., and Roulier, J. A. "An algorithm for computing a shape-preserving osculatory quadratic spline," *ACM Trans. Math. Softw.* 7, 331–347 (1981).

[80] McAllister, D. F., and Roulier, J. A. "Algorithm 574: Shape-preserving osculatory quadratic splines," *ACM Trans. Math. Softw.* 7, 384–386 (1981).

[81] McAllister, D. F., Passow, E., and Roulier, J. A. "Algorithms for computing shape preserving spline interpolations to data," *Math. Comput.* 31, 717–725 (1977).

[82] McCartin, B. J. "Applications of exponential splines in computational fluid dynamics," *AIAA J.* 21, 1059–1065 (1983).

[83] McConalogue, D. J. "A quasi-intrinsic scheme for passing a smooth curve through a discrete set of points," *Comp. J.* 13, 392–396 (1970).

[84] Meek, D. S., and Walton, D. J. "The use of Cornu spirals in drawing planar curves of controlled curvature," *J. Comput. Appl. Math.* 25, 69–78 (1989).

[85] Mehlum, E. "A curve-fitting method based on a variational criterion," *BIT* 4, 213–223 (1964).

[86] Meir, A., and Sharma, A. "Lacunary interpolation by splines," *SIAM J. Numer. Anal.* 10, 433–442 (1973).

[87] Mettke, H. "Quadratische Spline-Interpolation bei zusammenfallendem Interpolations- und Splinegitter," *Beiträge zur Num. Math.* 8, 113–119 (1980).

[88] Mettke, H. "Convex cubic HERMITE-spline interpolation," *J. Comput. Appl. Math.* 9, 205–211 (1983).

[89] Mettke, H. "Convex cubic HERMITE-spline interpolation (Corrigendum and addendum)," *J. Comput. Appl. Math.* 11, 377–378 (1984).

[90] Mettke, H., and Lingner, T. "Ein Verfahren zure konvexen kubischen Splineinterpolation," *Wiss. Zeitschr. der TU Dresden* 32, 77–80 (1983).

[91] Micula, G. "Theory and applications of spline functions," Preprint A89/1, Fachbereich Mathematik, Serie A, Freie Universität Berlin, 1989.

[92] Mishra, R. S., and Mathur, K. K. "Lacunary interpolation by splines, (0;0,2,3) and (0;0,2,4) cases," *Acta. Math. Acad. Sci. Hungar.* 36, 251–260 (1980).

[93] Morandi, R., and Costantini, P. "Piecewise monotone quadratic histosplines," *SIAM J. Sci. Stat. Comput.* 10, 397–406 (1989).

[94] Morandi, R., and Costantini, P. *PMQHS1: A code for computing piecewise monotone quadratic histosplines, Monografie di Software Matematico N. 46*, Instituto pe le Applicazioni del Calcolo, Roma, 1986.

[95] Mülthei, H. N., and Schorr, B. "Error analysis for a special X-spline," *Computing* 25, 253–267 (1980).

[96] Mund, E. H., Hallet, P., and Hennart, J. P. "An algorithm for the interpolation of functions using quintic splines," *J. Comput. Appl. Math.* 1, 279–288 (1975).

[97] Neumann, E. "Determination of an interpolating quadratic spline function," *Zastosowania Matematyki* 15, 245–250 (1976).

[98] Neumann, E. "Determination of a quadratic spline function with given values of the integrals in subintervals," *Zastosowania Matematyki* 16, 681–689 (1980).

[99] Neumann, E. "Cubic splines with given derivatives at the knots," *Comment. Math.* 11, 25–30 (1981).

[100] Neumann, E. "Cubic splines with given values of the second derivatives at the knots," *Demonstratio Mathematica* 14, 115–125 (1981).

[101] Neumann, E. "Determination of an interpolating quintic spline function with equally spaced and double knots," *Zastosowania Matematyki* 16, 133–141 (1977).

[102] Neumann, E. "Convex interpolating splines of arbitrary degree," *ISNM* 52, 211–222 (1980).

[103] Neumann, E. "Convex interpolating splines of arbitrary degree II," *BIT* 22, 331–338 (1982).

[104] Neumann, E. "Convex interpolating splines of arbitrary degree III," *BIT* 26, 527–536 (1986).

[105] Opfer, G., and Oberle, H. J. "The derivation of cubic splines with obstacles by methods of optimization and optimal control," *Numer. Math.* 52, 17–31 (1988).

[106] Passow, E., and Roulier, J. A. "Monotone and convex spline interpolation," *SIAM J. Numer. Anal.* 14, 904–909 (1977).

[107] Pence, D. D., "Computing nonlinear spline functions." In Cheney, W. (Ed.) *Approximation Theory III*, Academic Press, 1980.

[108] Prasad, J., and Varma, A. K. "Lacunary interpolation by quintic splines," *SIAM J. Numer. Anal.* 16, 1075–1079 (1979).

[109] Press, W. H. *et al. Numerical Recipes*, Cambridge University Press 1986.

[110] Pruess, S. "Alternatives to the exponential spline in tension," *Math. Comput.* 33, 1273–1281 (1979).

[111] Pruess, S. "Properties of splines in tension," *J. Approx. Th.* 17, 86–96 (1976).

[112] Pruess, S. "An algorithm for computing smoothing splines in tension," *Computing* 19, 365–373 (1978).

[113] Ramirez, V., and Lorente, J. "C^1 rational quadratic spline interpolation to convex data," *Appl. Numer. Math.* 2, 37–42 (1986).

[114] Reinsch, C. H. "Smoothing by spline functions," *Numer. Math.* 10, 177–183 (1967).

[115] Reinsch, C. H. "Smoothing by spline functions II," *Numer. Math.* 16, 451–454 (1971).

[116] Ren-Hong, W., and Shun-Tang, W. "On the rational spline functions," *J. Math. Res. Exposition* 4, 31–36 (1984).

[117] Renka, R. J. "Interpolatory tension splines with automatic selection of tension factors," *SIAM J. Sci. Stat. Comput.* 8, 393–415 (1987).

[118] Rentrop, P. "An algorithm for the computation fo the exponential spline," *Numer. Math.* 35, 81–93 (1980).

[119] Rentrop, P., and Wever, U. "Interpolation algorithms for the control of a sewing machine." In Neunzert, H., (Ed.) *Proc. Second European Symposium on Mathematics in Industry*, Teubner-Verlag, Stuttgart, 1988.

[120] Rentrop, P., and Wever, U. "Computational strategies for the tension parameters of the exponential spline." In Bulirsch, R. *et. al.* (Eds.) *Lecture Notes in Control and Information Science* 95, 122–134 (1987).

[121] Rice, J.R. *Numerical Methods, Software, and Analysis*, McGraw-Hill International Book Company, 1983.

[122] Sakai, M., and Silanes, M. C. L. de "A simple rational spline and its application to monotonic interpolation to monotonic data," *Numer. Math.* 50, 171–182 (1986).

[123] Sakai, M., and Usmani, R. A. "A shape preserving area true approximation of histogram by rational splines," *BIT* 28, 329–339 (1988).

[124] Sakai, M., and Usmani, R. A. "Exponential quadratic splines," *Proc. Japan Acad.* 60, Ser. A., 26–29 (1984).

[125] Sakai, M., and Usmani, R. A. "On exponential splines," *J. Approx. Th.* 47, 122–131 (1986).

[126] Sakai, M., and Schmidt, J. W. "Positive Interpolation with rational splines," *BIT* 29, 140–147 (1989).

[127] Sallam, S., and El-Tarazi, M. N. "Quadratic spline interpolation on uniform meshes," In Schmidt, J. W., and Späth, H. *Splines in Numerical Analysis*, Akademie-Verlag, Berlin, 1989.

[128] Schaback, R. "Spezielle rationale Splinefunktionen," *J. Approx. Th.* 7, 281–292 (1973).

[129] Schaback, R. "Interpolation mit nichtlinearen Klassen von Splinefunktionen," *J. Approx. Th.* 8, 173–188 (1973).

[130] Schaback, R. "Adaptive rational splines," *Constructive Approximation* 6, 167–179 (1990).

[131] Schmidt, J. W. "Convex interval interpolation with cubic splines," *BIT* 26, 377–387 (1986).

[132] Schmidt, J. W. "On convex cubic C^2-spline interpolation," *Intern. Series Numer. Math.*, vol. 81, 213–228 (1987).

[133] Schmidt, J. W. "An unconstrained dual program for computing convex C^1-spline approximants," *Computing* 39, 133–140 (1987).

[134] Schmidt, J. W. "On shape preserving spline interpolation: Existence theorems and determination of optimal splines," *Banach Center publ.* vol. 22, 377–388, Polish Scient. Publ. Warsaw, 1988.

[135] Schmidt, J. W. "Convex smoothing by splines and dualization," *Confer. Proceed.* "Numer. Methods Appl.," Sofia, (1988).

[136] Schmidt, J. W. "Results and problems in shape preserving interpolation and approximation with polynomial splines." In Schmidt, J. W., and Späth, H. (Eds.), *Splines in Numerical Analysis*, Akademie Verlag, Berlin, 1989.

[137] Schmidt, J. W., and Hess, W. "Positivity of cubic polynomials on intervals and positive spline interpolation," *BIT* 28, 340–352 (1988).

[138] Schmidt, J. W., and Hess, W. "Spline interpolation under two-sided restrictions on the derivatives," *Z. Angew. Math. Mech.* 69 (1989).

[139] Schmidt, J. W., and Hess, W. "Schwach verkoppelte Ungleichungssysteme und konvexe Spline-Interpolation," *Elem. Math.* 39, 85–95 (1984).

[140] Schmidt, J. W., and Hess, W. "Positive interpolation with rational quadratic splines," *Computing* 38, 261–267 (1987).

[141] Schmidt, J. W., and Hess, W. "Quadratic and related exponential splines in shape preserving interpolation," *J. Comput. Appl. Math.* 18, 321–329 (1987).

[142] Schmidt, J. W., and Mettke, H. "Konvergenz von quadratischen Interpolations- und Fächenabgleichsplines," *Computing* 19, 351–363 (1978).

[143] Schmidt, R. "Formerhaltende Interpolation," *Angew. Informatik* 29, 177–180 (1988).

[144] Schoenberg, I. J. "Splines and histograms." In *Spline Functions and Approx. Theory, Proc. Sympos. Univ. Alberta, Edmonton 1972*, ISNM 21, Basel, Switzerland, 1973.

[145] Schumaker, L. L. "On shape preserving quadratic spline interpolation," *SIAM J. Numer. Anal.* 20, 854–864 (1983).

[146] Schumaker, L. L. "On hyperbolic splines," *J. Approx. Th.* 38, 144–166 (1983).

[147] Schwarz, H. R. *Numerische Mathematik*, 2nd Edition, Teubner, Stuttgart, Germany, 1988.

[148] Schweikert, D. G. "The spline in tension (hyperbolic spline) and the reduction of extraneous inflection points," Thesis, Brown University, 1966.

[149] Schweikert, D. G. "An interpolating curve using a spline in tension," *J. Math. Physics* 45, 312–317 (1966).

[150] Schwetlick, H. *Numerische Lösung nichtlinearer Gleichungen*, R. Oldenbourg-Verlag, München, 1979.

[151] Seitelman, L. H. "Natural cubic splines are unnatural." In Wang, C. C. (Ed.), *Information Linkage between Applied Mathematics and Industry*, Academic Press, 1979.

[152] Söll, H. "Darstellung und Fortschreibung der monatlichen Haushalt-seinkommen mit Spline-Funktionen," *Splinefunktionen in der Statistik, Sonderh. z. Alleg. Stat. Arch., H.* 14, 43–65 (1978).

[153] Späth, H. *Algorithmen für elementare Ausgleichsmodelle*, R. Oldenbourg, 1973.

[154] Späth, H. *Spline-Algorithmen zur Konstruktion glatter Kurven und Flächen*, 4th Edition, R. Oldenbourg-Verlag, München, 1986.

[155] Späth, H. "Zur Glättung empirischer Häufigkeitsverteilungen," *Computing* 10, 353–357 (1972).

[156] Späth, H. "Interpolation by certain quintic splines," *Computer J.* 12, 292–293 (1969).

[157] Späth, H. "Die numerische Berechnung von interpolierenden Spline-Funktionen mit Blockunterrelaxation," Dissertation, Universität Karlsruhe 1969.

[158] Späth, H. "The numerical calculation of high degree Lidstone splines with equidistant knots by blockunderrelaxation," *Computing* 7, 65–74 (1971).

[159] Späth, H. "The numerical calculation of quintic splines by block under relaxation," *Computing* 7, 75–82 (1971).

[160] Späth, H. "Eine Vereinfachung des Newtonschen Verfahrens für Gleichungen mit homogenen Operatoren," *ZAMM* 45, 569 (1965).

[161] Späth, H. "Rationale Spline-Interpolation," *Angewandte Informatik*, 357–359 (1971).

[162] Späth, H. "Exponential spline interpolation," *Computing* 4, 225–233 (1969).

[163] Späth, H., and Meier, J. "Flexible smoothing with periodic cubic splines and fitting with closed curves," *Computing* 40, 293–300 (1988).

[164] Stoer, J. *Einführung in die Numerische Mathematik I*, 4th Edition, Springer-Verlag, Berlin, 1983.

[165] Stoer, J. "Curve fitting with clothoidal splines," *J. Res. Nat. Bur. Standards* 87, 317– 346 (1982).

[166] Varga, R. *Matrix Iterative Analysis*, Prentice Hall 1962.

[167] Varma, A. K. "Lacunary interpolation by splines I," *Acta. Math. Hung.* 31, 185–192 (1978).

[168] Varma, A. K. "Lacunary interpolation by splines II," *Acta. Math. Hung.* 31, 193–203 (1978).

[169] Werner, H. "Neuere Entwicklungen auf dem Gebiet der nichtlinearen Splinefunktionen," *ZAMM* 58, T86–T95 (1978).

[170] Werner, H. "An introduction to non-linear splines." In Sahney, B. N. (Ed.), *Polynomial and Spline Approximation*, 247–306 (1979).

[171] Werner, W. "Polynomial Interpolation: Lagrange versus Newton," *Math. of Computation* 43, 205–217 (1984).

[172] Wever, U. "Nonnegative exponential splines," *Computer-Aided Design* 20, 11–16 (1988).

[173] Woodford, C. H. "An algorithm for data smoothing using spline functions," *BIT* 10, 501–510 (1970).

[174] Woodford, C. H. "Smooth curve interpolation," *BIT* 9, 69–77 (1969).

[175] Xie, S. "Quadratic Spline Interpolation," *J. Approx. Th.* 40, 66–80 (1984).

Index